Fundamental Theories of Physics

Volume 204

The international monograph series “Fundamental Theories of Physics” aims to stretch the boundaries of mainstream physics by clarifying and developing the theoretical and conceptual framework of physics and by applying it to a wide range of interdisciplinary scientific fields. Original contributions in well-established fields such as Quantum Physics, Relativity Theory, Cosmology, Quantum Field Theory, Statistical Mechanics and Nonlinear Dynamics are welcome. The series also provides a forum for non-conventional approaches to these fields. Publications should present new and promising ideas, with prospects for their further development, and carefully show how they connect to conventional views of the topic. Although the aim of this series is to go beyond established mainstream physics, a high profile and open-minded Editorial Board will evaluate all contributions carefully to ensure a high scientific standard.

More information about this series at https://link.springer.com/bookseries/6001

Claus Kiefer
Editor

From Quantum to Classical

Essays in Honour of H.-Dieter Zeh

Springer

Editor
Claus Kiefer
Institute for Theoretical Physics
University of Cologne
Köln, Nordrhein-Westfalen, Germany

ISSN 0168-1222 ISSN 2365-6425 (electronic)
Fundamental Theories of Physics
ISBN 978-3-030-88780-3 ISBN 978-3-030-88781-0 (eBook)
https://doi.org/10.1007/978-3-030-88781-0

This Springer imprint is published by the registered company Springer Nature Switzerland AG
The registered company address is: Gewerbestrasse 11, 6330 Cham, Switzerland

H. Dieter Zeh in September 2005. Photo Rolf Kickuth

Preface

Why does a world described fundamentally by quantum theory look classical? Where does the observed arrow of time come from? Are these questions related? These are the questions that were at the heart of Dieter Zeh's thinking over many decades.

Dieter started his career as a theoretical physicist when working on nuclear physics in the sixties. His interests included the theory of α-decay, nucleosynthesis and the dynamics of deformed nuclei. It was his concrete research about the deformed nuclei that eventually led him, in 1968, to the discovery of what was later called decoherence.[1] Dieter was puzzled by the fact that individual nucleons could experience an orientation in spite of the whole nucleus being in a stationary quantum state. In his article on 'Roots and Fruits of Decoherence', he describes this as follows:

> If the nucleons in the deformed nucleus dynamically feel a definite orientation in spite of the global superposition, would an internal observer then not similarly have to become 'aware of' a certain measurement result?[2]

He then envisages a stationary quantum universe ('wave function of the universe') in which observers experience definite properties such as orientations in spite of the total quantum state being a superposition of all orientations. These observers are *decohered*, and this decoherence arises from their quantum correlation (entanglement) with many inaccessible degrees of freedom in the global superposition. Dieter identified this entanglement as the key to an understanding of the quantum-to-classical transition. This was not recognized by the physics community for many years. In his own words:

> I am indeed convinced that the importance of decoherence was overlooked for the first 60 years of quantum theory precisely because entanglement was misunderstood as no more than a statistical correlation between local objects.[3]

---

[1] His first paper on this subject could only appear in 1970, in the first volume of the newly created journal *Foundations of Physics*.

[2] H. D. Zeh, Roots and fruits of decoherence. In: *Quantum decoherence*, ed. by B. Duplantier, J.-M. Raimond, and V. Rivasseau (Birkhäuser, Boston, 2006), pp. 151–175.

[3] H. D. Zeh, *op. cit.*

Dieter always entertained a realistic interpretation of the wave function. Through this, he independently arrived at an Everett-type of interpretation without knowing about Everett's work. He was a proponent of this interpretation for the rest of his life, although he was not sympathetic to calling it a 'many-worlds interpretation', for the reason that we have only one quantum world (though with many quasi-classical branches). The only alternative to the Everett interpretation that Dieter accepted were theories with a dynamical collapse of the wave function, for the reason that these are realistic theories. But so far no empirical hint for collapse theories exists.

His later interests in the arrow of time and in quantum cosmology grew out of these early insights, because he recognized that the emergence of classical properties in quantum theory cannot be understood without understanding the irreversible nature of our Universe.[4] I had the pleasure to be a Ph.D. student under his supervision in the eighties and to apply his ideas on decoherence to quantum gravity and quantum cosmology. Even after my Ph.D., we stayed in close contact and participated in particular in a six-years project with six people organized in Heidelberg by Nucu Stamatescu, from 1990 to 1996, out of which our monograph on decoherence resulted.[5] During this project, we have met regularly for informal talks and discussion, more in the spirit of an eighteenth century academy than in the contemporary hectic and grant-driven academic atmosphere. It became evident there and elsewhere that Dieter was always interested in *understanding* things—in physics and beyond physics. He mentioned to me several times that he felt influenced by the Vienna Circle of Logical Empiricism about which he had heard for the first time in a radio talk by Ingeborg Bachmann in the fifties. Many of his general essays bear testimony of this influence.[6]

The present volume contains 13 articles in honour of Dieter Zeh and his work. They cover a wide range of topics. In the opening chapter, **Kristian Camilleri** focuses on the role Dieter Zeh played in the history of quantum mechanics and its interpretation. This is followed by **Wojciech Zurek**'s account on decoherence and, in particular, the important role of quantum Darwinism. The decoherence program is also the subject of **Maximilian Schlosshauer**'s contribution. He puts special emphasis on Zeh's commitment to a realistic interpretation of the wave function and discusses, in particular, the importance of experimental tests. Experimental decoherence in molecule interferometry is then discussed in detail in the contribution by **Markus Arndt, Stefan Gerlich** and **Klaus Hornberger**. This clearly demonstrates that decoherence has found its place in the centre of experimental physics, much different from the early years when Dieter had developed his ideas. Experiments also play a crucial role in **Markus Aspelmeyer**'s article. He focuses, in particular, on future experiments that involve superpositions of the gravitational field.

---

[4] See e.g. C. Kiefer, Obituary for Heinz-Dieter Zeh (1932--2018), in: *International Journal of Quantum Foundations*, **5**, 11–15 (2019).

[5] E. Joos, H. D. Zeh, D. Giulini, C. Kiefer, J. Kupsch, and I.-O. Stamatescu, *Decoherence and the Appearance of a Classical World in Quantum Theory*, second edition (Springer, Berlin, 2003).

[6] Dieter's former homepage, as he left it at the time of his death, can be found at http://www.thp.uni-koeln.de/gravitation/zeh/. It contains a collections of papers and essays.

Gravitation also plays a central role in the next contribution. **Andrei Barvinsky** and **Alexander Kamenshchik** discuss the meaning of the wave function of the Universe, in particular the role of the preferred basis for its interpretation. **Heinrich Päs**, in his article, focuses on Dieter's interpretation of the wave function and speculates on its relevance for developing and understanding a quantum theory of gravity. Time and gravity are also the subject of the next chapter written by **Henrique Gomes** and **Jeremy Butterfield**. They put emphasis on geometrodynamics and its connection with ideas of relational projects and shape dynamics.

The next three contributions focus on various aspects in which thermodynamics and statistical mechanics play a fundamental role. **Bei-Lok Hu** explains the connection between dissipation, fluctuations, noise, and decoherence in open quantum systems. **Lajos Diósi** presents various insights about quantum Brownian motion and its relation with decoherence and mentions important open problems for future research. **Domenico Giulini** discusses various subtle issues about the Second Law of thermodynamics and its statistical foundation on the basis of Ehrenfest's urn model. The next chapter by **Ion-Olimpiu Stamatescu** addresses issues in the theory of science, focusing on the role that computer simulations play for the understanding of physical theories. Last but not least, **Peter Byrne** presents a very personal account of Dieter and his commitment to the Everett interpretation of quantum theory.

I thank all the authors for their commitment and effort to produce such wonderful contributions. I am sure Dieter would have enjoyed to read them.

Cologne, Germany
August 2021

Claus Kiefer

Contents

Contributors

Markus Arndt Faculty of Physics, University of Vienna, Vienna, Austria

Markus Aspelmeyer Faculty of Physics, University of Vienna, Vienna, Austria; Institute for Quantum Optics and Quantum Information (IQOQI) Vienna, Austrian Academy of Sciences, Vienna, Austria

Andrei O. Barvinsky Theory Department, Lebedev Physics Institute, Moscow, Russia

Jeremy Butterfield Trinity College, University of Cambridge, Cambridge, UK

Peter Byrne Ink, Petaluma, CA, USA

Kristian Camilleri School of Historical and Philosophical Studies, University of Melbourne, Parkville, Australia

Lajos Diósi Wigner Research Centre for Physics, Budapest, Hungary

Stefan Gerlich Faculty of Physics, University of Vienna, Vienna, Austria

Domenico Giulini Institute for Theoretical Physics, Leibniz University Hannover, Hannover, Germany

Henrique Gomes Trinity College, University of Cambridge, Cambridge, UK

Klaus Hornberger Faculty of Physics, University of Duisburg-Essen, Duisburg, Germany

Bei-Lok Hu University of Maryland, College Park, MD, USA

Alexander Yu. Kamenshchik Dipartimento di Fisica e Astronomia, Università di Bologna and INFN, Bologna, Italy;
Landau Institute for Theoretical Physics, Russian Academy of Sciences, Moscow, Russia

Heinrich Päs Fakultät Für Physik, Technische Universität Dortmund, Dortmund, Germany

Maximilian Schlosshauer Department of Physics, University of Portland, Portland, OR, USA

Ion-Olimpiu Stamatescu Institut für Theoretische Physik, Universität Heidelberg, Heidelberg, Germany;
Forschungsstätte der Evangelischen Studiengemeinschaft, Heidelberg, Germany

Wojciech Hubert Zurek Theory Division, MS B213, LANL Los Alamos, Los Alamos, NM, USA

H. Dieter Zeh and the History of the Foundations of Quantum Mechanics

Kristian Camilleri

1 Introduction

In the prologue to his 2011 book *Elegance and Enigma*, Max Schlosshauer notes that "the last two or three decades have witnessed a stunning renaissance of quantum foundations. Today we find ourselves in the fortunate situation where occupation with quantum-foundational matters is no longer inconceivably far from reaching mainstream status" ([42], pp. 20–1). Things were not always so. In the 1960s those who concerned themselves with the foundations of quantum theory and challenged the prevailing orthodoxy in quantum mechanics found themselves marginalized by the physics community and with limited career prospects. H. Dieter Zeh (1932–2018) was one of the pioneering figures in the revival of the foundations of quantum mechanics. As Guido [1] points out: "The modern beginnings of decoherence as a subject in its own right are arguably the papers by H. D. Zeh of the early 1970s". He was, to use Olival Freire's provocative term, one of the "quantum dissidents", who dared to challenge the prevailing orthodoxy in quantum mechanics in the 1960s and 70s. As with many others who pursued this course of research at the time, this came

This chapter is based on a paper presented at the second international conference of the History of Quantum Physics in Utrecht on 17 July 2008, which was later published [10]. I would especially like to thank H. Dieter Zeh and Wojciech Zurek for their time and assistance in answering my questions through our email correspondence and for allowing me to quote from our correspondence. I would also like to thank H. Dieter Zeh and Fábio Freitas for granting me permission to use the transcript of the interview conducted in July 2008 in preparing this paper. Thanks must also go to Olival Freire for allowing me to use correspondence he obtained from the Wheeler papers at the American Philosophical Society, Philadelphia. I would also like to thank Max Schlosshauer for sharing with me his insights and considerable knowledge of decoherence during our many stimulating conversations.

K. Camilleri (✉)
School of Historical and Philosophical Studies, University of Melbourne, Parkville 3010, Australia
e-mail: kcam@unimelb.edu.au

C. Kiefer (ed.), *From Quantum to Classical*, Fundamental Theories of Physics 204,
https://doi.org/10.1007/978-3-030-88781-0_1

at considerable cost. His ground-breaking work on decoherence would go largely unrecognized for more than a decade before the theoretical and experimental study of decoherence began to attract more widespread attention of physicists in the 1980s and 90s.

Zeh's early work on the foundations of quantum mechanics and his struggle for recognition provides an instructive case study in the history of the re-emergence of the foundations of quantum mechanics in the latter decades of the twentieth century. In this chapter I examine Zeh's early work on decoherence and the obstacles he faced in Heidelberg in winning support for his controversial ideas. Zeh's commitment to the Everett interpretation undoubtedly served as a major obstacle to the wider recognition of his early theoretical contributions to the understanding of decoherence. It was only through the divorcing of interpretation from the study of the dynamics of environment-induced decoherence in the 1980s that Zeh's work began to attract renewed attention from a number of physicists. This occurred largely through the work of Wojciech Zurek in the 1980s and 90s. However, Zeh's struggles were not simply the result of his unorthodox views on interpretation. Unlike Zurek, Zeh did not enjoy the luxury of the backing and support of senior colleagues who were sympathetic to his theoretical interests. Whereas Zurek benefitted immensely from his association with John Wheeler, whose reputation gave him the freedom to pursue foundational questions, Zeh found himself increasingly marginalized in Heidelberg with little prospect of professional advancement. The story of Zeh's early struggles thus constitute an important chapter in the history of the foundations of quantum mechanics.

2 H. Dieter Zeh and the Origins of Decoherence

The problem of explaining how the essentially 'classical' features of the macro-world emerge from a quantum theory of the micro-world has been one of the central problems for the interpretation of quantum mechanics since the 1930s. In the 1980s the study of the entanglement between quantum systems and their environments began to shed new light on the emergence of 'classicality' in quantum mechanics. As Erich Joos explains: "The interaction with the environment which is in principle present for all systems seems to play a decisive role for an understanding of classical properties in the framework of quantum theory" ([26], p. 12). This process, known as 'environment-induced decoherence' or simply 'decoherence', has been the subject of much recent interest and provides a new insight into explaining why quantum superpositions effectively disappear in the macroscopic world—in other words why we don't encounter 'Schrödinger cats'. While there continues to be much debate and disagreement about the extent to which decoherence can explain the appearance of the 'classical' world, the delocalization of interference resulting from quantum entanglement between a local subsystem and its environment is now an experimentally well-confirmed process [5, 12, 23, 34, 38, 39].

In 1970 Zeh proposed that it is precisely the openness of quantum systems, i.e. their interaction with the environment, which is *essential* to explaining why in a unitary dynamics of the Schrödinger equation, the interference terms approximately vanish, in other words, why the 'superposition' of states effectively disappears in the macroscopic world. Zeh's work on the dynamics of quantum entanglement began from the assumption of the universality of the wave function, which invariably led him to an Everett-like relative-state interpretation. Thus, decoherence, as it would later become known, was for Zeh, from its inception an integral part of the Everett picture. As will become clear, this was one of the reasons that Zeh's early papers remained largely ignored by physicists for over a decade. In the early 1980s Wojciech Zurek published two important papers in which he derived the preferred basis and the superselection rules on the basis on environment-induced decoherence. While Zurek's work arose independently of Zeh's and differed in its formal approach, it represented a renewed effort to engage in some of the key physical ideas which had appeared in Zeh's early papers. In spite of the similarities in their approaches to decoherence, Zurek's programmatic aims for decoherence and its relationship to interpretational issues have diverged in certain crucial respects from those of Zeh.

Zeh completed his PhD on nuclear physics at Heidelberg University, and after spending time at Berkeley, Cal Tech, and La Jolla, he returned to Heidelberg in 1966 to submit his *Habilitationsschrift* and take up a position there as a lecturer. In 1970 Zeh published a paper, 'On the interpretation of measurement in quantum theory', in which he discussed the consequences of the entanglement of a system with its environment, anticipating a number of the important insights behind decoherence. As the title of the paper suggests, this work was originally intended as a contribution to the discussion of the measurement problem. Zeh proposed that the difficulties which arise in this standard von Neumann treatment of measurement stem from the "invalid" assumption that a macroscopic system (in this case the quantum system interacting with the measuring apparatus) can be completely 'isolated' from its environment. The aim of the paper was to show that by taking into account the interaction with the environment—the "remainder of universe"—the dynamics would show why we do not observe superpositions of quantum states in macroscopic systems [55]. Proceeding from the assumption of the "universal validity of quantum theory", Zeh argued that strictly speaking "the only 'closed system' is the universe as a whole" ([55], p. 73).

Zeh proposed that by taking a realistic interpretation of the universality of the Schrödinger equation, one could derive certain 'superselection rules', which had been postulated to preclude the existence of certain superpositions which are quantum-mechanically possible, but which are in practice never observed (e.g. the superposition of negative and positive charges). Here he argued that the superposition of different charge numbers is in practice never observed because such a superposition is dynamically unstable as a result of its interaction with the environment ([55], p. 75). This, he argued, is also the case in the chiral states (left-handedness or right-handedness) of sugar molecules, which do not appear in a superposition of quantum states. Through their interaction with the environment, the interference terms practically vanish, effectively leaving two dynamically stable, 'decoupled' (quasi-classical) states. In an early version of his paper, in the form of a typescript entitled '*Probleme der Quantentheorie*' written in German some time around 1966–7, Zeh spelled out this basic idea:

> The realistic interpretation of a universally valid quantum theory leads therefore to the superposition of these states becoming unobservable, if the difference of its interactions with the "remainder of the world" is effective enough to fix the state macroscopically. In particular, states that differ through the charge number, have a sufficiently different interaction with the environment in order to become always "automatically measured". That explains the superselection rule which forbids superposition between such states ([53], p. 10).

This notion, which later became one of the cornerstone of the decoherence program in later years, was pursued in more rigorous fashion in a paper Zeh published with Kübler in 1973 entitled, 'The Dynamics of Quantum Correlations'. Here the authors proposed that by taking as their starting point the universal validity of the Schrödinger wave equation, it was possible to use a density matrix to represent the state of a subsystem (which itself does not obey a von Neumann equation) in deriving the superselection rules as a dynamical consequence of universal entanglement [31].

Zeh recognised that the assumption of a universality of the superposition principle led to rather "unusual consequences". Here Zeh proposed, in a manner strikingly similar to Everett, that in the case where we have "two different components", (such as in the case of a spin state of an electron) only one of which is observed, the two final states "represent two completely *decoupled worlds*". Because of the entanglement with the environment, the two angular momentum states would no longer be in a quantum superposition. Zeh argued that while "the 'other' component cannot be observed any more", it is necessary to posit its existence if "only to save the consistency of the theory" ([55], p. 74). He acknowledged that while the question of "whether the other components "exist" after the measurement" might well be deemed meaningless from the positivist standpoint, "the question of whether or not the assumption of this existence (i.e. of an objective world) leads to a contradiction" *is* indeed a meaningful question. As Zeh later recalled, "From the beginning "decoherence" (uncontrollable entanglement) was for me an argument for the Everett interpretation (whose papers I did not know when I completed the first draft)" [51]. The concluding section of the early version of the paper (the 1967 typescript '*Probleme der Quantentheorie*') is particularly illuminating:

> But how can one interpret this superposition? If (e.g. through a measurement) a superposition of (to simplify matters) two vector elements [*Zeigerstellungen*] has evolved ($\Psi = c_1\Psi_1 + c_2\Psi_2$), then the states Ψ_1 and Ψ_2 will become practically independently from one another after that, as the matrix element $<\Psi_1|H|\Psi_2>$ becomes negligibly small. One has to deal with two "worlds" which are practically independent from one another after the measurement. It appears impossible to escape this consequence as long as one accepts the universal validity of quantum theory. Both worlds differ macroscopically – including the observing organism. One must accept it however as an experience that "the" consciousness is only realized in one of these worlds at a time. This statement is similar to the one which tells us that the consciousness is only realized in one person at a time. One can describe the state of this partial-world however in very good approximation through Ψ_1 or Ψ_2 alone. That is the "reduction of the wave function"! In this sense, Schrödinger's experiment with the cat would split the world into a continuum of worlds, in every one of which the cat is killed at a different time ([53], p. 9).

Zeh recalls that he only 'discovered' Everett after completing the preliminary draft of the paper. Having finished writing the paper, he went down to the library to do something "completely different", whereupon he stumbled upon an article by Bryce DeWitt on the quantum theory of gravity which made use of the universal wave function proposed by Everett [52].[1] Zeh subsequently referred to Everett's relative-state interpretation in the version of the paper published in *Foundations of Physics*, though he now added that his own work marked an advance upon Everett's in providing a basis for "the dynamical stability conditions" of the branches [55], p. 74).

Zeh's work can be seen as forming part of the renewed interest in the foundations of quantum mechanics which had emerged by the late 1960s. Zeh recalls, "I always had an interest in foundations, of course, and I knew there was something not understood in quantum theory" [52]. In the introductory section of the paper, Zeh discussed the contributions to "the problem of measurement and the related problem of how to describe classical phenomena in the framework of quantum theory" which had "received increased attention during recent years", citing the work of Wigner, Jauch, Shimony, Yansee, Ludwig, Daneiri, Prosperi, Loinger, Bell, Bohm and Bub [55], p. 69).

The genesis of Zeh's ideas on the interpretation of quantum mechanics can be traced to his research on low energy nuclear physics which formed the subject of his *Habilitationsschrift* in 1966 [54]. Zeh was at this time concerned with how to explain the collective degrees of freedom of nuclei. The standard method was to use what is called a Hartree–Fock approximation to describe 'stationary states' of heavy nuclei by means of determinants of time-dependent single-nucleon wave functions. In the case of deformed nuclei, however, each of the nuclei are in angular momentum eigenstates, but the nucleus as a whole will not be in an angular momentum eigenstate. Zeh felt that the key to the problem lay with accepting the fundamental role that entanglement plays in quantum theory, not merely as "a statistical correlation between local objects", but as a feature of the underlying 'reality' ([52], [61], pp. 2–3).[2] It dawned on Zeh that the "feeling" of a (fixed) orientation is a formal consequence" of the symmetry violation that occurs in "a strong entanglement between many nuclei". In other words, "the individual nucleons "observe" an apparent asymmetry in spite of the symmetric global superposition of all orientations". Zeh recalls that this situation led him to reflect on the possibility that the dynamical consequences of entanglement might well hold the key to a solution of the measurement problem ([61], p. 4). In this way, he speculated about the possibility of "a nucleus that is big enough to contain a complex subsystem which may resemble a registration device or even a conscious observer. It/she/he would then become entangled with its/her/his "relative world",

[1] As Osnaghi, Freitas and Freire point out in their paper on the history of the Everett interpretation: "The paper in which DeWitt presented his famous Wheeler-DeWitt equation relies on Everett's approach in order to provide an interpretive framework for 'the state functional of the actual universe'" [37].

[2] Zeh recalls that one important argument in convincing him of the 'reality' of the wave function was the "explanation of parity violation by means of the superposition of a K-meson and its antiparticle forming a new "real" particle" [51].

such as with a definite orientation". Here the external observer is incorporated into quantum theory by becoming "*part of the environment* of the observed object" thereby becoming "entangled, too, with the property he is observing". As a consequence the observer "thus feels part of a much bigger "nucleus" or closed system: the quantum universe" ([61], p. 5). Yet these early ideas, as Zeh says himself, "went beyond what is now called decoherence".

As a student Zeh had entertained an interest in philosophy, particularly epistemology, and had in fact contemplated studying philosophy at university [51]. Though Zeh never undertook the study of philosophy in any systematic fashion, his concern with fundamental issues in physics and his interest in epistemological questions was evident in his defense of the Everett interpretation, reflecting his familiarity with logical positivism and perhaps more importantly, the work of the neo-Kantian philosopher Hans Vaihinger. Vaihinger had argued that the underlying reality of the world remains unknowable, but we behave "as if" the constructions of physics such as electrons, protons and electromagnetic waves exist, and to this extent such 'heuristic fictions' constitute our reality.[3] In this way, Zeh argued that the universal wave function may be employed as a 'heuristic fiction', but it is no less 'real' than the entities posited by other physical theories (e.g. quarks), the existence of which is routinely taken for granted.[4] In his 1967 manuscript, he wrote: "Because of their unobservability naturally one can deny the existence of the remaining [unobserved] partial worlds", but we may assume they "exist in the same sense as do all 'heuristic fictions' of physics" ([53], p. 10). Elaborating on this point a few years later, he explained, we customarily "regard those things as 'existing in reality' which are extrapolated from the observed by means of established laws" ([57], p. 109). From this point of view, Zeh maintained that one can regard the unobserved branching states (worlds) of the universal state function as 'real' since they are an extrapolation of the superposition principle which plays a central role in the dynamics of quantum entanglement. Vaihinger's epistemology thus plays a part in Zeh's defence of the Everett interpretation [59].

3 Early Reception and the Dark Ages of Decoherence

Zeh initially submitted an early version of his paper to several journals, among them *Die Naturwissenschaften* and *Nuovo Cimento B*, only to have it repeatedly rejected.[5] Indeed, in the late 1960s, it was known that a number of prominent journals

[3] Vaihinger also argued that we act "as if" other minds exist as well, and take this to be part of psychic reality. Vaihinger's *Die Philosophie des Als Ob* was first published in 1913, though it was probably written some years earlier [46, 47].

[4] Vaihinger is cited in Zeh's paper 'On the Problem of Conscious Observation in Quantum Mechanical Description' circulated in the *Epistemological Letters* in 1981, which was later revised slightly and published in *Physics Foundation letters* ([59], p. 4).

[5] The reason cited was that "quantum theory does not apply to macroscopic objects". Indeed according to Zeh, one reviewer commented that "This paper is completely senseless" [51].

adopted a strict editorial policy with regard to submissions dealing with foundations of quantum mechanics.[6] John Clauser recalled that during the 1960s "a very powerful secondary stigma began to develop within the physics community towards anyone who sacrilegiously was critical of quantum theory's fundamentals… Any student who questioned the theory's foundations, or, God forbid, considered studying the associated problems as a legitimate pursuit in physics was sternly advised that he would ruin his career by doing so" ([13], p. 72). This was precisely the advice that Zeh received. Zeh [62] later recalled, "An influential Heidelberg Nobel Prize winner frankly informed me that any further activities on this subject would end my academic career!" The Nobel Prize winner was J. Hans D. Jensen.[7] As Zeh later confirmed in an interview in 2008:

> When I wrote this paper … [Jensen] said he did not understand that, and he sent a copy unfortunately to Rosenfeld in Copenhagen… He wrote a letter to Jensen which Jensen never showed me where he must have been very cynical about what I had said, and I remember that Jensen told that to some other colleagues, then when I noticed they were talking about them, they were chuckling. But he never told me precisely what was in this letter… Then Jensen told me that I should not continue this work, and so then our relationship deteriorated [52].

The letter from Rosenfeld to Jensen that Zeh mentions here was recently discovered in the Rosenfeld Papers contained in the Niels Bohr Archive. It is dated February 14, 1968.

> I established a rule in my life never to step on anybody's toe, but a preprint written by a certain Dr 'Toe' [Zeh, in German] from your institute that I have received makes me deviate from this rule. I have all the reasons in the world to assume that such a concentrate of wildest nonsense is not being distributed around the world with your blessing, and I think to be of service to you by directing your attention to this calamity.[8]

---

[6] As Clauser recalls, "even as late as the early and mid seventies, whenever a manuscript discussing the foundations of quantum mechanics (and especially one discussing hidden variables) was submitted to either the *Physical Review* or *Physical Review Letters*, editor Samuel Goudsmit would, in turn, enclose a one-page APS policy statement along with it to the manuscript's referee. That policy, in essence, urged the referee summarily to reject any paper on this subject, unless the paper was both mathematically based *and* gave new quantitative experimental predictions. Bohr's response to EPR certainly could not have been published under Goudsmit's stated criteria" ([13], p. 72).

[7] In 1963 Jensen shared the Nobel prize with Maria Göppert-Mayer for their work on the shell nuclear model.

[8] The original German reads: "Ich mach es zu einer Lebensregel, so weit vermeidlich auf keinen Zeh zu treten, aber der Empfang eines von einem gewissen Dr. Zeh aus Ihrem Institut verfassten preprint veranlasst mich von dieser Regel abzuweichen. Ich habe allen Grund anzunehmen, dass ein solches Konzentrat wildesten Unsinnes nicht mit Ihrem Segen in die Welt verbreitet ist, und ich glaube Ihnen von Dienst zu sein, indem ich Ihre Aufmerksamkeit auf dieses Unglück richte". L. Rosenfeld to J. H. D. Jensen, 14 Feb 1968. A series of further letters were exchanged between Rosenfeld and Jensen on the matter: Jensen to Rosenfeld, 1 March 1968; Rosenfeld to Jensen, 6 March 1968; Jensen to Rosenfeld, 10 Apr 1968; 9 May 1968; Rosenfeld to Jensen, 25 Apr 1968. Rosenfeld Papers, Niels Bohr Archive, Copenhagen. Anja Jacobsen and Felicity Pors recovered these letters The German translation is mine. Quoted in Freire [19], p. 306.

This was certainly not the first time Rosenfeld had meddled in such matters by writing to the head of an Institute regarding the pretensions of a younger colleague. Just two years earlier, Rosenfeld had written to Abdus Salam, who was the Head of the International Centre of Theoretical Physics in Trieste, regarding a preprint written by a young Austrian physicist, K. S. Tausk that had come into his hands on the measurement problem.[9] Rosenfeld could be merciless on such occasions, and by the 1960s he seems to have taken it upon himself to police the orthodoxy. Zeh was certainly not the only physicist during this period to incur the wrath of Rosenfeld.

After sending a preliminary version of the paper to Eugene Wigner for comment, Wigner suggested that he should try to publish the paper with some revisions in the newly formed journal *Foundations of Physics* which would be more receptive to alternative approaches to quantum physics.[10] The paper was eventually accepted with some revisions and published in the inaugural volume of *Foundations of Physics*. It was later included among the seminal papers in the foundations of quantum mechanics that appeared in Wheeler and Zurek's *Quantum Theory and Measurement*, published in 1983. As Olival Freire has pointed out, Wigner was instrumental in encouraging further work in different directions on the foundations of quantum mechanics [17, 18]. Wigner also saw to it that Zeh was invited to a conference on the Foundations of Quantum Mechanics at the International School of Physics "Enrico Fermi" in Varenna in the summer of 1970, which was organised by Bernard d'Espagnat [15].[11] There Zeh presented a paper on the quantum theory of measurement based on his earlier considerations. Here he again emphasised that the description of a measurement by means of the Schrödinger wave equation is "unrealistic" because "it assumes a macroscopic system can be isolated" ([56], p. 269). Zeh was explicitly critical of the Copenhagen interpretation and reiterated his own version of the Everett interpretation, according to which "consciousness appears to be connected with dynamically stable branches of the universal wave function" ([56], p. 270). DeWitt, who also presented a paper on the many worlds view at the same conference, welcomed Zeh's work as an important contribution to the Everett interpretation.

Throughout the 1970s Wigner remained open to the possibility of a fundamental revision of quantum mechanics, and he became increasingly convinced that "far more

[9] Rosenfeld described Tausk's preprint, entitled on 'Relation of Measurement with Ergodicity, Macroscopic Systems, Information and Conservation Laws', as "such incredible thrash [sic] that I hardly could believe my eyes when I read it. I feel that I ought to write you about it in the event that (as I hope) this masterpiece has just escaped your attention.... The author is, I suppose, very young and inexperienced; one good turn you could do him, since you presumably know him better than I do, would be to represent that before blandly assuming that the trivialities which fill his paper could have been overlooked by such people as Niels Bohr and Heisenberg, he might perhaps reflect that he could be the one who misses the point" (Rosenfeld to Salam, Copenhagen, 20 Sep 20 1966. Quoted in Freire [19], p. 182).

[10] The editorial board was made up of physicists with a range of differing views on quantum mechanics including Henry Margenau, Louis de Broglie, David Bohm, Wigner himself, and Vladimir Fock, many of whom were critical of the orthodox view.

[11] The participants of the conference presented a wide range of views on the interpretation of quantum mechanics. Among them were Wigner, Jauch, Yanase, d'Espagnat, Prosperi, Bell, Shimony, DeWitt, Zeh, Selleri and Bohm.

fundamental changes will be necessary" (Wigner to Shimony, 12 Oct 1977 in Freire [19], p. 167). He encouraged other physicists to pursue a range of alternative solutions to the measurement problem in the 1960s and 70 s, and in doing so, "helped to legitimize heterodoxy on this subject" ([19], p. 167). Abner Shimony, who completed his Ph.D. in physics under Wigner's supervision in Princeton, later remarked that "Wigner's authority as one of the great pioneers and masters of quantum mechanics" and his encouragement and support was instrumental in his own "decision to devote much research effort to these problems" ([43], p. xii). Wigner was critical of some aspects of Zeh's work, and objected to the "extrapolation of the superposition principle to the entire universe", on the grounds that experimental evidence for the principle was limited almost entirely to microscopic systems. Yet, he was adamant that Zeh was "unquestionably correct" in his observation "that it is practically impossible to isolate a macroscopic object from its surroundings" ([49], p. 380).[12] Wigner seems to have been acutely aware of the difficulties faced by younger researchers who had an interest in foundations. As he put it in a letter to Joseph Jauch, "I am less concerned about myself than about other people who are much younger than I am and whose future careers such statements may hurt".[13]

Relations between Zeh and Jensen deteriorated in Heidelberg, and Zeh effectively became a *persona non grata*.[14] Yet he continued to pursue his work on quantum mechanics, publishing another article in *Foundations of Physics* in 1973 entitled 'Towards a Quantum Theory of Observation'. Elaborating on his earlier work, Zeh defended the Everett interpretation, and argued that it may even be possible that there may be dynamical consequences of this interpretation which differ from collapse models. Zeh investigated the possibility that in certain cases, the world components (or Everett branches) which are not observed (but which we may assume still *exist*), could interact with one another in the global superposition, which in turn would have dynamical consequences for the time dependent behaviour of the "observed component" [57]. Physicists, for the most part, tended to ignore Zeh's work. *Foundations of Physics* did not enjoy a reputation as a 'respectable' physics journal, and while there was some interest in new approaches to quantum mechanics, this remained a rather marginalised area within the broader physics community.

Even when physicists did engage in debate over the interpretation of quantum mechanics in the 1970s, these discussions typically focused on hidden variables or proposals to modify the unitary Schrödinger dynamics with physical collapse models. As Zeh recalls, many of those physicists 'who were interested in the foundation of quantum theory saw in decoherence an unwelcome competition to their own ideas,

[12] At the 1970 Varenna conference Wigner included Zeh's proposal among the six approaches to the measurement problem ([48], pp. 17–18).

[13] Eugene Wigner to Josef M. Jauch, 6 September 1966. Wigner Papers, box 94, folder 7. Quoted in Freire [19], p. 250.

[14] Zeh recalls that whenever his attention turned to 'fundamental questions' in Heidelberg, Jensen told him to 'let sleeping dogs lie' [52]. The transcript of the interview contains the phrase, 'Don't wake up sleeping dogs', which is a direct translation of the German idiom '*Schlafende Hunde soll man nicht wecken*' but I have taken the liberty of rephrasing this in the more commonly used expression in English.

since they usually wanted to *change* quantum theory, while we maintained there is no reason to do so if you accept Everett' [51]. Having decided that his career prospects had been irreparably damaged, Zeh decided to continue to work on the foundations of quantum mechanics in relative isolation, and ceased publishing in physics journals. He instead chose to circulate his ideas by means of 'unofficial publications', notably *The Epistemological Letters* of the Ferdinand-Gonseth Association in Biel (Switzerland), which provided a discussion forum for many prominent dissenting voices in the quantum debate, among them Bell, Bohm and Popper.[15]

Zeh would later label this period 'the dark ages of decoherence' ([61], p. 10). This period only came to end with the publication of two important papers by Wojciech Zurek in *Physical Review*, which treated the problem of the preferred basis and the derivation of superselection rules from the perspective of environment-induced decoherence [64, 65]. In his 1981 paper Zurek spelt out that "interaction with the environment is the key feature that distinguishes the here-proposed model of the apparatus from the manifestly quantum systems" as a result of which "the apparatus cannot be observed in a superposition of the pointer-basis states because its state vector is being continuously collapsed" ([64], p. 1522). These papers provided the most important impetus for the renewed interest in the dynamical consequences of environmental entanglement for the emergence of 'classicality' in quantum systems. Zurek was careful to point that one could not answer the "insoluble question" of "what causes the collapse of the apparatus-environment combined wave function" ([64], p. 1517). However, "the question 'what mixture does the wave packet collapse into?' acquires a definite answer" ([64], p. 1522). This also provides a dynamical explanation of superselection rules, which are usually postulated for the purpose of eliminating superpositions of the pointer-basis states ([65, 70], pp. 87–89). Zurek coined the term 'environment-induced superselection' for this feature of quantum decoherence, which he later termed 'einselection'.

As noted earlier, Zeh's work had found very little support from colleagues in Heidelberg. By contrast, Zurek's early interest in decoherence, while he was in Austin and then later at Caltech, was encouraged and supported by John Wheeler, under whose direction Zurek undertook postdoctoral research in 1979–1981. During this period Wheeler engaged in many stimulating discussions on quantum theory with Zurek. As Zurek recalls, "Wheeler was absolutely essential in defining the problem, or rather, the whole set of problems" which were critical in the way he approached environment-induced decoherence [63]. We find Wheeler frequently acknowledged in many of Zurek's papers on decoherence spanning more than two decades.[16]

This brings to light a crucially important difference in the social and intellectual contexts in which Zeh and Zurek had formulated their early ideas on decoherence.

---

[15] In 1981 Zeh informally circulated a paper on 'The Problem of Conscious Observation in Quantum Mechanical Description', in which he presented a 'multi-consciousness interpretation' which appears to be one of the earliest formulations of the many-minds interpretations of quantum mechanics.

[16] It is quite likely that he was also important in ensuring that Zurek was invited to participate in conferences where his ideas on decoherence could be discussed, in an environment when foundational questions in quantum mechanics were usually reserved for more senior colleagues.

As Zurek put it in a letter to Fritz Rohrlich in 1982, "I credit John Wheeler with my interest in the fundamental issues raised by the quantum theory of measurement" (Zurek to Fritz Rohrlich, 11 November 1982. Quoted in Kaiser [30], p. 313). After a particularly inspiring visit to Wheeler's summer cottage in Maine, Zurek wrote to Wheeler "Every day I come back to our discussions on physics and philosophy" (Wocjek Zurek to John Wheeler, 10 August 1979. Quoted in Kaiser [30], p. 313). This was a privilege not afforded to Zeh. As David Kaiser explains:

> Wheeler's reputation helped to shield the young students from other professors' disdain for such philosophical patter. Aside from Wheeler's strong backing, Zurek recalls the general attitude in the hallways at that time: graduate students like himself and Wootters had received a very obvious and loud message that thinking seriously about foundations was a waste of time and a detriment to one's career. Nonetheless Wheeler and his small circle of students soldiered on. Wheeler brought in a steady stream of visiting scholars to keep discussion fresh. He also organized a brand new seminar on quantum measurement … an entire semester spent puzzling over paradoxes like Schrödinger's cat… The following year, PhD in hand, Zurek helped Wheeler coteach the seminar ([30], pp. 216–7).[17]

This underscores the crucial importance of the support of senior colleagues for younger researchers wishing to pursue research in the foundations of quantum mechanics. Zurek recalled that after his 1981 pointer basis paper, he was "pleasantly surprised" to receive several invitations to attend meetings, where he encountered was an "atmosphere of intellectual fellowships and excitement", which he had not been expecting ([42], p. 37).

4 Decoherence and Its Interpretation

There was, however, a further important difference in the way Zeh and Zurek each approached the foundations of quantum mechanics. Their different attitudes to interpretational problems reflect the different origins, motivations and philosophical perspectives that guided their early work on decoherence. Zeh's early papers originated in his early interest in the dynamical consequences of the Everett interpretation.[18] Decoherence, as he would often later insist, does "not by itself solve the measurement problem" [62]. Zurek, on the other hand, steered away from 'interpretational' issues in his early papers. In the dynamical models developed by Zurek, Caldeira, and Leggett in the 1980s, the study of decoherence was divorced from the interpretational context in which Zeh had first considered it. As Zeh remarked: "The emphasis on 'decoherence by itself' … became a concept only during the '80s" [51]. Thus, in contrast to Zeh's views that decoherence forces a radical departure from

[17] This readings set for the class would eventually serve as the basis for the 800 + page edited volume *Quantum Theory and Measurement*, which Zurek edited with Wheeler.

[18] In his *Philosophy of Quantum Mechanics* published in 1974, Max Jammer cited [55] paper as one of the few contributions to the measurement problem sympathetic to the many-worlds interpretation of quantum mechanics ([24], p. 519).

the Copenhagen orthodoxy, Zurek advanced the view that decoherence is compatible with, or even provides a confirmation of, many of Bohr's insights into quantum mechanics. By attending to the different interpretive frameworks in which Zeh and Zurek presented their work on decoherence that we can better understand why Zeh's work was largely ignored, while Zurek's work attracted considerable interest. Put simply, in the case of decoherence, interpretation influenced reception.

Zurek's early thinking on the role of the environment in decoherence can be traced back to a term paper he had written as a doctoral student for Wheeler's class on the Einstein-Bohr debate involving the double-slit experiment. At the 1927 Solvay conference Einstein argued that by measuring the recoil imparted to the measuring apparatus by the photon passing through one the slits, one could determine the trajectory of the individual photon in the construction of the interference pattern. Bohr responded by pointing out that Einstein had failed to take into account the indeterminacy in the momentum of the measuring apparatus which would preclude a precise determination of the path of the photon ([4], pp. 211–224). Zurek thus set about "to examine Bohr's idea in detail" by determining "exactly what interference pattern is produced when we attempt to determine the slit through which each photon passes" ([50], p. 474). Encouraged by Wheeler, the paper was revised in collaboration with Bill Wootters and submitted to the journal *Physical Review* where it was published in 1979. The title of the paper: 'Complementarity in the double-slit experiment: quantum nonseparability and a quantitative statement of Bohr's principle', makes clear the extent to which the authors saw their work as an extension of Bohr's views.

However, whereas Bohr had treated the measuring apparatus *classically*, albeit with the limitations imposed by the uncertainty relations, Wooters and Zurek represented "the plate (with recoil) as a *quantum–mechanical* harmonic oscillator. The photon and the plate, having once interacted, become *nonseparable parts of a single quantum–mechanical system*. This forces us to consider the effect of our measurement of the plate on the photon wave function, and consequently on the interference pattern produced by these photons" ([50], p. 474 emphasis added). This allowed a quantitative analysis of the two-slit experiment. In the concluding section of the paper, the authors proposed a reformulation of Bohr's principle of complementarity "in terms of an inequality which sets the limit on the amount of retrievable information about the photon's paths (photon-particle) for an assumed sharpness of the interference pattern (photon-wave)" ([50], p. 474). In providing a quantum–mechanical treatment of the measuring instrument, Wooters and Zurek departed from Bohr's insistence on the primacy of classical concepts. However in a different sense, the paper can also be read as a *confirmation* of Bohr's intuition that only through the interaction with the measuring apparatus does the quantum system become describable by means of classical concepts. Far from constituting an attempt to contradict Bohr, for Zurek, the analysis of complementarity formed an important point of departure in his thinking about the role of the environment in the decoherence of quantum systems.

Zurek's approach to quantum entanglement thus differed from Zeh's. Zurek concentrated on the local dynamics of the reduced density matrix rather than sweeping interpretational questions. In a symposium on the foundations of quantum mechanics in 1981, Zurek explained that the "problem of the preferred basis which arises in the context of quantum theory of measurement" is "virtually independent of the interpretation of quantum theory" ([70], p. 85). Zurek's approach was to focus on dynamical problems, leaving interpretational issues largely to one side. It was his hope that through a deeper understanding of the process of decoherence in quantum mechanics, one could eventually shed important new light on some of the foundational questions in the theory. As he would later put it:

> I think the whole point of the paper (and more broadly, of my approach to decoherence) was that I could say things that were relevant to foundational questions and that followed directly from quantum theory, without any interpretational baggage attached. (I think that was one difference with Dieter Zeh, who was more interested in interpretational aspects.) And that – using decoherence as a vantage point – it was possible to do calculations that had implications for experiments, and make statements that were "interpretationally non-committal" [63].

Note the contrast with Zeh's view:

> I am indeed surprised about the indifference of most physicists regarding the potential consequences of decoherence [for the interpretation of quantum mechanics] …, since this concept arose as a by-product of arguments favouring … an Everett-type interpretation. In contrast to the Copenhagen interpretation, which insists on fundamental classical concepts … [decoherence rests on an interpretation of] the wave function as a complete and universal representation of reality ([61], p. 1).

As these passages show, the attitudes of both physicists to the relevance of decoherence for the interpretation of quantum mechanics reveal their own individual experiences of the history of decoherence and its conceptual development. For Zeh, decoherence had emerged from considerations in nuclear physics which led him to explore the dynamical consequences of the assumption of a universal wave function (essentially an Everett-type interpretation), whereas Zurek's work on decoherence had originated in his treatment of the two-slit experiment and the dynamical consequences of the interaction of a measuring apparatus and the environment. The different attitudes to the interpretation of decoherence are, to some extent, expressions of the 'hidden historicity'—to use Ulrich Röseberg's term—of the development of their theoretical insights [40].

The 1980s witnessed a renewed interest in the dynamics of entanglement and its consequences for decoherence of quantum systems. In particular, the papers of Caldeira and Leggett were instrumental in the development of new models based on quantum Brownian motion, for the study of the effect of the environment in dissipating quantum interference [7–9, 32].[19] In 1984 Zurek widely circulated a paper, 'On the Reduction of the Wavepacket: How Long does it Take?", in which

[19] These models typically used a master equation to describe the dynamics of a particle moving in one dimension, interacting with an environment represented by independent harmonic oscillators in thermal equilibrium. Such models could be used to formulate a dynamical equation for the reduced density matrix representing the local quantum sub-system. While these early models marked important theoretical progress in the field, they were applicable only in highly idealized situations, such as the short wavelength limit which set an upper bound to the decoherence rate, and the high temperature limit. It was not until the later papers by Unruh and Zurek [45] and Hu et al. [22], which modelled the environment as a massless scalar field, that a generalized master equation for deriving the decoherence rate of a superposition of wave packets for a single particle was formulated.

he drew on the Caldeira-Leggett master-equation to calculate the time rate of the vanishing of interference terms due to decoherence effects [66]. The results showed that decoherence occurs incredibly rapidly—of the order of 10^{40} times faster than the 'relaxation time (the time a system in interaction with its environment takes to reach thermal equilibrium)! For all practical purposes, the decoherence effect is "virtually instantaneous".

The appearance of Zurek's papers in the early 1980s also reignited Zeh's interest in pursuing the decoherence program, and sensing that the time was now right "to improve things" he agreed to take on Erich Joos as a doctoral student [51]. The subject of Joos' thesis was a study of the dynamics of decoherence, specifically through a comparison of master equations with the Zeno effect, which had up until that time been claimed to require a collapse of the wave function.[20] In 1985 Joos and Zeh published a seminal paper expanding on the theoretical work Joos had carried out in his dissertation, in which they showed that the scattering of photons and air molecules leads to the rapid decoherence, which continuously 'localises' a wave packet, thereby preventing it from dispersing as predicted by the Schrödinger equation for a single particle [28]. In this way the authors drew the conclusion that "the 'classical' properties of macroscopic systems' have "their origin in the nonlocal character of quantum states" (p. 223). The paper presented one of the first models of scattering-induced decoherence using master equations derived from the density matrix. As Joos would put it later: "'Particles' appear localised in space not because there are particles, but because the environment continually measures position. The concept of a particle seems to be derivable from the quantum concept of state" ([26], p. 12). However, Zeh and Joos steered clear of any interpretational statements in their joint paper. Indeed Zeh recalls that he advised Joos to concentrate on the dynamical part of the problem "without ever talking about Everett", and focus his attention on the consequences for the reduced density matrix [52]. Indeed after Zurek's papers appeared in 1981/2, Zeh deliberately refrained from mentioning Everett, partly so as not to jeopardize Joos' academic career [51].

Zurek's classic article 'Decoherence and the Transition from Quantum to Classical' in the October 1991 in *Physics Today* is often seen as marking a turning point in the recognition of decoherence among the wider community of quantum physicists. There Zurek took a somewhat cautious, if not ambivalent, stance on the question of interpretation, arguing that: "Decoherence is of use within the framework of either of the two interpretations: it can supply a definition of the branches in Everett's Many Worlds Interpretation, but it can also delineate the border that is so central to Bohr's point of view" ([67], p. 44). According to Zurek: "The key feature of the Copenhagen Interpretation is the dividing line between quantum and classical. Bohr emphasized that the border must be mobile" ([67], p. 36). Decoherence provides a dynamical answer to why this is so. As he later put it:

[20] The 'quantum Zeno effect' was the term first coined by George Sudarshan and Baidyanaith Misra for the situation in which an unstable particle, if observed continuously, does not undergo unitary dynamical evolution but remains 'frozen'. The term now denotes the situation where time evolution is suppressed by interactions with the environment, scattering of particles, stochastic fields, etc.

> The role of decoherence is to establish a boundary between quantum and classical. The boundary is in principle moveable, but in practice largely immobilized by the irreversibility of the process of decoherence… The equivalence between 'macroscopic' and 'classical' is then validated by the decoherence considerations, but only as a consequence of the practical impossibility of keeping objects which are macroscopic perfectly isolated ([68], p. 311).

In this way, for Zurek contended that decoherence was able to 'explain' the classical-quantum divide, in a manner which "occurs in complete accord with Bohr's 'Copenhagen Interpretation'" ([68], p. 311). But in same breath, he contended that the "interpretation that emerges from these considerations is obviously consistent with Everett's 'Relative State' point of view" ([68], p. 311). This rapprochement of Bohr and Everett would become characteristic of Zurek's work. As he wrote in another paper: "*In spite of the Everett-like framework of this discussion, the picture that emerges in the end—when described from the point of view of the observer—is very much in accord with the views of Bohr*" ([69], pp. 89–90 emphasis mine). For Zeh, this nod to Bohr was little more than "political correctness". In his view, Bohr's insistence of the primacy of classical concepts was diametrically opposed to the very possibility of a dynamical understanding of the emergence of classicality. In the end, disagreements of this kind may amount to nothing more than differing ways of reading Bohr.

Nevertheless, as the history of quantum mechanics shows, different readings of Bohr were often not mere hermeneutic exercises, but could have important consequences. By aligning oneself with the so-called "Copenhagen interpretation", one could avoid the retributions of defenders of the orthodoxy. Indeed such branding strategies could have dramatic implications for the reception of ideas. H. D. Zeh's early work on decoherence was ignored in the 1970s, largely because of his unapologetic commitment to an Everett-type interpretation. By contrast, Zurek's early work on decoherence, which had the backing of Wheeler, was carried out with a far more conciliatory rhetorical approach to the "Copenhagen interpretation". This, I would suggest, contributed to its greater acceptance. Labelling oneself as "pro-" or "anti-Copenhagen" was thus strategically important for physicists in certain institutional and cultural contexts. The term 'Copenhagen' carried important connotations in the foundations of quantum mechanics, which could make or break a career in physics.

5 Decoherence and the 'New Orthodoxy'

By the 1990s the idea that decoherence can be regarded as a justification or completion of the Copenhagen interpretation began to find favour among a number of physicists. In his book *The Interpretation of Quantum Mechanics* published in 1994 Roland Omnès presented the newly emerging views on decoherence as signifying a new period of "renewal of the conventional interpretation" which "began in the period 1975–1982 with the discovery and understanding of the decoherence effect" has now arisen ([35], p. xii). Such a revival, made possible by the recent developments in theory and experiment, as Omnès explained, mark a crucial "transition between a period when Bell's ideas and questions concerning hidden variables were dominant and a return to the origins—the basic problems first envisioned by Bohr, Heisenberg and Pauli" ([36], p. 69). Here Zeh's contribution is effectively written out of the history of decoherence.[21] Thus for Omnès, after a period of challenges to the orthodox viewpoint, emerging largely from an interest in hidden variables theories and Bell's work in the 1950s and 1960s, "a new effort aiming … at a clarification and a justification of Bohr's interpretation has now followed" ([35], xii). A somewhat similar view is expressed in Jeffrey Bub's *Interpreting the Quantum World* in which he devotes a chapter to what he terms the 'new orthodoxy' in quantum mechanics.

> The 'new orthodoxy' appears to centre now on the idea that the original Copenhagen interpretation has been vindicated by the recent technical results on environmental decoherence. Sophisticated versions of this view are formulated in terms of 'consistent histories' or 'decoherent histories' and trade on features of Everett's 'relative state' interpretation… There seems to be a growing consensus that a modern, definitive version of the Copenhagen interpretation has emerged, in terms of which the Bohr-Einstein debate can be seen as a rather old-fashioned way of dealing with issues that are now more clearly understood. This 'new orthodoxy' weaves together several strands: [including] the physical phenomenon of environment-induced decoherence ([6], pp. 6, 212).

Whether such a consensus did indeed emerge in the 1990s now seems doubtful. It may appear remarkable that an interpretation that purports to weave together elements of decoherence, the consistent histories approach and the Everett interpretation could be called a 'modification', or even a 'vindication' of Bohr's views ([6], p. 212). Joos and Zeh remained adamant that, far from vindicating Bohr, these developments reflect a decisive break with the "orthodoxy of the Copenhagen school" and a "desire to achieve a better understanding of the quantum–classical relation" ([27], p. 54). Murray Gell-Mann and James Hartle also saw these developments in decoherence and the consistent histories approach as pointing to an "extension, clarification, and completion of the Everett interpretation" ([20], p. 306). Yet, it is clear that many physicists now see those very same developments as having contributed decisively to a "modernized version of the interpretation first proposed by Bohr in the early days of quantum mechanics" ([35], p. 498).

Later interpretations of decoherence have in some cases led to clear divisions in the programmatic aims and the interpretational commitments of decoherence theorists. This is clearly evident in the revised second edition of *Decoherence and the Appearance of a Classical World in Quantum Theory* which includes chapters from Joos and Zeh, as well contributions from Claus Kiefer, Domenico Giulini, Joachim Kupsch, and Ion-Olimpiu Stametescu. The book arose from a series of seminars on the foundations of quantum theory organised by Stamatescu, which were held at the *Forschungstätte der Evangelischen Studiengemeinschaft* (The Protestant Institute for Interdisciplinary Research) in Heidelberg, outside of the university. In the Preface

[21] While the Joos-Zeh 1985 paper on the dynamical treatment of scattering and the timescales for decoherence, noteworthy for its avoidance of any strong interpretational commitments, is cited by Omnès in his 1994 book *The Interpretation of Quantum Mechanics*, none of Zeh's early papers are mentioned. Omnès does, however, cite Zeh's early work in his later book *Understanding Quantum Mechanics* ([36], pp. 74, 290).

to the 2003s edition, the authors explained that differences of opinion had arisen regarding the basic conceptions, aims and motivations of decoherence, leading to minor revision:

> For this reason we have rearranged the order of the authors: they now appear in the same order as the chapters, such that *those most closely related to the "early" and most ambitious concept of decoherence* are listed first. The first three authors (Joos, Zeh, Kiefer) agree with one another that decoherence (in contradistinction to the Copenhagen interpretation) allows one to eliminate primary classical concepts, thus neither relying on an axiomatic concept of observables nor on a probability interpretation of the wave function in terms of classical concepts. While the fourth and fifth authors (Giulini and Kupsch) still regard the probability interpretation (expressed by means of expectation values, for example) as basic for an interpretation of the formalism, the sixth author (Stamatescu) critically reviews collapse models, which put all quantum probabilistic aspects into their (novel) dynamics [29], v emphasis added).

The 1990s saw a number of important experimental breakthroughs which confirmed many of the basic assumptions and theoretical predictions made using decoherence models. In 1996 a team at Boulder, Colorado generated a superposition of two spatially separated but localised wave packets of a single trapped $^{9}Be^{+}$ ion and measure the way the superposition was gradually destroyed through its interaction with a external radiation field serving as the 'environment' [34]. In 1997 Zeh commented that reality of the wave function "is now strongly supported by the beautiful experiments with individual atoms in activities (Schrödinger cat states, quantum engineering, phase space tomography etc.)" ([58], p. 449). Yet the experiments on decoherence did not prove as decisive on this point as Zeh had hoped.

In 1996, the experimental group at the École Normale Supérieure in Paris led by Michel Brune, Serge Haroche and Jean-Michel Raimond, published a preliminary letter in which they proposed to study the "disappearance of decoherence" by an experiment involving the manipulation of electromagnetic fields into a Schrödinger cat-like superposition using rubidium atoms [14]. The authors cited the early papers by Zeh and Zurek, as well the more recent publications from Zurek and Unruh. However, in the subsequent papers, which provided further confirmation of the predictions of decoherence models, the references to Zeh disappeared, whereas Zurek's publications continued to be cited extensively along with the [35] book [3, 33, 38, 39].

In 1997, Raimond, Brune and Haroche proposed an experiment to measure the 'reversible decoherence of a mesoscopic superposition of field states'. The authors stated that "the proposed experiment would demonstrate *the essential role of complementarity* in the decoherence process". Further, it was claimed that as the onset of decoherence "becomes increasingly faster with the size of the system" in such a way that "can easily be *interpreted in terms of complementarity*" ([38], p. 1964). Zeh recalls that he was surprised at this turn of events: "At first I expected that these results … would be seen as a confirmation of universal entanglement, and hence Everett, or at least as a severe argument against Copenhagen. But it turned out to be just the opposite—they later regarded entanglement as a form of complementarity!" [51]. When decoherence was taken up by experimentalists in the 1990s, it was often interpreted as providing further confirmation of Bohr's views, or at least, consistent with them.

Zeh was often bemused by these attempts to forge a connection between the dynamics of entanglement and Bohr's original viewpoint. "In order to remain 'politically correct'", he wrote in 2003, "some authors have even *re-defined* complementarity in terms of entanglement, although the latter has never been a crucial element in the Copenhagen interpretation" ([60], p. 28). In a letter to Wheeler in 1980, Zeh had expressed his deep frustration at the tendency of Bohr's followers to see in his utterances, anticipations of later developments: "I expect the Copenhagen interpretation will sometime be called the greatest sophism in the history of science, but I would consider it a terrible injustice if—when someday a solution should be found—some people claim that this is of course what Bohr always meant, only because he was sufficiently vague" (Zeh to Wheeler, 30 October, 1980, [25]). Yet precisely "what Bohr really meant" and what properly constitutes the "Copenhagen interpretation" has been the subject of endless debate and disagreement.[22]

Zeh's pioneering work on the foundations of quantum mechanics, and his contributions to the understanding of decoherence have received increasing recognition in recent years. Fifty years after his first paper on 'On the interpretation of measurement in quantum theory', decoherence has become a thriving area of theoretical and experimental research in the foundations of quantum mechanics, and there is a proliferation of competing interpretations of quantum mechanics. The Everett interpretation has undergone something of a revival in the last thirty years. But Zeh's work and its early reception stands today as a reminder of the difficulties faced by those quantum dissidents who dared to challenge the prevailing orthodoxy in the 1960s and 70s, and the intellectual and social context with which they had to contend.

References

1. G. Bacciagaluppi, The role of decoherence in quantum mechanics, in *Stanford Encyclopedia of Philosophy* ed. by E.N. Zalte. First published Nov 3, 2003; substantive revision Aug 23, 2007 (2007). http://plato.stanford.edu/entries/qm-decoherence/
2. M. Beller, *Quantum dialogue: the making of a revolution* (University of Chicago Press, Chicago, 1999)
3. P. Bertet, S. Osnaghi, A. Rauschenbeutel, G. Nogues, A. Auffeves, M. Brune et al., A complementarity experiment with an interferometer at the quantum-classical boundary. Nature **411**, 166–179 (2001)

[22] Those familiar with the extensive literature which has attempted to make sense of the Copenhagen interpretation will point out that the term 'Copenhagen interpretation' has historically come to denote a range of different interpretations, some of which bear little if any connection with Bohr's original view of complementarity. See Beller [2], Camilleri [11], Faye and Folse [16], Howard [21], Scheibe [41] and Stapp [44]. We should recall that the term 'Copenhagen interpretation' only appeared in the 1950s in connection with the Soviet Marxist critique and emergence of a new wave of alternative approaches to quantum mechanics in the work of physicists like Bohm, de Broglie, Janossy and Vigier and Everett.

4. N. Bohr, Discussion with Einstein on the epistemological problems in atomic physics, in *Albert Einstein: Philosopher-scientist*, vol. 7, ed. by P.A. Schilpp (Open Court, La Salle, 1949), pp. 201–241
5. M. Brune, E. Hagley, J. Dreyer, X. Maitre, A. Maali, C. Wunderlich et al., Observing the progressive decoherence of the "meter" in a quantum measurement. Phys. Rev. Lett. **77**, 4887–4890 (1996)
6. J. Bub, *Interpreting the Quantum World* (Cambridge University Press, Cambridge; New York, 1997).
7. A.O. Caldeira, A.J. Leggett, Influence of dissipation on quantum tunnelling in macroscopic systems. Phys. Rev. Lett. **46**, 211–214 (1981)
8. A.O. Caldeira, A.J. Leggett, Path integral approach to quantum Brownian motion. Physica A **121**, 587–616 (1983)
9. A.O. Caldeira, A.J. Leggett, Influence of damping on quantum interference: an exactly soluble model. Phys. Rev. A **31**, 1059–1066 (1985)
10. K. Camilleri, A history of entanglement: decoherence and the interpretation problem. Stud. Hist. Philos. Mod. Phys. **40**, 290–302 (2009)
11. K. Camilleri, Constructing the myth of the Copenhagen interpretation. Perspect. Sci. **17**, 26–57 (2009)
12. I. Chiorescu, Y. Nakamura, C.J. Harmans, J.E. Mooij, Coherent quantum dynamics of a superconducting flux qubit. Science **21**, 1869–1871 (2003)
13. J. Clauser, Early history of Bell's theorem, in *Quantum [Un]speakables: From Bell to Quantum Information*. ed. by R.A. Bertlmann, A. Zeilinger (Springer, Berlin, 2002), pp. 61–98
14. L. Davidovich, M. Brune, J.M. Raimond, S. Haroche, Mesoscopic quantum coherences in cavity QED: Preparation and decoherence monitoring schemes. Phys. Rev. A **53**, 1295–1309 (1996)
15. B. d'Espagnat, *Foundations of Quantum Mechanics. Proceeding of the International School of Physics "Enrico Fermi"*. (Academic Press, New York, London, 1971)
16. J. Faye, H.J. Folse (eds.), *Niels Bohr and Contemporary Philosophy* (Kluwer Academic Publishers, Dordrecht; Boston, 1994)
17. O. Freire, The historical roots of "foundations of quantum physics" as a field of research (1950–1970). Found. Phys. **34**, 1741–1760 (2004)
18. O. Freire, Orthodoxy and heterodoxy in the research on the foundations of quantum physics: E. P. Wigner's case, in *Cognitive Justice in a Global World. Prudent Knowledge for a Decent Life*, ed. by S. Santos. (Lexington Books, Lanham, MD, 2007), pp. 203–224
19. O. Freire, *The Quantum Dissidents: Rebuilding the Foundations of Quantum Mechanics (1950–1990)* (Springer, Berlin, 2015)
20. M. Gell-Mann, J. Hartle, Quantum mechanics in the light of cosmology, in *Murray Gell-Mann: selected papers*, ed. by H. Fritsch (World Scientific, New Jersey, 2010), pp. 303–325
21. D. Howard, Who invented the "Copenhagen Interpretation"? A study in mythology. Philos. Sci. **71**, 669–682 (2004)
22. B.L. Hu, J.P. Paz, Y. Zhang, Quantum Brownian motion in a general environment: exact master equation with nonlocal dissipation and coloured noise. Phys. Rev. D **45**, 2843–2861 (1992)
23. E. Il'ichev, N. Oukhanski, A. Izmalkov, T. Wagner, M. Grajcar, H.-G. Meyer, A.Y. Smirnov, A.M. van den Brink, M.H. Amin, Continuous monitoring of Rabi oscillations in a Josephson flux. Phys. Rev. Lett. **91**, 097906.1–097906.4 (2003)
24. M. Jammer, *The philosophy of quantum mechanics: the interpretations of quantum mechanics in historical perspective* (Wiley & Sons, New York, 1974)
25. A. John, Wheeler Papers. Series II, Box Wo-Ze, folder Zeh. American Philosophical Society, Philadelphia.
26. E. Joos, Quantum theory and the appearance of the classical world, in *New Techniques and Ideas in Quantum Measurement Theory*. ed. by D. Greenberger (New York Academy of Sciences, New York, 1986), pp. 6–13
27. E. Joos, The emergence of classicality from quantum theory, in *The re-emergence of emergence: the emergentist hypothesis from science to religion*, ed. by P. Davies, P. Clayton (Oxford University Press, 2006), pp. 53–77

28. E. Joos, H.D. Zeh, The emergence of classical properties through interaction with the environment. Zeitschrift für Physik B Condensed Matter **59**, 223–243 (1985)
29. E. Joos, H.D. Zeh, C. Kiefer, D. Giulini, J. Kupsch, I.O. Stametescu, *Decoherence and the Appearance of a Classical World in Quantum Theory*, 2nd ed. (Springer, Berlin; New York, 2003).
30. D. Kaiser, *How the Hippies Saved Physics: Science, Counterculture, and the Quantum Revival* (Norton, W. W, 2011)
31. O. Kübler, H.D. Zeh, Dynamics of quantum correlations. Ann. Phys. **76**, 405–418 (1973)
32. A.J. Leggett, The superposition principle in a macroscopic system, in *Foundations of Quantum Mechanics in the Light of New Technology. Selected Papers from the Proceedings of the First Through Fourth International Symposia on Foundations of Quantum Mechanics*, ed. by S. Nakajima, Y. Murayama, A. Tonomura (World Scientific, Singapore, New Jersey, London, Hong Kong, 1996), pp. 27–35
33. X. Maitre, E. Hagley, J. Dreyer, A. Maali, C.W.M. Brune, J.M. Raimond et al., An experimental study of a Schrödinger cat decoherence with atoms and cavities. J. Mod. Opt. **44**, 2023–2032 (1997)
34. C. Munro, D.M. Meekhof, B.E. King, D.J.A. Wineland, "Schrödinger's cat" superposition state of an atom. Science **272**, 1131–1136 (1996)
35. R. Omnès, *The Interpretation of Quantum Mechanics* (Princeton University Press, Princeton, N.J., 1994)
36. R. Omnès, *Understanding Quantum Mechanics* (Princeton University Press, Princeton, N.J., 1999)
37. S. Osnaghi, F. Freitas, O. Freire Jr., The origin of the Everettian heresy. Stud. Hist. Philos. Mod. Phys. **40**, 97–123 (2009)
38. J.M. Raimond, M. Brune, S. Haroche, Reversible decoherence of a mesoscopic superposition of field states. Phys. Rev. Lett. **79**, 1964–1967 (1997)
39. J.M. Raimond, M. Brune, S. Haroche, Manipulating quantum entanglement with atoms and photons in a cavity. Rev. Mod. Phys. **73**, 565–582 (2001)
40. U. Röseberg, Hidden historicity: the challenge of Bohr's philosophical thought, in *Niels Bohr and Contemporary Philosophy*. ed. by J. Faye, H. Folse (Kluwer, Dordrecht, 1994), pp. 325–344
41. E. Scheibe, *The Logical Analysis of Quantum Mechanics* (Pergamon Press, Oxford; New York, 1973).
42. M. Schlosshauer (ed.), *Elegance and Enigma: The Quantum Interviews* (Springer, Berlin, 2011)
43. A. Shimony, *Search for a Naturalistic World View* (Cambridge University Press, Cambridge, 1993)
44. H. Stapp, The Copenhagen Interpretation. Am. J. Phys. **40**, 1098–1116 (1972)
45. W.G. Unruh, W.H. Zurek, Reduction of a wavepacket in quantum Brownian motion. Phys. Rev. D **40**, 1071–1094 (1989)
46. H. Vaihinger, *Die Philosophie des Als Ob. System der Theoretischen, Praktischen und Religiosen Fiktionen der Menschheit auf Grund Eines Idealistischen Positivismus, mit einem Anhang über Kant und Nietzsche*. Leipzig: F. Meiner (1924)
47. H. Vaihinger, *The Philosophy of 'As If': A System of the Theoretical, Practical and Religious Fictions of Mankind* (C. K. Ogden, Trans.). (Kegan Paul Trench Trubner, London, 1924)
48. E. Wigner, The subject of our discussions, in *Foundations of Quantum Measurement*. ed. by B. d'Espagnat (Academic Press, New York; London, 1971), pp. 1–19
49. E. Wigner, Epistemological perspective on quantum theory, in *Contemporary Research in the Foundations and Philosophy of Quantum Theory*. ed. by C.A. Hooker (Reidel, Dordrecht, 1973), pp. 369–385
50. W.K. Wootters, H.D. Zurek, Complementarity in the double-slit experiment: quantum nonseparability and a quantitative statement of Bohr's principle. Phys. Rev. D **19**, 473–484 (1979)
51. H.D. Zeh, Personal communication (22, 24 May, 2, 24 June, 21 November, 2008)
52. H.D. Zeh, Interview with Zeh conducted by Fábio Freitas, 25 July 2008. Audio recording and transcript deposited at the American Institute of Physics.

53. H.D. Zeh, *Probleme der Quantentheorie.* Unpublished manuscript. (1967). https://protect-au.mimecast.com/s/KLq2CJypvAfqPDkj1cVnxwE?domain=thp.uni-koeln.de
54. H.D. Zeh, Symmetrieverletzende Modellzustände und kollektive Bewegungen. Zeitschrift für Physik A Hadrons and Nuclei **202**, 38–48 (1967)
55. H.D. Zeh, On the interpretation of measurement in quantum theory. Found. Phys. **1**, 69–76 (1970)
56. H.D. Zeh, On the irreversibility of time and observation in quantum theory, in *Foundations of Quantum Mechanics*. ed. by B. d'Espagnat (Academic Press, New York; London, 1971), pp. 263–273
57. H.D. Zeh, Toward a quantum theory of observation. Found. Phys. **3**, 109–116 (1973)
58. H.D. Zeh, What is achieved by decoherence? in *New Developments on Fundamental Problems in Quantum Physics*, ed. by M. Fererro, A. van der Merwe. (Kluwer, Dordrecht, 1997), pp. 441–451. https://protect-au.mimecast.com/s/KLq2CJypvAfqPDkj1cVnxwE?domain=thp.uni-koeln.de
59. H.D. Zeh, The problem of conscious observation in quantum mechanical description. Found. Phys. Lett. **13**, 221–233 (2000). (A early version of this paper was originally circulated in 1981 by means of the Epistemological letters of the Ferdinand-Gosneth Association in Biel (Switzerland). http://arxiv.org/abs/quant-ph/9908084)
60. H.D. Zeh, Basic concepts and their interpretation, in *Decoherence and the Appearance of a Classical World in Quantum Theory*, ed. by E. Joos, H.D. Zeh, O. Kiefer, D. Giulini, J. Kupsch, I.O. Stamatescu (Springer, Berlin; New York, 2003), pp. 7–40
61. H.D. Zeh, Roots and fruits of decoherence, in *Quantum Decoherence*, ed. by B. Duplantier, J. M. Raimond, V. Riveasseau. (Birkhäuser, 2006), pp. 151–175. http://arxiv:quant-ph/0512078v2
62. H.D. Zeh, How decoherence can solve the measurement problem (2007). https://protect-au.mimecast.com/s/KLq2CJypvAfqPDkj1cVnxwE?domain=thp.uni-koeln.de.
63. W.H. Zurek, Personal communication (4 July 2008)
64. W.H. Zurek, Pointer basis of quantum apparatus: into what mixture does the wave packet collapse? Phys. Rev. D **24**, 1516–1525 (1981)
65. W.H. Zurek, Environment-induced superselection rules. Phys. Rev. D **26**, 1862–1880 (1982)
66. W.H. Zurek, Reduction of the wave packet: how long does it take? in *Frontiers of Nonequillibrium Statistical Mechanics*, ed. by G.T. More, M.O. Scully. (Plenum Press, New York, 1986), pp. 145–149. http://arxiv.quantum-ph/0302044v1.
67. W.H. Zurek, Decoherence and the transition from quantum to classical. Phys. Today **44**(October), 36–44 (1991)
68. W.H. Zurek, Preferred states, predictability, classicality and the environment-induced decoherence. Progress Theoret. Phys. **89**, 281–312 (1993)
69. W.H. Zurek, Zurek replies: negotiating the tricky border between quantum and classical. Phys. Today **46**(April), 84–90 (1993)
70. W.H. Zurek, Pointer basis, and inhibition of quantum tunnelling by environment-induced superselection, in *Foundations of Quantum Mechanics in the Light of New Technology. Selected Papers from the Proceedings of the First Through Fourth International Symposia on Foundations of Quantum Mechanics*, ed. by S. Nakajima, Y. Murayama, A. Tonomura. (World Scientific, Singapore, New Jersey, London, Hong Kong, 1996), pp. 85–93

Emergence of the Classical World from Within Our Quantum Universe

Wojciech Hubert Zurek

I met Dieter Zeh for the first time at a mini-conference organized by John Wheeler in Austin, Texas, circa 1983. I was then at Caltech, and I was curious about Zeh. I found out about him through Eugene Wigner who visited Wheeler in Texas around the time I submitted the manuscript of my "pointer basis" paper to *Physical Review*.

My inspiration for the pointer basis came from the papers on non-demolition measurements involved in gravity wave detection (see e.g. Caves et al. [1]), but Wigner sensed affinity with Zeh's views, and suggested I send him my manuscript. Dieter sent back a copy of Zeh [2] with a note "look at page 268" scribbled in pencil (I still have it). I read it (p. 268 starts with a discussion of the loss of the interference pattern due to scattered radiation based on the discussion of the double slit experiment in Feynman's "Lectures") and I also read his 1970 *Foundations of Physics* paper. What came through was (in addition to Zeh's allegiance to Everett) conviction that macroscopic systems are difficult to isolate, and so Schrödinger equation is not valid. I cited Zeh's Fermi School lecture and his *Foundations of Physics* paper [3] in my early work. (The citation to Zeh [2] was added "in proofs" to my 1981 paper that introduced pointer basis.)

I remember that on our first meeting Zeh noted, jokingly, that if we ever wrote a paper together, he would be, alphabetically, the first author. I was always sorry we never did, but I also have a feeling this might have been a frustrating experience for both of us: Dieter was a committed Everettian, and I was resolved to remain open minded about the interpretation. I felt Copenhagen and Everett's interpretations were both incomplete, and more calculating and thinking unconstrained by the allegiance to either was the best way to make progress.

Dieter drew attention to the fact that macroscopic systems can never be isolated from the environment, and realized that, therefore, one should not expect them to follow unitary dynamics generated by Schrödinger equation. This was a valuable insight.

Given Zeh's adherence to Everett's views (obvious in his earliest papers), I was surprised he did not address the gaping hole in Everett's interpretation—the selec-

W. H. Zurek (✉)
Theory Division, MS B213, LANL Los Alamos, Los Alamos, NM 87545, USA

C. Kiefer (ed.), *From Quantum to Classical*, Fundamental Theories of Physics 204,
https://doi.org/10.1007/978-3-030-88781-0_2

tion of the preferred states (or the origin of the "branches"). He mentioned it on occasion [4], and even realized that stability should be a criterion. However, in the end he seemed content with the eigenstates of the density matrix as candidates for classicality, and it is clear (and was eventually also accepted by Dieter) that they are unsuitable.

I have however recently re-read Everett's writings, and I am now less surprised—this issue is never pointed out in Everett's paper or in his "long thesis" [5, 6]. Indeed, it appears Zeh (or other Everettians, e.g. DeWitt [7, 8]) did not fully appreciate its importance. This point is, however, very much appreciated by the present-day leading followers of Everett (see e.g. Kuypers and Deutsch [9], Saunders et al. [10], Wallace [11, 12] and references therein).

Dieter and I have met, over the years, several more times. He was at the meeting in Santa Fe [13]. He also contributed [14] to the proceedings of the meeting in Spain [15] I helped organize. We have both lectured at the Poincaré seminar in 2005 in Paris (for proceedings see Raimond and Rivasseau [16]). Dieter talked there about "roots and fruits of decoherence" [17].

The encounters I remember most fondly happened when I spent half a year, month at a time, in Heidelberg under the auspices of the Humboldt Foundation. I had an office on Philosophenweg, in the same building where Dieter had his office before he retired from the University. I met there his colleagues and some of the co-authors of the collection [18]. There were lively discussions.

The most pleasant and memorable encounters were the dinners at Zeh's home in Waldhilsbach near Heidelberg. When the weather was good, we dined in his garden. During and especially after dinner the conversation turned to physics. There was always good wine, and when Dieter refilled my glass, he usually emphasized (with a twinkle in his eyes) that the chances of encountering police that late at night when I drove back to Heidelberg were minimal.

I think we have both consciously stayed away from the issues of interpretation. Dieter was an enthusiastic Everettian, and I thought that both of the prevailing interpretations were really unfinished projects. Decoherence was the first (and excellent) step rather than the final word on the subject of the quantum-to-classical transition, but I saw it as an interpretation-neutral development, and more a likely beginning of more insights rather than the end. He obviously disagreed, at least about the interpretation-neutral development part.

We have nevertheless agreed about the importance of decoherence for interpretation, but, again, I thought more could be done and calculated. I was, in particular, convinced that the information theoretic tools can be useful in his task, while he seemed skeptical, and thought that information is often brought into quantum foundational discussion as a *deus ex machina*. One subject on which we emphatically agreed (and where information mattered) was the connection between decoherence and the second law in quantum chaotic systems where it yields the entropy production at the Boltzmann-Kolmogorov-Sinai rate [19–21].

Quantum Darwinism was too early in its development for me to seriously engage Zeh on this subject. I have a feeling he would have been initially skeptical (this was generally the pattern, even with pointer states and einselection), but I also think he might have welcomed it as he eventually welcomed pointer states, even though—long time after they were defined [22, 23]—he was still favoring the eigenstates of the reduced density matrix as preferred basis (see Zeh [13]). He has also, over time, warmed up to the envariant derivation of Born's rule [24, 25], probably in part because it fulfilled one of the original goals of Everett—derivation of probabilities that fit well within the relative state approach—even though he was also happy to accept Born's rule as an additional postulate.

Part of the problem in discussing these matters with Dieter was also the difference in the attitude we had towards the interpretation of quantum theory. I think he was focused on showing how decoherence turns superpositions into mixtures, and believed that was enough. I think we both agreed that this did not solve all of the "measurement problem", but Dieter was happy with the post-decoherence *status quo*, and (armed with the Everettian point of view) felt no need to do more. I—on the other hand—had a feeling that the process responsible for the acquisition of information is crucial in how the classical world we perceive emerges from the quantum substrate, and there is more to understand.

Pointer states and environment induced superselection were the first step. Quantum Darwinism is a further important step in this direction. It clarifies issues left open by decoherence and affirms the importance of einselection and of the pointer states.

I mentioned quantum Darwinism briefly in one of our encounters. As expected, there was interest but also resistance. I was therefore surprised when—reviewing Zeh's papers Ref. [26–29] while preparing this contribution—I discovered that the key idea of quantum Darwinism is acknowledged (however briefly) in his most recent writings [30].

I think of this paper as in a way an (obviously belated) attempt to appraise Dieter of the progress on quantum Darwinism and explain my view of its role in the interpretation of quantum theory (including its role in the perception of the "collapse").

1 Introduction

Quantum Darwinism [31–39] builds on decoherence theory [18, 40–43]. It shows how perception of classical reality arises via selective amplification and spreading of information in our fundamentally quantum Universe. Quantum Darwinism (QD) depends on the information transfer from the system to the environment, as does decoherence. However, QD goes beyond decoherence as it recognizes that many copies of the system's pointer states are imprinted on the environment. Agents acquire data indirectly, intercepting environment fragments (rather than measuring systems of interest *per se*). Thus, data disseminated through the environment provide us with shared information about stable, effectively classical pointer states. (Humans use primarily photon environment: we see "objects of interest" by intercepting tiny fractions of photons that contributed to decoherence.)

Quantum Darwinism recognizes that the objective classical reality we perceive and we believe in is, in the end, a model constructed by observers whose consciousness relies on indirect means of detecting objects of interest. The confidence we seem to have in this classical reality ignores, to first approximation, (or at the very least ignored for a long time) the means of its apprehension. One can recall here "It from Bit": John Wheeler [44] emphasized that we construct "it", the world "out there", from the bits of information (see also Fig. 7, on p. 195 of Ref. [45]). In the case of quantum Darwinism, this information is delivered to us by the decohering environment.

Indeed, the solid and objective reality we all believe exists is a construct, devised by our consciousness, and based on the second hand information eavesdropped by us from the environment that has also decohered the objects of interest. That environment has two functions: It helps determine the states in which the objects out there can exist, and it delivers to us information about the preferred states of the very objects it has decohered—about their einselected, stable, hence, effectively classical pointer states [22, 23].

Quantum Darwinism is not really "hypothetical"—it is a "fact of life". Its central tenet—the indirect acquisition of information by observers—is what actually happens in our quantum Universe. It is surprising that the significance of the indirect means of acquisition of information for the emergence of classical reality was not recognized earlier. Part of the reason may be that decoherence and its role was not identified until about half a century after the interpretation of quantum theory was hotly debated in the wake of the introduction of the Schrödinger equation. By then the subject of interpretation was considered settled, and, hence, uninteresting, or at the very least more a domain of philosophy rather than physics.

But perhaps an even more important reason for the delayed appreciation of the importance of the information transmitted by the decohering environment is that information theory—a crucial tool in the analysis of information flows and in quantifying redundancy (that is a consequence of amplification which in turn follows as a natural consequence of decoherence)—was formulated (by Shannon, in 1948) over twenty years after the advent of modern quantum mechanics. As the discussions of the interpretation of quantum theory by Bohr, Heisenberg, Schrödinger, and others started immediately after Schrödinger's equation was introduced, the interpreters at the time (including von Neumann, Dirac, London and Bauer, and others, in addition to the forefathers mentioned above) did not have the information-theoretic tools needed to take into account what appears to be, in retrospect, evident—that the information we have about what exists is crucial for what we believe exists.

It is nevertheless interesting that even these early discussions appealed (at least occasionally) to the "irreversible act of amplification". Quantum Darwinism recognizes that amplification is a natural and an almost inevitable consequence of decoherence, and that it does not need to involve special arrangements present in Geiger counters, photographic plates, or cloud chambers that were invoked in the early discussions.

The inevitable byproduct of decoherence is then, typically, abundance of the copies of information about the preferred states in the environment. This information is often (but not always) available to the observers. When it is available, fractions of the environment can deliver to agents multiple copies of the data about the pointer states. Agents who receive this indirect information from distinct environment fragments will agree on the objective reality of the states they infer from their data.

Not all decohering environments are equally useful as communication channels. Light excels, and we, humans, rely largely on photons. Other senses can also provide us with useful information. One can—in all of the examples of perception—point out redundancies in the information carried by the environments that are employed, although the corresponding communication channels may not be as straightforward to quantify as is the case for light [46–48].

2 Solving the Measurement Problem

The literature on the foundations of quantum theory abounds in categorical statements to the effect that an idea (e.g. Everett interpretation or decoherence) does or does not solve the measurement problem. Such claims are only rarely accompanied by an attempt to explain what is the measurement problem. One has a feeling that even presenting its definition would be controversial as there are different reasons to feel discomfort with what happens in quantum measurements, and, more generally, with quantum physics.

Nevertheless, and at the risk of inciting a controversy, I now undertake the task of defining the measurement problem, as we need to describe it before we discuss the extent to which consciousness is implicated. As is widely appreciated, the classical ideal of a single unique preexisting state that, for composite systems, has a Cartesian structure (allowing separately for a definite state of the classical system S and the classical apparatus or agent A) is inconsistent with the unitarity of quantum evolutions and with the structure of entangled states typically resulting from the interaction of a quantum $\mathcal{S}$ and $\mathcal{A}$.

What is then left to do, if this goal of recovering a fundamentally classical world is out of reach? The obvious answer is the analysis of perceptions. Would they look any different from what we experience in a Universe that is fundamentally quantum?

We emphasize, in the previous paragraph, the phrase 'such as ours'. It is an important caveat, a sign of departure from complete generality and towards explaining perception of reality in *our* rather than some generic Universe. A different hypothetical Universe may not have the properties that we take for granted in our Universe and that will matter for the ensuing considerations. Similarly, abilities of the agents should be constrained by the assumptions about their means of perception before we attempt to understand what they perceive.

We shall make, in particular, three important assumptions. To begin with, we shall assume that, at energies relevant for the measurement we carry out, one can identify systems (such as the system $\mathcal{S}$ or the apparatus $\mathcal{A}$) that maintain their identity over timescales relevant for measurements or perceptions.

Existence of systems is essential in stating the measurement problem, as without systems the state of the whole $\mathcal{SA}$ evolves unitarily and deterministically, and the issue of the definite measurement outcomes does not arise—there is no need (and no way) to ask about the outcomes when there is just a single state of the whole indivisible $\mathcal{SA}$. It follows that we shall also not apologize for assuming existence of other systems (such as the environment $\mathcal{E}$, with its subsystems) essential for the discussion of decoherence and quantum Darwinism.

Nevertheless, this assumption raises the question—how are the systems defined? It was noted earlier [49] and is still an open questions. We shall not attempt to address it here.

For the purpose of our discussion we shall assume that systems exist and that what they are depends to at least some on the past of the part of Universe where they exist (e.g. a chair was made, a cat was born and grew up, etc.). We can (as a consequence of quantum Darwinism) relax and make more precise what is expected of systems: Basically, we expect "systems of interest" to make imprints of their states on the environment.

This is a strong assumption, but both photons and air would qualify here as environments, so we make it in recognition of the particular features of the Universe we inhabit.

The second assumption we shall make is that the Hamiltonians present in the Universe are local. This is certainly consistent with what we know about our Universe. Locality is essential if causality is to be respected, and depends on the existence of space (hence, presumably, time). Locality of interactions does not preclude nonlocality of states of composite quantum systems.

Last not least, and as a third assumption, we assume that the agents that control measuring devices and acquire information are a bit like us. To start with the obvious, agents should inhabit—should be embedded—in the Universe rather than study its evolution "from the outside". They should be also capable of making measurements using local interactions we have already assumed, so their senses should be somewhat like ours. And they should be able to make records of the outcomes and process information in these records using logical circuits not unlike our neurons or classical computers.

In fact, we shall make an even stronger assumption that the agent's senses work much like ours. That is, we assume that agents utilize excitations in the environments (like we use photons and phonons) to gather information.

We shall focus on photons as they deliver vast majority of our data, and serve as the principal communication channel through which we perceive our world, including the outcomes of the measurements. Other channels (except perhaps hearing, which relies on sound waves that, because of their wave nature, have properties similar to light) are somewhat different and less relevant for the problem at hand, as taste, smell, or touch rarely arise in discussions of quantum measurements. Nevertheless, the conclusions we shall arrive at can be reasonably easily (*mutatis mutandis*, and in presence of the other assumptions, such as the locality of interactions) applied to these other channels.

These assumptions are adopted to simplify the discussion. They are likely stronger than necessary, and it is an interesting question to see whether they can be relaxed and what would be the consequences of their modifications. This is an intriguing area of research that is only in its infancy but is already leading to interesting insights (see e.g. Brandão et al. [50], Knott et al. [51], Qi and Ranard [52], Baldijão et al. [53]).

With these preliminaries out of the way, we are now ready to tackle the issue of perception of being conscious of a classical world while embedded within a quantum Universe. We end this section by drawing a sharp distinction between the two similar but, in fact, quite different ways of enquiring about the origin of the "everyday classical reality". Thus, one can either ask "Why do we *perceive* our world as classical?", or "Why our world *is* classical?".

We shall pursue the answer to the first of these two questions, understood as the question about the perception of objective classical reality—effectively classical states and correlations. The second question also has a definite answer—our Universe is *not* fundamentally classical, even on the macroscopic level that includes observers and measuring devices. This point was in effect conceded even by Bohr, when in his famous double slit debate he has defended consistency of quantum mechanics by invoking Heisenberg's indeterminacy for the movable slit in the apparatus proposed by Einstein, even though that slit was an integral part of measuring equipment (hence, according to the Copenhagen interpretation, it should have been *ab initio* classical).

3 Observers in the Classical and in the Quantum Universe

In the classical Newtonian setting it was possible to imagine that states are an exclusive property of systems, and that observers perceive them "as they are", that they see what really exists. Thus, one could envisage such a fundamental ontic state of a system existing independently of the epistemic state—of the imprecise information observer has about it. Epistemic states could differ, depending on what and how much observers know about the same fundamental ontic state. For instance, one could distinguish between insiders and outsiders—observers who knew that state more or less precisely. In classical settings one could also imagine that the only limit on how well an insider knew the fundamental state had to do with the resolution of measurements used to determine it or perhaps with the memory capacity of the observer (as knowing the state more precisely generally involves more data, hence, a lengthier description). Hence, one could think of a fundamental ontic state as a limit approached via a sequence of epistemic states corresponding to increasingly accurate measurements.

In our quantum Universe einselection determines preferred states of the objects of interest, but it does not constrain observers to measure exclusively these preferred states. Thus, one might be concerned that einselection is inconclusive by pointing out that, in principle, agents could measure observables other than the pointer observable.

If the aim of measurement is to predict, this would make little sense, but such concerns about the answers offered by decoherence have been raised on occasion (see e.g. Sánchez-Cañizares [54]). Quantum Darwinism puts them conclusively to rest.

This danger of disagreement between observers is nevertheless, in view of the preceding remarks about the preferences imposed by the einselection—a somewhat fanciful motivation. It is in observer's interest to measure pointer observables. So, why do we really need to investigate quantum Darwinism?

The answer is simple: In our Universe environments really act as communication channels. Quantum Darwinism actually happens, and how it happens settles the issue of the alignment of measurements between different observers (and of the emergence of objective classical reality agents can be aware of) much more decisively than decoherence alone.

Decoherence inevitably implies information acquisition by the environment, so its transmission is a natural consequence. Quantum Darwinism recognizes that the environments do not just destroy quantum coherence, but that systems—while they are being monitored—imprint multiple copies of information about their preferred pointer states onto the subsystems of that environment. Our senses do not couple objects of our interest directly with our information storage repositories. Rather, we eavesdrop on the information imprinted, in many copies, in the environment. This constrains what we can perceive and be conscious of.

Einselection turns out to be just one of the two functions of the "environmental monitoring and advertising agency". Over and above suppression of the uncomfortably quantum information, monitoring by the environment produces multiple copies of the data about states that can survive repeated copying—that is, preferred pointer states. Some (but not all) environments disseminate this information intact, putting it within the reach of agents, who can then get it indirectly (that is, without the danger of disrupting the system with direct measurements).

3.1 *Seeing is Believing*

The requirement of redundancy appears to be built into our senses, and in particular, our eyesight: The wiring of the nerves that pass on the signals from the rods in the eye—cells that detect light when illumination is marginal, and that appear sensitive to individual photons [55]—tends to dismiss cases when fewer than ~ 7 neighboring rods fire simultaneously [56]. Thus, while there is evidence that even individual photons can be (occasionally and unreliably) detected by humans [57], redundancy (more than one photon) is usually needed to pass the signal onto the brain.

This makes evolutionary sense—rods can misfire, so such built-in veto threshold suppresses false alarms. Frogs and toads that "make their living" by hunting small, poorly illuminated prey, have apparently lower veto thresholds, possibly because ejecting their tongue at a non-existent target is worth the risk, while missing a meal is not. Moreover, amphibians are cold-blooded, so they may not need to contend with as much noise, as thermal excitation of rods appears to be the main source of "false positives".

Quantum Darwinism relies on repeatability. As observers perceive outcomes of measurements indirectly—e.g. by looking at the pointer of the apparatus or at the photographic plate that was used in a double-slit experiment—they will depend, for their perceptions, on redundant copies of photons that are scattered from (or absorbed by) the apparatus pointer or the blackened grains of emulsion.

Thus, repeatability is not just a convenient assumption of a theorist: This hallmark of quantum Darwinism is built into our senses. And the discreteness of the possible measurement outcomes—possible perceptions—follows from the distinguishability of the preferred states that can be redundantly recorded in the environment [58, 59].

What we are conscious of is then—it appears—based on redundant evidence. Quantum Darwinist update to the existential interpretation is to assert that states exist providing that one can acquire redundant evidence about them indirectly—from the environment. This of course presumes stability in spite of decoherence [60, 61] so there is no conflict with the existential interpretation (that was originally formulated primarily on the basis of decoherence, Ref. [49, 62]) but the threshold for objective existence is nevertheless raised.

3.2 Decoherence of Records and Information Processing

Decoherence affects record keeping and information processing hardware (and, hence, information retention and processing abilities) of observers. Hence, it is relevant for our consciousness. As was already noted some time ago by Tegmark [63], individual neurons decohere on a timescale very short compared to e.g. the "clock time" on which human brain operates or other relevant timescales. Therefore, even if somehow one could initiate our information processing hardware in a superposition of its pointer states (which would open up a possibility of being conscious of superpositions), it would decohere almost instantly. The same argument applies to the present day classical computers. Thus, even if information that is explicitly quantum (that is, involves superpositions or entanglement) was inscribed in computer memory, it would decohere and (at best) become classical. (More likely, it would become random nonsense.)

It is a separate and intriguing question whether a robot equipped with a quantum computer could "do better" and perceive quantumness we are bound to miss. However, if such a robot relied (as we do) on the fragments of the environment for the information about the system of interest, it could access only the same information we can access. This information is classical, and quantum information processing capabilities would not help—only pointer states can be accessed through this communication channel.

Existential interpretation—as defined originally (e.g., Ref. [62])—relied primarily on decoherence. Decoherence of systems immersed in their environments leads to einselection: Only preferred pointer states of a system are stable, so only they can persist. Moreover, when both systems of interest and agent's means of perception and information storage are subject to decoherence, only correlations of the pointer states of the measured systems and the corresponding pointer states of agent's memory that store the outcomes are stable—only such Cartesian correlations (where each object can have a definite if unknown state) can persist.

Decoherence, one might say, "strikes twice": It selects preferred states of the systems, thus defining what can persists, hence, exist. It also limits correlations that can persist, so that observer's memory or the apparatus pointer will only preserve correlations with the einselected states of the measured system when they are recorded in the einselected memory states.

This is really the already familiar discussion of the quantum measurement problem with one additional twist: Not just the apparatus, but also the systems of everyday interest to observers are subject of decoherence. Thus, the post-measurement correlations (when investigated using discord) must now be "classical-classical", [64], while in quantum measurements only the apparatus side was guaranteed to be einselected (hence, certifiably classical).

4 Quantum Darwinism, Classical Reality, and Objective Existence

Monitoring of the system by the environment (the process responsible for decoherence) will typically leave behind multiple copies of its pointer states in $\mathcal{E}$. Pointer states are favored—only states that can survive decoherence can produce information theoretic progeny in this manner [58, 59]. Therefore, only information about pointer states can be recorded redundantly. States that can survive decoherence can use the same interactions that are responsible for einselection to proliferate information about themselves throughout the environment.

Quantum Darwinism [37, 38, 43] recognizes that observers use the environment as a communication channel to acquire information about pointer states indirectly, leaving the system of interest untouched and its state unperturbed. Observers can find out the state of the system without endangering its existence (which would be inevitable in direct measurements). Indeed, readers of this text are—at this very moment—intercepting a tiny fraction of the photon environment by their eyes to gather information.

This is how virtually all of our information is acquired. A direct measurement is not what we do. Rather, we count on redundancy, and settle for information that exists in many copies. This is how objective existence—the cornerstone of classical reality—arises in the quantum world.

4.1 Proliferation of Information and Einselection

Quantum Darwinism was introduced relatively recently. Previous studies of the records "kept" by the environment were focused on its effect on the state of the system, and not on their utility. Decoherence is a case in point, as are some of the studies of the decoherent histories approach [65, 66].

The exploration of quantum Darwinism in specific models has started at the beginning of this millennium [31–35]. I do not intend to review all of the results of this ongoing research. The basic conclusion of these studies is, however, that the dynamics responsible for decoherence is also capable of imprinting multiple copies of the pointer basis on the environment. Moreover, while decoherence is always implied by quantum Darwinism, the reverse need not be true. For instance, when the environment is completely mixed, it cannot be used as a communication channel, but it will still suppress quantum coherence in the system and lead to einselection of pointer states.

Let us consider a simple example of quantum Darwinism: For many subsystems, $\mathcal{E} = \bigotimes_k \mathcal{E}^{(k)}$, the initial pre-decoherence state of the system and the environment, $(\alpha|\uparrow\rangle + \beta|\downarrow\rangle)|\varepsilon_0^{(1)}\varepsilon_0^{(2)}\varepsilon_0^{(3)}...\rangle$ evolves into a "branching state";

$$|\Upsilon_{\mathcal{SE}}\rangle = \alpha|\uparrow\rangle|\varepsilon_\uparrow^{(1)}\varepsilon_\uparrow^{(2)}\varepsilon_\uparrow^{(3)}...\rangle + \beta|\downarrow\rangle|\varepsilon_\downarrow^{(1)}\varepsilon_\downarrow^{(2)}\varepsilon_\downarrow^{(3)}...\rangle \tag{1}$$

The state $|\Upsilon_{\mathcal{SE}}\rangle$ represents many records inscribed in its fragments, collections of subsystems of $\mathcal{E}$ (Fig. 1). This means that the state of $\mathcal{S}$ can be found out by many, independently, and indirectly—hence, without disturbing $\mathcal{S}$. This is how evidence of objective existence arises in our quantum world.

Linearity assures that all the branches persist: collapse to one outcome in Eq. (1) is not in the cards. However, large $\mathcal{E}$ can disseminate information about the system.

4.2 Mutual Information in Quantum Correlations

To develop theory of quantum Darwinism we need to quantify information between fragments of the environment and the system. Quantum mutual information is a convenient tool we shall use for this purpose.

The mutual information between the system $\mathcal{S}$ and a fragment $\mathcal{F}$ (that will play the role of the apparatus $\mathcal{A}$) can be computed using their von Neumann entropy:

$$H_X = -\mathrm{Tr}\rho_X \lg \rho_X \ . \tag{2}$$

Given the density matrices of the system $\mathcal{S}$ and a fragment $\mathcal{F}$ mutual information is simply;

$$I(\mathcal{S}:\mathcal{F}) = H_{\mathcal{S}} + H_{\mathcal{F}} - H_{\mathcal{S},\mathcal{F}} \ . \tag{3}$$

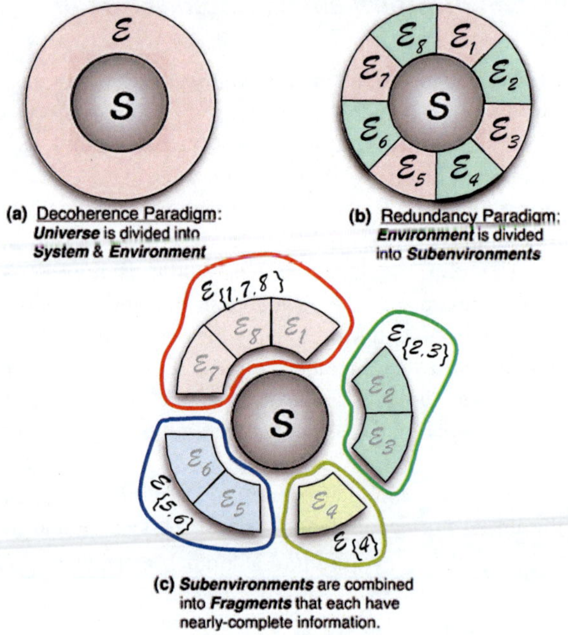

Fig. 1 Quantum Darwinism and the structure of the environment. The decoherence paradigm distinguishes between a system ($\mathcal{S}$) and its environment ($\mathcal{E}$) as in (a), but makes no further recognition of the structure of $\mathcal{E}$; it could as well be monolithic. In the environment-as-a-witness paradigm, we recognize subdivision of $\mathcal{E}$ into subenvironments—its natural subsystems, as in (b). The only essential requirement for a subsystem is that it should be individually accessible to measurements; observables corresponding to different subenvironments commute. To obtain information about the system $\mathcal{S}$ from its environment $\mathcal{E}$ one can then carry out measurements of fragments $\mathcal{F}$ of the environment—non-overlapping collections of its subsystems. Sufficiently large fragments of the environment that has monitored (and, therefore, decohered) $\mathcal{S}$ can often provide enough information to infer the state of $\mathcal{S}$, by combining subenvironments as in (c). There are then many copies of the information about $\mathcal{S}$ in $\mathcal{E}$, which implies that some information about the "fittest" observables that survived monitoring by $\mathcal{E}$ has proliferated throughout $\mathcal{E}$, leaving multiple copies—qmemes — their (quantum) informational offspring. This proliferation of the information about the fittest states defines quantum Darwinism. Multiple qmemes allow many observers to find out the state of $\mathcal{S}$: Environment becomes a reliable witness with redundant copies of information about preferred observables, which leads to objective existence of preferred pointer states

We already noted the special role of the pointer observable. It is stable and, hence, it leaves behind information-theoretic progeny—multiple imprints, copies of the pointer states—in the environment. By contrast, complementary observables are destroyed by the interaction with a single subsystem of $\mathcal{E}$. They can in principle still be accessed, but only if *all* of the environment is measured. Indeed, because we are dealing with quantum systems, things are much worse than that: The environment must be measured in precisely the right (typically global) basis along with $\mathcal{S}$ to allow for such a reconstruction. Otherwise, the accumulation of errors over multiple measurements will lead to an incorrect conclusion and re-prepare the environment, so that it is no longer a record of the pre-decoherence state of $\mathcal{S}$, and phase information is irretrievably lost.

When each environment qubit is a perfect copy of the state of $\mathcal{S}$, redundancy in this simple example is given by the number of fragments—that is, in this case, by the number of the environment qubits—that have complete information about $\mathcal{S}$. In this simple case there is no reason to define redundancy in a more sophisticated manner. Such a need arises in more realistic cases when the records in individual subsystems of $\mathcal{E}$ are imperfect.

4.3 *Objective Reality from Redundant Information*

An environment fragment $\mathcal{F}$ can act as apparatus with a (possibly incomplete) record of $\mathcal{S}$. When $\mathcal{E}_{\backslash\mathcal{F}}$ ('the rest of the environment') is traced out, $\mathcal{SF}$ decoheres, and the reduced density matrix describing joint state of $\mathcal{S}$ and $\mathcal{F}$ is:

$$\begin{aligned} \rho_{\mathcal{SF}} &= \mathrm{Tr}_{\mathcal{E}\backslash\mathcal{F}} |\Psi_{\mathcal{SE}}\rangle\langle\Psi_{\mathcal{SE}}| = \\ &= |\alpha|^2 |\uparrow\rangle\langle\uparrow| |F_\uparrow\rangle\langle F_\uparrow| + |\beta|^2 |\downarrow\rangle\langle\downarrow| |F_\downarrow\rangle\langle F_\downarrow| \end{aligned} \tag{4}$$

When $\langle F_\uparrow | F_\downarrow\rangle = 0$, $\mathcal{F}$ contains perfect record of the preferred states of the system. In principle, each subsystem of $\mathcal{E}$ may be enough to reveal its state, but this is unlikely. Typically, one must collect many subsystems of $\mathcal{E}$ into $\mathcal{F}$ to find out about $\mathcal{S}$.

The redundancy of the data about pointer states in $\mathcal{E}$ determines how many times the same information can be independently extracted—it is a measure of objectivity. The key question of quantum Darwinism is then: *How many subsystems of $\mathcal{E}$—what fraction of $\mathcal{E}$—does one need to find out about $\mathcal{S}$?*. The answer is provided by the mutual information

$$I(\mathcal{S}:\mathcal{F}_f) = H_\mathcal{S} + H_{\mathcal{F}_f} - H_{\mathcal{SF}_f} \,,$$

the information about $\mathcal{S}$ available from $\mathcal{F}_f$, that can be obtained from the fraction $f = \frac{\sharp\mathcal{F}}{\sharp\mathcal{E}}$ of $\mathcal{E}$ (where $\sharp\mathcal{F}$ and $\sharp\mathcal{E}$ are the numbers of subsystems).

In case of perfect correlation a single subsystem of $\mathcal{E}$ would suffice, as $I(\mathcal{S}:\mathcal{F}_f)$ jumps to $H_\mathcal{S}$ at $f = \frac{1}{\sharp\mathcal{E}}$. The data in additional subsystems of $\mathcal{E}$ are then redundant. Usually, however, larger fragments of $\mathcal{E}$ are needed to find out enough about $\mathcal{S}$. Red plot in Fig. 2 illustrates this: $I(\mathcal{S}:\mathcal{F}_f)$ still approaches $H_\mathcal{S}$, but only gradually. The length of this plateau can be measured in units of f_δ, the initial rising portion of $I(\mathcal{S}:\mathcal{F}_f)$. It is defined with the help of the *information deficit* δ observers tolerate:

$$I(\mathcal{S}:\mathcal{F}_{f_\delta}) \geq (1-\delta)H_\mathcal{S} \tag{5}$$

Redundancy is the number of such records of $\mathcal{S}$ in $\mathcal{E}$:

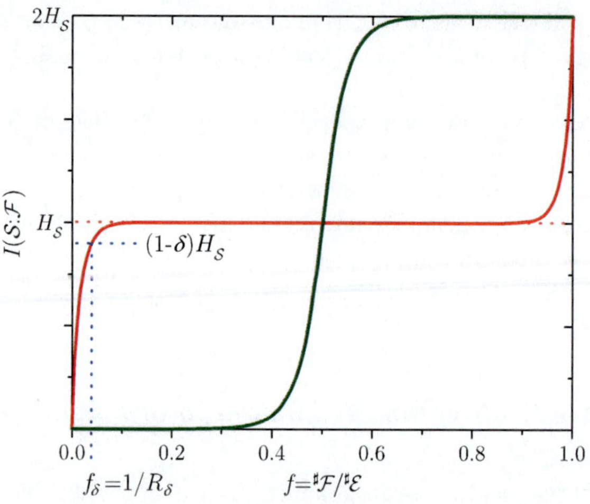

Fig. 2 Information about the system contained in a fraction f of the environment. The red plot shows a typical $I(\mathcal{S} : \mathcal{F}_f)$ established by decoherence. Rapid rise means that nearly all classically accessible information is revealed by a small fraction of $\mathcal{E}$. It is followed by a plateau: additional fragments only confirm what is already known. Redundancy $\mathcal{R}_\delta = 1/f_\delta$ is the number of such independent fractions. Green plot shows $I(\mathcal{S} : \mathcal{F}_f)$ for a random state in the composite system $\mathcal{SE}$. In this case, almost no information is revealed until nearly half of the environment is measured

$$\mathcal{R}_\delta = 1/f_\delta \tag{6}$$

$\mathcal{R}_\delta$ sets the upper limit on how many observers can find out the state of $\mathcal{S}$ from $\mathcal{E}$ independently and indirectly. In models, and especially in the photon scattering analyzed extending the decoherence model of Joos and Zeh [67] redundancy $\mathcal{R}_\delta$ is huge [46–48] and depends on δ only weakly (logarithmically).

This is 'quantum spam': $\mathcal{R}_\delta$ imprints of the pointer states are broadcast throughout the environment. Many observers can access them independently and indirectly, assuring objectivity of pointer states of $\mathcal{S}$. Repeatability is key: States must survive copying to produce many imprints.

Insights into the nature of the constraints imposed by unitarity in the process of copying [58, 59] show when, in spite of the no-cloning theorem repeated transfers of information are possible. Distinguishability of the "originals", the states that can be repeatably copied, is essential. Discrete preferred states set the stage for quantum jumps. Copying yields branches of records inscribed in subsystems of $\mathcal{E}$. Initial superposition yields superposition of branches, Eq. (1), so there is no literal collapse. However, fragments of $\mathcal{E}$ can reveal only decohered branches to the observer (and cannot reveal their superposition). Such evidence will suggest 'quantum jump' from superposition to a single outcome.

4.4 Environment as a Witness

Not all environments are good in this role of a witness. Photons excel: They do not interact with the air or with each other, faithfully passing on information. Small fraction of photon environment usually reveals all we need to know. Scattering of sunlight quickly builds up redundancy: a 1 μm dielectric sphere in a superposition of 1 μm size increases $\mathcal{R}_{\delta=0.1}$ by $\sim 10^8$ every microsecond [46, 47]. A mutual information plot illustrating this case is shown in Fig. 3.

Air is also good in decohering, but its molecules interact, scrambling acquired data. Objects of interest scatter both air and photons, so both acquire information about position, and favor similar localized pointer states. Moreover, only photons are willing to reveal what was that pointer state.

Quantum Darwinism shows why it is so hard to undo decoherence [68]. Plots of mutual information $I(\mathcal{S} : \mathcal{F}_f)$ for initially pure $\mathcal{S}$ and $\mathcal{E}$ are antisymmetric (see Figs. 2 and 3) around $f = \frac{1}{2}$ and $H_{\mathcal{S}}$ [31]. Hence, a counterpoint of the initial quick rise at $f \leq f_\delta$ is a quick rise at $f \geq 1 - f_\delta$, as last few subsystems of $\mathcal{E}$ are included in the fragment $\mathcal{F}$ that by now contains nearly all $\mathcal{E}$. This is because an initially pure

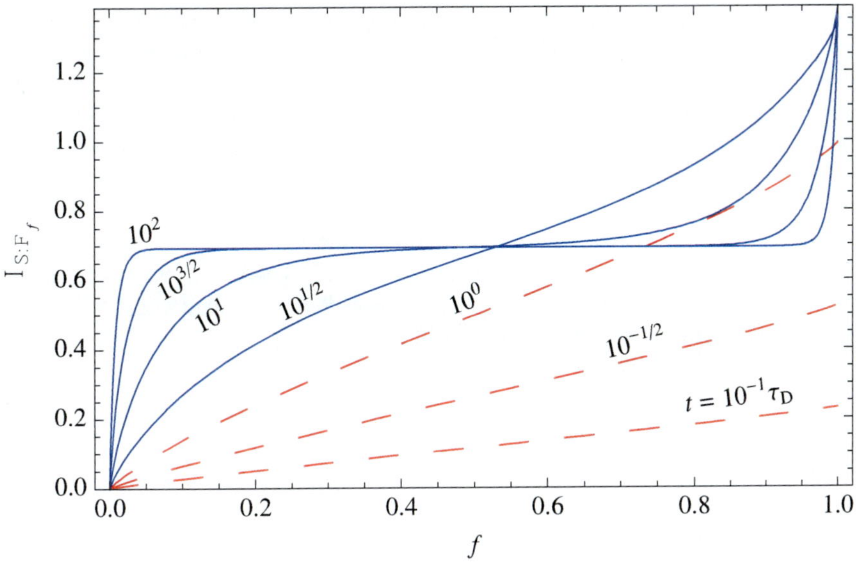

Fig. 3 The quantum mutual information $I(\mathcal{S} : \mathcal{F}_f)$ vs. fragment size f at different elapsed times for an object illuminated by point-source black-body radiation. For point-source illumination, individual curves are labeled by the time t in units of the decoherence time τ_D. For $t \leq \tau_D$ (red dashed lines), the information about the system available in the environment is low. The linearity in f means each piece of the environment contains new, independent information. For $t > \tau_D$ (blue solid lines), the shape of the partial information plot indicates redundancy; the first few pieces of the environment increase the information, but additional pieces only confirm what is already known [47]

$\mathcal{SE}$ remains pure under unitary evolution, so $H_{\mathcal{SE}} = 0$, and $I(\mathcal{S} : \mathcal{F}_f)|_{f=1}$ must reach $2H_{\mathcal{S}}$. Thus, a measurement on *all* of $\mathcal{SE}$ could confirm its purity in spite of decoherence caused by $\mathcal{E}\backslash\mathcal{F}$ for all $f \leq 1 - f_\delta$. However, to verify this one has to intercept and measure all of $\mathcal{SE}$ in a way that reveals pure state $|\Upsilon_{\mathcal{SE}}\rangle$, Eq. (1). Other measurements destroy phase information. So, undoing decoherence is in principle possible, but the required resources and foresight preclude it.

In quantum Darwinism decohering environment acts as an amplifier, inducing branch structure of $|\Upsilon_{\mathcal{SE}}\rangle$ distinct from typical states in the Hilbert space of $\mathcal{SE}$: $I(\mathcal{S} : \mathcal{F}_f)$ of a random state is given by the green plot in Fig. 2, with no plateau or redundancy. Antisymmetry [31] means that $I(\mathcal{S} : \mathcal{F}_f)$ 'jumps' at $f = \frac{1}{2}$ to $2H_{\mathcal{S}}$.

Environments that decohere $\mathcal{S}$, but scramble information because of interactions between subsystems (e.g. air) eventually approach such random states. Quantum Darwinism is possible only when information about $\mathcal{S}$ is preserved in fragments of $\mathcal{E}$, so that it can be recovered by observers. There is no need for perfection: Partially mixed environments or imperfect measurements correspond to noisy communication channels—their capacity is depleted, but we can still get the message [69, 70].

Quantum Darwinism settles the issue of the origin of classical reality by accounting for all of the operational symptoms objective existence in our quantum Universe: A single quantum state cannot be found out through a direct measurement. However, pointer states usually leave multiple records in the environment. Observers can use these records to find out the (pointer) state of the system of interest. Observers can afford to destroy photons while reading the evidence—the existence of multiple copies implies that many can access the information about the system indirectly and independently, and that they will all agree about the outcome. This is how objective existence arises in our quantum world.

There has been significant progress in the study of the acquisition and dissemination of the information by the environments and the consequences of quantum Darwinism. We point interested readers to recent overview [39] as well as to experimental efforts (see Unden et al. [71] and references therein), and to a popular account [72] as well as to the recent more technical advance [36].

5 Discussion

Our Universe is quantum, as experimental confirmations of quantum superpositions on increasingly large scales and as studies entanglement demonstrate. Yet, the world we inhabit appears classical to us. It seems devoid of superpositions or entanglement.

Emergence of classicality from the quantum substrate was a long-standing mystery. Decoherence accounts for the environment - induced superselection. Einselection leads to preferred pointer states. It posits that objects acquire effective classicality because they are in effect monitored by their environments that destroy superpositions of selected quasi-classical pointer states that are immune to decoherence, as they do not entangle with the environment.

Quantum Darwinism extends the role of the environment: It is motivated by the insight that we acquire our data about the Universe indirectly, by intercepting fractions of the environment that have caused decoherence. Thus, only information reproduced, by decoherence, in many copies, can be regarded as objective: It can be accessed by many who will agree about what they perceive, and who can re-confirm their data by measuring additional environment fragments.

Decohering environment destabilizes all superpositions—it preserves intact only the einselected pointer states. Einselection is focused on the persistence of states in spite of the environment. It poses and answers a different question than quantum Darwinism which is focused on how the information about the objects we perceive reaches our senses, and, ultimately, our consciousness. However, the answer to both questions revolves around pointer states—the states that are einselected are also the only states that can be imprinted in many copies on the environment.

Quantum Darwinism recognizes environment's role as a medium, as a communication channel through which information reaches observers.[1] The ability to obtain information about systems of interest indirectly, by eavesdropping on their environments (rather than through direct measurement) safeguards their pointer states from the disturbance that would be otherwise inflicted by the acquisition of information. Pointer states are selected by the interaction with the environment for their immunity to decoherence. The information about them is inscribed, in multiple copies, on the subsystems of that environment. Only the data about the pointer states is then readily available to observers.

This is a tradeoff: Observers can acquire information that has predictive value (as pointer states persist in spite of decoherence) but they lose the choice of what to measure—only the information about the observables selected by the interaction with the environment is within reach [34, 35, 39].

This dual role of the environment (its "censoring" of the Hilbert space and its broadcasting of the information about the einselected states) supports and extends the *existential interpretation*. Persistence is a precondition of existence. The original focus [62] of the existential interpretation was preservation of pointer states in presence of decoherence. Zeh [75] acknowledged the role of the existential interpretation at the time when the emphasis was shifting from einselection to the quantum Darwinian, "environment as a witness" point of view.

Quantum Darwinism adds a new and crucial element to the existential interpretation: It explains how pointer states can be found out without getting disrupted. Thus, the concern about the fragility of quantum states can be—for macroscopic systems that decohere and that comprise our everyday reality—put aside. We can find out about them without disrupting their (pointer) states.

Crucial advance in appreciating the role of the environment as a witness came when the scattering of radiation by a nonlocal "Schrödinger cat" superposition of

[1] Similar ideas (albeit with a somewhat different emphasis) are pursued by others, largely within the quantum information community. Recent discussions of the variations on the theme of quantum Darwinism including relevant references can be found in e.g. in Korbicz [73], or in Le and Olaya-Castro [74].

dielectric sphere was shown [46, 47] to produce copious redundancy of the information about its location (see Fig. 3). This is an accurate model of what actually happens: We gain most of our information about macroscopic objects in our Universe by detecting a small fractions of photons they scatter or emit. This is how we find out what exists, by employing the environment as both a witness to the "crime" of decohering superpositions and as a guilty party (after all, it is responsible for decoherence in the first place).

Quantum Darwinism is increasingly recognized as key to emergence of the familiar classical reality from within our quantum Universe. Its implications are independent of the interpretational stance, although it does rely on the universal applicability of quantum theory (see, however, Baldijão et al. [53]). It is clearly compatible with the Everett's relative states. It is also compatible with a non-dogmatic reading of Bohr's views, as is decoherence and einselection, as discussed by Camilleri and Schlosshauer [76].

Quantum Darwinism answers the key question at the core of the interpretational discussions: How is it possible that fragile quantum states can give rise to robust reality that appears to be objective, undisturbed by the acquisition of information about it by observers? The answer is straightforward—we use the very same environment that is responsible for decoherence and hence for einselection as a communication channel. Thus, we find out relevant information without interacting directly with the objects of interest.

This millennium has witnessed a significant progress on the interpretational questions that were posed (or at least should have been posed) at the advent of quantum physics. These advances—decoherence, einselection of the pointer states, and quantum Darwinism—resolve issues that seemed either impossible to address or remained unnoticed only a few decades ago. These answers are not "interpretations". They rely on quantum theory *per se*, and do not call on any additional ingredients apart from these needed to pose the questions in the first place. Interpretations can be 'adjusted' to accommodate these advances, but it is the interpretations (Copenhagen or Everett's) that have to adjust: Quantum theory is the basis of decoherence, einselection of the pointer states, and quantum Darwinism, and that basis is non-negotiable—it does not need to be and cannot be adjusted.

Will—in view of these advances—the measurement problem be regarded as "solved"? I hope it might be (or, at the very least, such a truly significant advance might be 'widely acknowledged') but—to be honest—I am far from certain. I sense that a large fraction of these who are aware of the measurement problem are also rooting for it to remain a problem.

Thus, any advance in this area (like decoherence, einselection, or quantum Darwinism) is more often than not greeted with the disappointed "...but this is just quantum theory...".

That is hard to argue with—it is quantum theory. Indeed, there are authorities with well-deserved stellar reputations whose reaction to the above advances is (at least approximately) "...but this is just quantum theory...".

There are also others who take a more constructive approach and suggest ways of modifying quantum theory that would—in their opinion—make it more palatable

[77–86]. The explored possibilities range from a variety of probabilistic approaches (see e.g. references in Baldijão et al. [53]) to superdeterminism (see e.g. Hossenfelder [87]). Some of these proposals may lead to experimental tests. So far, there is however no experimental evidence for anything other than quantum theory.

There are of course also physicists who acknowledge the importance of the advances discussed here, and many have contributed to both theory and experimental tests of decoherence, einselection, and quantum Darwinism.

Nevertheless, the question can be posed: Will quantum mechanics be replaced by a more palatable theory...? Personally, I would be delighted if there was a testable alternative, or if it turned out that quantum theory can be deduced from a more fundamental set of simple principles, or if a chink in its armor was found by an experiment, a chink that would hint about how to arrive at such a deeper theory. I applaud efforts to look beyond quantum theory, 'as we know it'.

However, I also strongly suspect that such search is futile, at least in the sense that—even if quantum theory were to be supplanted by a still deeper theory—quantum weirdness will not go away as a result: After all, it was the experiment that forced us to accept the inevitability quantum weirdness, and the principle of superposition or entanglement will not go away. I admit that the resistance of general relativity to quantization is intriguing, and heroic attempts to test superpositions of larger and heavier objects (by Arndt, Aspelmeyer, Bouwmeester, Zeilinger, and others) are well worth the effort, but I also admit some of my enthusiasm may well be self-serving, based on the conviction that these and other tests of quantum mechanics will continue to reveal more about decoherence and einselection—that is, the real origin of our everyday classical reality.

I strongly suspect that the ultimate message of quantum theory is that the separation between what exists and what is known to exist—between the epistemic and the ontic—must be abolished. It is hard to argue with "wanting more", but wanting more need not imply replacing or abolishing quantum theory.

Quantum Darwinism, einselection, pointer states, decoherence, etc., are also a result of "wanting more". Better understanding of quantum physics and of its relation to the classical world we inhabit led to insights that deepened our appreciation of what quantum theory is about. Almost a century ago, when quantum mechanics was still in its infancy, dissolution of the strict distinction between what exists and what is known, between ontic and epistemic—characteristic feature of the classical Newtonian physics—was difficult to envision. Moreover, in the absence of the theory of information—it was impossible to quantify.

Surprisingly, as the nascent quantum theory matured, quantum awkwardness did not go away. The lesson of the developments that started with decoherence, and led, through einselection and pointer states to quantum Darwinism and related insights, is that quantum states are epiontic. It now appears that the distinction between what is and what is known about what exists—something we take for granted in our everyday world—is not there at the fundamental level: It emerges only as a consequence of the separation between the systems of interest (e.g. macroscopic bodies such as planets, billiard balls, and sometimes perhaps even buckyballs) and the carriers of information about them (such as photons).

Acknowledgements I have benefited from discussions with Charles Bennett, Robin Blume-Kohout, Sebastian Deffner, Davide Girolami, Jonathan Halliwell, James Hartle, Paweł Horodecki, Ryszard Horodecki, Claus Kiefer, Harold Ollivier, Juan Pablo Paz, David Poulin, Hai-Tao Quan, Jess Riedel, Akram Touil, Bin Yan, Michael Zwolak, and, of course, Dieter Zeh. This research was supported by DoE through the LDRD program at Los Alamos, and in part by the Foundational Questions Institute's Consciousness in the Physical World program, in partnership with the Fetzer Franklin Fund.

References

1. C.M. Caves et al., Rev. Mod. Phys. **52**, 341–392 (1980)
2. H.D. Zeh, *Foundations of Quantum Mechanics*, ed. by B. D'Espagnat (Academic, New York, 1971), pp. 263–273
3. H.D. Zeh, Found. Phys. **1**, 69 (1970)
4. H.D. Zeh, Found. Phys. **3**, 109–116 (1973)
5. H. Everett III., Rev. Mod. Phys. **29**, 454 (1957)
6. H. Everett III, Ph. D. Dissertation, Princeton University, reprinted in DeWitt and Graham, (1973)(1957)
7. B.S. DeWitt, *Foundations of Quantum Mechanics*, ed. by B. d'Espagnat (Academic Press, New York, 1971)
8. B.S. DeWitt, N. Graham (eds.), *The Many—Worlds Interpretation of Quantum Mechanics* (Princeton University Press, Princeton, 1973)
9. S. Kuypers, D. Deutsch, Everettian relative states in the Heisenberg picture (2020). arXiv:2008.02328
10. S. Saunders, J. Barrett, A. Kent, D. Wallace, *Many Worlds?: Everett, Quantum Theory, & Reality* (Oxford University Press, 2010)
11. D. Wallace, Decoherence and its role in the modern measurement problem. Phil. Trans. Roy. Soc. A **370**, 4576–4593 (2012)
12. D. Wallace, *The Emergent Multiverse: Quantum Theory according to the Everett Interpretation* (Oxford University Press, Oxford, 2012)
13. H.D. Zeh, *Complexity, Entropy, and the Physics of Information*, ed. by W.H. Zurek (Addison Wesley, Redwood City, 1990), p. 405
14. H.D. Zeh, *Physical Origins of Time Asymmetry*, ed. by J.J. Halliwell et al. (Cambridge University Press, 1994), pp. 390–404
15. J.J. Halliwell, J. Pérez-Mercader, W.H. Zurek, *Physical Origins of Time Asymmetry* (Cambridge University Press, Cambridge, 1994)
16. J.-M. Raimond, V. Rivasseau (eds.), *Quantum Decoherence (Poincaré Seminar 2005)* (Birkhaüser Verlag, Basel, 2006)
17. H.D. Zeh, *Quantum Decoherence*, in Raimond and Rivasseau (2006), pp. 151–175
18. E. Joos, H.D. Zeh, C. Kiefer, D. Giulini, J. Kupsch, I.-O. Stamatescu, *Decoherence and the Appearance of a Classical World in Quantum Theory* (Springer, Berlin, 2003)
19. H.D. Zeh, *The Physical Basis of the Direction of Time* (Springer, Heidelberg, 2007)
20. W.H. Zurek, Physica Scripta **T76**, 186 (1998)
21. W.H. Zurek, J.-P. Paz, Phys. Rev. Lett. **72**, 2508 (1994)
22. W.H. Zurek, Phys. Rev. D **24**, 1516 (1981)
23. W.H. Zurek, Phys. Rev. D **26**, 1862 (1982)
24. W.H. Zurek, Phys. Rev. Lett. **90**, 120404 (2003)
25. W.H. Zurek, Phys. Rev. A **71**, 052105 (2005)
26. H.D. Zeh, Phys. Lett. A **172**(4), 189–192 (1993)
27. H.D. Zeh, Found. Phys. Lett. **13**(3), 221–233 (2000)

28. H.D. Zeh, *Decoherence: Theoretical, Experimental, and Conceptual Problems*, ed. by P. Blanchard, D. Giulini, E. Joos, C. Kiefer, I.-O. Stamatescu. Lecture Notes in Physics (Springer, Berlin, 2000), pp. 19–42
29. H.D. Zeh, Phys. Lett. **309**, 329–334 (2003)
30. H.D. Zeh, Z. Naturforsch. A **71**(3), 195–212 (2016)
31. R. Blume-Kohout, W.H. Zurek, Found. Phys. **35**, 1857 (2005)
32. R. Blume-Kohout, W.H. Zurek, Phys. Rev. A **73**, 062310 (2006)
33. R. Blume-Kohout, W.H. Zurek, Phys. Rev. Lett. **101**, 240405 (2008)
34. H. Ollivier et al., Phys. Rev. Lett. **93**, 220401 (2004)
35. H. Ollivier et al., Phys. Rev. A **72**, 423113 (2005)
36. A. Touil et al., Eavesdropping on the decohering environment: quantum darwinism, amplification, and the origin of objective classical reality (2021). arXiv:2107.00035
37. W.H. Zurek, Nat. Phys. **5**, 181 (2009)
38. W.H. Zurek, Phys. Today **67**(10), 44 (2014)
39. W.H. Zurek, Phil. Trans. R. Soc. A **376**(2123), 20180107 (2018)
40. M. Schlosshauer, Rev. Mod. Phys. **76**, 1267 (2004)
41. M. Schlosshauer, *Decoherence and the Quantum to Classical Transition* (Springer, Berlin, 2007)
42. M. Schlosshauer, Phys. Rep. **831**, 1–57 (2019)
43. W.H. Zurek, Rev. Mod. Phys. **75**, 715 (2003)
44. J.A. Wheeler, *Complexity, Entropy, and the Physics of Information*, ed. by W.H. Zurek(Addison Wesley, Redwood City, 1990), p. 3
45. J.A. Wheeler, W.H. Zurek, *Quantum Theory and Measurement* (Princeton University Press, Princeton, 1983)
46. C.J. Riedel, W.H. Zurek, Phys. Rev. Lett. **105**(2), 020404 (2010)
47. C.J. Riedel, W.H. Zurek, N. J. Phys. **13**, 073038 (2011)
48. M. Zwolak, C.J. Riedel, W.H. Zurek, Phys. Rev. Lett. **112**, 140406 (2014)
49. W.H. Zurek, Phil. Trans. R. Soc. Lond. Ser. A **356**, 1793 (1998)
50. F.G.S.L. Brandão, M. Piani, P. Horodecki, Nat. Comm. **6**, 7908 (2015)
51. P.A. Knott et al., Phys. Rev. Lett. **121**, 160401 (2018)
52. X.-L. Qi, D. Ranard, arXiv:2001.01507
53. R.D. Baldijão et al., Quantum Darwinism and the spreading of classical information in non-classical theories (2020). arxiv.org:/2012.06559
54. J. Sánchez-Cañizares, Foundations of Science **24**, 275–285 (2019)
55. N.M. Phan, M.F. Cheng, D.A. Bessarab, L.A. Krivitsky, Phys. Rev. Lett. **112**(21), 213601 (2014)
56. F. Rieke, D.A. Baylor, Rev. Mod. Phys. **70**, 1027 (1998)
57. J. Tinsley, M. Molodtsov, R. Prevedel et al., Nat Comm. **7**, 12172 (2016). https://doi.org/10.1038/ncomms12172
58. W.H. Zurek, Phys. Rev. A **76**, 052110 (2007)
59. W.H. Zurek, Phys. Rev. A **87**, 052111 (2013)
60. J.-P. Paz, W.H. Zurek, *Coherent Atomic Matter Waves, Les Houches Lectures*, ed. by R. Kaiser, C. Westbrook, F. David (Springer, Berlin, 2001), p. 533
61. W.H. Zurek, Phys. Today **44**, 36 (1991). (see also an 'update', quant-ph/0306072)
62. W.H. Zurek, Prog. Theor. Phys. **89**, 281 (1993)
63. M. Tegmark, Phys. Rev. E **61**, 4194–4206 (2000)
64. M. Piani, P. Horodecki, R. Horodecki, Phys. Rev. Lett. **100**, 090502 (2008)
65. M. Gell-Mann, J.B. Hartle, *Proceedings of the 4th Drexel Conference on Quantum Non-Integrability: The Quantum-Classical Correspondence*, ed. by D.-H. Feng, B.-L. Hu (International Press of Boston, Hong Kong, 1998)
66. J.J. Halliwell, Phys. Rev. D **60**, 105031 (1999)
67. E. Joos, H.D. Zeh, Z. Phys. B **59**, 223 (1985)
68. M. Zwolak, W.H. Zurek, Sci. Rep. **3**, 1729 (2013)
69. M. Zwolak, H.-T. Quan, W.H. Zurek, Phys. Rev. Lett. **103**, 110402 (2009)

70. M. Zwolak, H.-T. Quan, Zurek, Phys. Rev. A **81**, 062110 (2010)
71. T. Unden et al., Phys. Rev. Lett. **123**(14), 140402 (2019)
72. P. Ball, *Beyond Weird* (Chicago University Press, Chicago, 2018)
73. J.K. Korbicz, Roads to objectivity: Quantum Darwinism, Spectrum Broadcast Structures, and Strong Quantum Darwinism (2020). arxiv:2007.04276
74. T.P. Le, A. Olaya-Castro, Quantum. Sci. Technol. **5**, 045012 (2020)
75. H.D. Zeh, Found. Phys. Lett. **12**(2), 197–200 (1999)
76. K. Camilleri, M. Schlosshauer, Stud. Hist. Phil. Mod. Phys. **49**, 73–83 (2015)
77. S.L. Adler, Why decoherence has not solved the measurement problem: a response to P. W. Anderson. Stud. Hist. Philos. Mod. Phys. **34**, 135–142 (2003)
78. S.L. Adler, *Quantum Theory as an Emergent Phenomenon: The Statistical Mechanics of Matrix Models as the Precursor of Quantum Field Theory* (Cambridge University Press, Cambridge, 2004)
79. A.J. Leggett, Macroscopic quantum systems and the quantum theory of measurement. Supp. Prog. Theor. Phys. **69**, 80–100 (1980)
80. A.J. Leggett, Testing the limits of quantum mechanics: motivation, state of play, prospects. J. Phys. Condens. Matter **14**, R415–R451 (2002)
81. R. Penrose, Gravity and state vector reduction, in *Quantum Concepts in Space and Time*, ed. by C.J. Isham, R. Penrose (Clarendon Press, 1986), pp. 129–146
82. R. Penrose, *The Emperors New Mind* (OxfordUniversity Press, Oxford, 1989)
83. R. Penrose, On Gravitys role in Quantum State Reduction. Gen. Relat. Gravit. **28**, 581–600 (1996)
84. G. 't Hooft, *The Cellular Automaton Interpretation of Quantum Mechanics* (2014). arXiv:1405.1548
85. S.W. Weinberg, Precision tests of quantum mechanics. Phys. Rev. Lett. **62**, 485–488 (1989)
86. S.W. Weinberg, Collapse of the state vector. Phys. Rev. A **85**, 062116 (2012)
87. S. Hossenfelder, Superdeterminism: A Guide for the Perplexed (2020) arXiv:2010.01324

Decoherence: From Interpretation to Experiment

Maximilian Schlosshauer

1 Introduction

I never had the opportunity to meet Dieter Zeh in person, but over the years we maintained a fairly regular correspondence. The last email I received from him dates from April 1, 2015, by which time our exchanges had become rather infrequent. Because this email happens to be an excellent representation of his inimitable style and thoughtfully articulated views, and because it touches on some points to which I shall return later in this essay, I will take the liberty to quote it in full below. Zeh had sent it to me in response to a paper [1] that Kristian Camilleri and I had written on the relation between decoherence and the views of Bohr and Heisenberg. There, we had suggested that Bohr's doctrine of "classical concepts" should be understood epistemologically. As we put it,

> Bohr's fundamental point was that any interpretation of quantum mechanics must in the end fall back on the use of classical concepts, because such concepts play an indispensible role in experimental contexts in which we acquire empirical knowledge of the world.

Thus, we saw Bohr's insistence on the primacy of classical concepts as motivated by his emphasis on the functional role of experiments, and we therefore suggested that this understanding of classical concepts should be distinguished from the kind of classicality obtained from dynamical solutions to the quantum–classical problem, such as the account given by decoherence.

"This argument may be acceptable," Zeh wrote to me, "if it is to be understood merely from a historical point of view":

> In this case, Bohr's classical concepts would be akin to Wittgenstein's famous ladder that one may afterward discard. But I'm sure that this is not what Bohr had in mind. I say this because,

M. Schlosshauer (✉)
Department of Physics, University of Portland, 5000 North Willamette Boulevard, 97203 Portland, OR, USA
e-mail: schlossh@up.edu

C. Kiefer (ed.), *From Quantum to Classical*, Fundamental Theories of Physics 204,
https://doi.org/10.1007/978-3-030-88781-0_3

> on such a reading, the axiomatic view of quantum theory would amount to nothing more than, for example, Sommerfeld's axiomatic introduction of Maxwell's equations (as presented in his textbook)—that is to say, without taking a historical or operational route. (Of course, one must not forget the empirical justification, in contrast with modern cosmological theories!) Bohr, however, has always categorically rejected a dynamical analysis of the measurement process, and certainly he would have never applied the wave function to the "environment."
>
> Personally, I do not care much for the historical development, only for the outcome of this development. But one should require that this outcome be conceptually complete. In particular, usually I have explicitly used the term "quasiclassical" concepts, although later on I realized that Gell-Mann and Hartle had used this term to refer to something completely different from wave packets. Needless to say, in this context the wave function, or a superposition, are to be understood ontically (what else?); the use of "observables" appears to be entirely out of context in such a framework.[1]

In just these two paragraphs, Zeh takes us on a tour de force, giving not only his own assessment of Bohr's views, but also finding the opportunity to mention Wittgenstein's ladder, Maxwell's equations, Sommerfeld's textbook, cosmological theories, and the distinction between axiomatic, operational, and empirical approaches to theories. It is a feat of compression, and a wonderful example of the breadth and depth of his thinking that was always on display. Most of his emails (and papers) read like this.

In this essay, I will share a few observations on the decoherence program and the role played by Zeh in its development. I will do so through two different lenses. First, in Sect. 2, I will recount Zeh's interpretive stance toward quantum mechanics and how it shaped his conception and understanding of decoherence. Second, in Sect. 3, I will discuss some of the recent experiments on decoherence and their interplay with theoretical models, using, in particular, the example of collisional decoherence to illustrate an effort that started with Zeh's formative ideas and continues to this date. I will offer concluding remarks in Sect. 4.

2 Interpretation

Zeh's discovery [2] of the crucial importance of the environment for the description of quantum systems was made in virtual isolation and dismissed for years afterward [3–5]. Zeh concluded at the time that his early work on decoherence had all but destroyed his mainstream academic career [3, 5], and the 1970s and early 1980s constituted what he has called the "dark ages of decoherence" [6]. In his 2008 interview with Freitas [3], he recounted:

> Except for Wigner, nobody seemed to be interested in my suggestion, and so it was just neglected. This changed not until 1981, when Zurek's paper appeared. (...) I was very happy that it had appeared in the *Physical Review*. I expected that the idea would now soon become better known, which turned out to be true—slowly.

[1] My translation.

Camilleri [4] (see also his contribution to this volume) has given an analysis of the early history of decoherence, showing how it was shaped by Zeh's and Zurek's contrasting interpretive commitments. He traces one reason for the slow reception to Zeh's association of decoherence with an Everett-style "many-minds" interpretation, which Zeh had independently formulated [2, 7] when he first conceived of the ideas that later became the theory of decoherence. Zeh continued to argue in favor of this interpretation for the rest of this life [8], writing, for example, in 2006 [6]:

> A consistent interpretation of [quantum] theory in accordance with the observed world requires a *novel and nontrivial identification of observers* with appropriate quantum states of local systems which exist only in certain, dynamically autonomous *components* of the global wave function. Accordingly, it is the observer who "splits" indeterministically—not the (quantum) world. [Emphasis in the original.]

He called this interpretation "consistent" (one of his favorite terms) because, in his view, it required no separation into quantum and classical, be it at the level of terminology or physics. He suggested that by "consistently using this global unitary description, all those much discussed 'absurdities' of quantum theory can be explained" [8], and that [9]

> the Heisenberg–Bohr picture of quantum mechanics can now be claimed dead. Neither classical concepts, nor any uncertainty relations, complementarity, observables, quantum logic, quantum statistics, or quantum jumps have to be introduced on a fundamental level.

Elsewhere, he wrote that "this interpretation is an attempt to replace the 'pragmatic irrationalism' that is common in quantum theory textbooks (complementarity, dualism, fundamental uncertainty etc.) by a consistent application of just those concepts which are actually, and without exception *successfully*, used when the theory is applied" [6].

Yet it seems to me that Zeh's attachment to the Everett view was never primary. Rather, it appears to have been a *consequence* of his much deeper conviction that, as he put it in his email, "the wave function, or a superposition, are to be understood ontically (what else?)," together with the purely empirically grounded assumption of the universal validity of the Schrödinger evolution. In his interview with Freitas [3], he explained his interpretation of the wave function:

> I do believe that the wave function describes the real world. I don't mean that we may not find something entirely new some day, but from all we know now, it's the only consistent description. (...) I think it is untenable to claim that the wave function is only a tool to calculate probabilities. There are many individual (real) physical properties that have to be described by nonlocal superpositions. For example, you have to describe total spin eigenstates of a couple of particles as superpositions of product states by means of Clebsch–Gordan coefficients. Even if you take the particles apart, they have to stay in this superposition in order to conserve angular momentum. This total spin can be confirmed individually if you bring the particles back together again, or statistically by means of Bell type correlations. So the observable (hence real) total spin is defined by the nonlocal quantum state. I don't see any other way to formulate this. That's why I believe that the wave function describes reality.

He went on to elaborate:

What I mean by saying that a concept is real (or describes reality) is that it can be used consistently to describe our observations. So I would never use this word in a framework where you need complementary concepts, or a conceptual dualism, that is, where you have to switch between different concepts. So Bohr had to conclude that there is no microscopic reality in his view (and he had even to assume uncertainty relations to apply to the classical properties of macroscopic objects). I think this question came up in the beginning of quantum theory when people started asking: "If the electron behaves sometimes like a particle, and sometimes like a wave, then what is it really?" It's a question for consistency. But I think that you can consistently derive all apparent particle aspects from a universal wave function. Consider Mott's description of particle tracks, for example. So in this respect, I think that the wave function can be regarded as a concept that represents reality.

Indeed, he appeared to be open to any interpretation or version of quantum theory in which the quantum state was considered real. In this vein, his main objection to dynamical collapse theories [10] was simply that there was no experimental evidence of collapse-induced effects. As he noted in his interview with Freitas [3]:

Dynamical collapse theories are very honest and serious suggestions, but they must lead to observable consequences. One has to demonstrate experimentally where and how the global Schrödinger equation breaks down, and as long as that has not be done, it's very speculative to suggest such things. So I prefer to assume the opposite, which necessarily leads to Everett.

He was open to Bohmian mechanics, too, since it was realistic (a physical pilot wave guiding physical particles), though he found the role of particles as distinguishing the "relevant" branch of the wave function, while leaving an infinity of "empty" branches, to contribute unnecessary clutter [11]. Everett, then, seemed to him the most appealing interpretation arising from the assumption of the physical reality of the wave function, because, in his view, it was minimalistic, in the sense that it required only one physical entity (the wave function), one kinematical concept (the superposition principle), and one dynamical law (the Schrödinger equation). Faced with the ontological extravagance of such an interpretation, with its myriad parallel (but, thanks to decoherence, dynamically decoupled) branches, he argued that [9]

the existence of "other" components (with their separate conscious versions of ourselves) is a heuristic fiction, based on the assumption of a general validity of dynamical laws that have always been confirmed when tested.

His concept of a "heuristic fiction," a term he frequently used when discussing the question of the reality of the branches, was in reference to the philosophy of Vaihinger [12], who had suggested that we may treat concepts and entities "as if" they were real if they can be extrapolated from consistently confirmed physical laws.[2] Note that in Zeh's statement, the assumption of the reality of the wave function is implicit, for he makes reference only to "dynamical laws" (namely, the Schrödinger equation) as giving rise to the "heuristic fiction" of coexisting branches.

[2] When Freitas, in his interview with Zeh [3], posed the question of the reality of the "unobserved branches of the wave function," Zeh did not mention Vaihinger's concept. Instead, his response was simply that this reality "is the consequence, but I think it's also a matter of definition what you call 'real.' In any normal definition of the word, I would call [the branches] real."

Interestingly, he never warmed to attempts to derive the Born rule from within the Everett interpretation, even though this problem—and, more generally, the meaning of probability—are often thought to be the true Achilles heel of the Everett interpretation [13, 14]. He was content to say that the probability weights had to be postulated [8], emphasizing that [3]

> the only consequence you can derive from decoherence is that the branches become dynamically autonomous: separate "worlds." What remains open is only the precise probability weight, which means, for example, that not all branches are equally probable, so we do not end up in one of these so-called "maverick" components, where frequencies of measurement results do not agree with the Born rule.

Just as the Everett view of quantum mechanics was to Zeh a consequence of a global, universal wave function governed by universal unitary dynamics, so was ubiquitous entanglement and the resulting decoherence. This is why he never stopped insisting that, as was noted above, decoherence must not be separated from the interpretation of quantum mechanics. It also meant that decoherence, by itself, was unsurprising—or, as he put it [15], that it is

> a normal consequence of interacting quantum mechanical systems. It can hardly be denied to occur—but it cannot explain anything that could not have been explained before. Remarkable is only its quantitative (realistic) aspect that seems to have been overlooked for long.

This is not to imply that Zeh considered his discovery of decoherence trivial or obvious, though Joos may well be correct when he considers the comparably late recognition of the consequences of environmental entanglement a "historical accident" [16, p. 13].[3] Rather, it is to suggest that what Zeh might have ultimately found more relevant was the notion of the reality of the wave function that had led him to his discovery in the first place. As he explained in 2006 [6], the concept of decoherence

> arose as a by-product of arguments favoring either a collapse of the wave function as part of its dynamics, or an Everett-type interpretation. In contrast to the Copenhagen interpretation, which insists on fundamental classical concepts, both these interpretations regard the wave function as a complete and universal representation of reality.

Perhaps this may also help explain why Zeh did become little involved in the intense experimental and theoretical research efforts on decoherence we have seen over the past two or three decades (though he was always a close and critical observer). He may have felt that, in the way he had originally formulated decoherence, what mattered was already fundamentally in place. I was under the impression that he thought that many of the recent developments—be it experimental confirmation, realistic modeling, fresh terminology, or applications in quantum information science—did not add much that was genuinely new, and that, to the contrary, they often diluted

---

[3] As for the prehistory of decoherence, there are some remarks one can find, for example, in the writings of Heisenberg that hint at the role of the environment, albeit without the all-important reference to entanglement [1]. For instance, Heisenberg suggested that "the interference terms are ... removed by the partly undefined interactions of the measuring apparatus, with the system and with the rest of the world" [17, p. 23].

or even misrepresented the purity and universality of his conception of decoherence. One of his concerns in this context was the dissociation of decoherence from the kind of realistic, Everettian view he had been championing. Thus, when the first cavity QED experiments [18–20] allowed for the observation of a controlled decoherence process and provided an impressive confirmation of his insights about the role of the environment, he was surprised to read the authors' interpretation [20] of their observations:

> At first I expected that these results would be seen as a confirmation of universal entanglement, and hence Everett, or at least as a severe argument against Copenhagen. But it turned out to be just the opposite—they later regarded entanglement as a form of complementarity! (Quoted in Ref. [4].)

Comments such as this suggest that Zeh saw the insights brought about by decoherence theory not only as improving upon the kind of interpretive framework constructed by the Copenhagen view, but as actually being at odds with it [4]. (Of course, given the variety of viewpoints that have been lumped together under the heading of the Copenhagen interpretation [21], one needs to assess claims of this kind in the context of the specific views of individuals such as Bohr and Heisenberg [1].) As early as in his first paper on decoherence [2], Zeh was also always careful (and correct) to stress that decoherence does not solve the measurement problem, because the reduced density matrices obtained from the trace over the environment are improper and therefore not ignorance-interpretable (see also his Refs. [6, 22, 23]). This, indeed, is now the universally acknowledged position.

Another one of Zeh's concerns was with the proper definition of the term "decoherence." Zeh emphasized that decoherence should refer only to a process of uncontrollable entanglement with a large number of degrees of freedom [6]. In particular, it should be carefully distinguished from the phenomenon of "ensemble dephasing" obtained by averaging over many noisy realizations of the evolution for an ensemble of systems [22, 24]. In the literature, this distinction is often blurred, with the term "decoherence" quite loosely applied to any process that leads to a diminishing of coherence. While at the level of the reduced density matrix of the system, loss of coherence due to environmental entanglement and loss of coherence due to classical noise may manifest themselves in a similar manner, the two processes are physically distinguished in the sense that the effects of noise could in principle be reversed by local operations on the system (akin to the spin-echo technique in NMR), while undoing the effect of environmental entanglement requires measurements on the environment to recover information that has been transferred from the system [25, 26]. Because one does not usually have full control over the environment, the entanglement and resulting leakage of information from system to environment become effectively irreversible, and it is this irreversibility that Zeh emphasized as an essential property of the decoherence process [6]. (Incidentally, one of his other major research interests was the arrow of time, a topic on which he published a book [27].) Such uncontrollable entanglement is in contrast with the controlled entanglement we may use as a resource, for example, in a Bell experiment or in quantum information processing, and decoherence is typically detrimental to the presence and exploitation of such controlled entanglement [28–30].[4]

Despite Zeh's efforts to tie decoherence theory to a realistic interpretation of the quantum state, its wider reception by the physics community has been largely characterized by indifference to the attachment of a particular interpretation of quantum mechanics. This may be no surprise, as decoherence is simply an application of the standard quantum formalism to open quantum systems, and therefore its consequences can be couched in the language of any interpretation of quantum mechanics [35, 36]. In some of these interpretations, however, the concepts and insights of decoherence have come to play an essential role. For example, decoherence is a crucial ingredient in making the Everett interpretation work, since it leads to a dynamical definition of the branches and therefore addresses the preferred-basis problem that had posed a serious challenge to this type of interpretation [37–41]. A relative-state view thoroughly informed by decoherence is at the heart of Zurek's "existential interpretation" [37, 42, 43], which later came to include further results derived from system–environment entanglement, such as quantum Darwinism [44–46] and a proposed derivation of Born's rule on the basis of what Zurek has termed environment-assisted invariance [43, 47–49]. In modal interpretations [50], decoherence can specify the values to be made definite [51, 52] and identify problematic rules for value assignments [53]. In the consistent-histories interpretation [54–56], decoherence is used to dynamically define consistent, quasiclassical histories [42, 57–60] and records [26, 42, 57–59, 61], with the stability of such records explained by their redundant environmental encoding [26, 42, 48, 57, 62]. Ideas associated with the views of Bohr and his followers, such as the importance of amplification, irreversibility, and communication, have been illuminated and supported by reference to the role of the environment [44, 48, 49, 63–72]. In some interpretations, on the other hand, decoherence appears to play no important role. For example, in QBism [73], an interpretation in which quantum states encode an observer's personal probabilistic expectations associated with his future measurement interactions, decoherence would simply reflect the observer's adjustment of these expectations in light of the presence of an environment.[5]

At its most basic, interpretation-neutral level, decoherence simply addresses a consistency problem, by giving a dynamical account of how and when the quantum probability distributions approach the classically expected distributions. (This, surely, is a view of decoherence Zeh would have considered incomplete at best.) It seems that decoherence has proven plainly too useful and relevant a concept to experimenters and theoreticians alike to be claimed by any particular interpretive stance.

[4] Under very specific, carefully designed conditions, however, decoherence may sometimes act as a generator of controlled entanglement [31–34].

[5] Unsurprisingly, Zeh's own assessment of QBism was negative [8]. He saw it as an interpretation that "replaces the whole physical world by a black box, representing an abstract concept of 'information' about inconsistent classical variables, and assumed to be available to vaguely defined 'agents' rather than to observers who may be consistent parts of the physical world to be described."

Likewise, it has been found to fit too well the pragmatic, information-theoretic view of quantum mechanics suggested by modern quantum information science to be left to any particular ontological or metaphysical camp.

3 Experiments

The field of decoherence research has seen astonishing growth since the 1980s, and today decoherence has become one of the central paradigms of modern quantum mechanics [26, 74–79]. There have been at least two significant catalysts for this development. The first is the dramatic progress that has been made in the experimental realization and control of quantum systems and quantum phenomena. The other catalyst has been the influential role played by quantum information science. Both are areas for which a deep and detailed understanding of the physical sources and dynamics of decoherence is essential, and, accordingly, they have been a driving factor for research into decoherence. This has led to ever-more realistic decoherence models (below we will discuss the example of models for collisional decoherence). It has also resulted in the development of a variety of methods for the effective control and mitigation of decoherence, a task of acute importance in any implementation of quantum computing devices. Such methods include decoherence-free subspaces [80, 81], reservoir engineering [25, 82–85], dynamical decoupling [86–91], and quantum error correction [92–96]. In turn, these methods have nourished further experimental advances. Decoherence-free subspaces, for example, have been experimentally implemented in quantum systems such as photons [97, 98], trapped ions [99–102], and NMR qubits [103]. They have been experimentally shown to improve the performance of quantum algorithms [104] and fault-tolerant quantum key distribution [105], and have even found use in applications beyond quantum information, for example, in protecting a neutron interferometer from mechanical vibrations [106].

More generally, the information-theoretic view of quantum mechanics has proved helpful in understanding and formalizing decoherence as a process of environmental monitoring that results in a transfer of information from the system into quantum correlations between system and environment [44, 78, 79]. This perspective has also inspired new ways of applying the ideas of decoherence, for example, by studying the redundant storage of information in the environment, as explored in the program of quantum Darwinism [44–46].

There are numerous experiments in which the observation and control of decoherence processes are important components. Examples include atom–photon interactions in a cavity [18, 19, 107–110], matter-wave interferometry [111–115], superconducting systems [116–119], ion traps [25, 84, 120–128], quantum dots [129, 129–134], and micromechanical resonators [135–139, 139]. Some of these experiments chiefly aim to minimize decoherence so that the desired superposition states of mesoscopically or macroscopically distinguishable states can be prepared and maintained, often with an eye toward the implementation of devices for quantum information processing. Other experiments are expressedly designed to study the

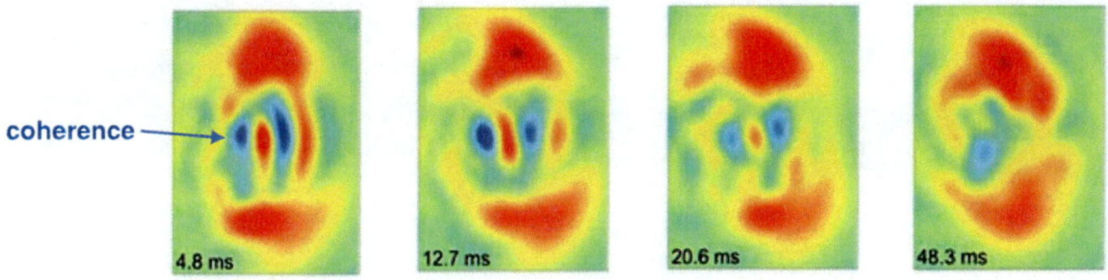

Fig. 1 Observation of decoherence in the cavity QED experiment of Ref. [110]. A superposition of two coherent photon fields with mean photon number $\bar{n} = 3.5$ is prepared and its gradual decoherence is monitored. The figure shows the 2D Wigner representation of the photon state as a function of the time. Oscillations in the region between the two peaks represent coherence between the components of the superposition. Figure adapted with permission from Ref. [110]

dynamics of decoherence itself, by making possible the observation of its gradual action, as well as by exerting appropriate control over the strength and structure of the interaction with the environment such that different decoherence timescales and preferred bases can be realized.

Two classic classes of experiments in this latter category are the cavity QED experiments of Refs. [18, 19, 107–110] and the matter-wave interferometry experiments with large molecules of Refs. [111–115]. In the cavity QED experiments, nonclassical superpositions of two coherent-field states $\left|\alpha e^{\mathrm{i}\chi}\right\rangle$ and $\left|\alpha e^{-\mathrm{i}\chi}\right\rangle$ were generated inside the cavity, followed by the observation of their controlled decoherence. Specifically, in the experiment of Ref. [110], the mean photon number was $\bar{n} = |\alpha|^2 = 3.5$ and the phase $\chi = 0.37\pi$, resulting in a very small overlap of the two state components (given by $\left|\langle\alpha e^{\mathrm{i}\chi}|\alpha e^{-\mathrm{i}\chi}\rangle\right|^2 = e^{-4|\alpha|^2 \sin^2\chi} \approx 8 \times 10^{-6}$). The gradual decoherence of the superposition was measured by reconstructing the photon state at different moments in time (see Fig. 1). The theoretically predicted decoherence time [109, 110, 145, 146] was found to be in good agreement with the experimentally measured value. In the matter-wave interferometry experiments [111–115], interference fringes were observed for large molecules in a Talbot–Lau interferometer, and controlled decay of the visibility of interference fringes due to collisions with background gas molecules [112, 113], as well as due to thermal decoherence [114], was demonstrated (see Fig. 2). For both collisional and thermal decoherence, which-path information carried away by environmental particles—in the form of gas molecules scattered by the fullerenes, and photons emitted from the heated fullerenes—led to a decrease in spatial coherence between wave-function components corresponding to passage through different slits in the diffraction grating. By adjusting the pressure of the background gas, the strength of the collisional decoherence could be tuned. Likewise, the amount of thermal decoherence was varied by heating the fullerenes to different temperatures.

A detailed understanding of the sources and dynamics of decoherence has also helped realize and observe increasingly large nonclassical superposition states [147]. For example, interference fringes for molecules consisting of up to 2,000 atoms and with masses in excess of 25,000 Da have been reported [115]. In cavity QED,

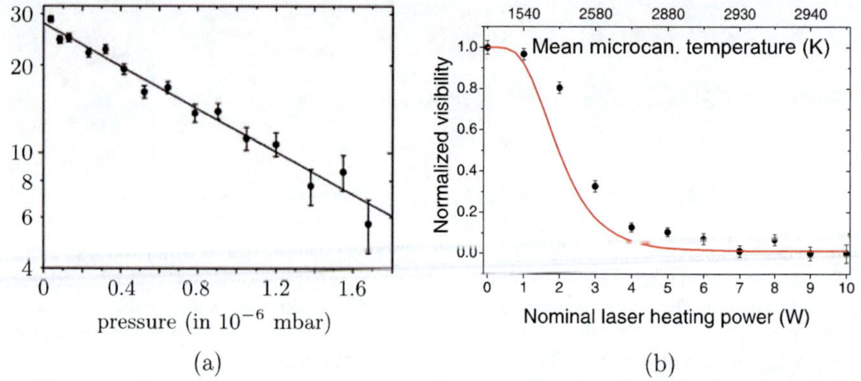

Fig. 2 Reduction of the visibility of interference fringes for C_{70} fullerenes in the presence of decoherence, as measured in the experiments of Refs. [112–114]. **a** Decoherence due to collisions with a background gas, shown as a function of gas pressure [112, 113]. The solid line represents theoretical predictions [140–142]. Figure adapted with permission from Ref. [113]. **b** Decoherence due to emission of thermal radiation from heated molecules [114]. The solid line shows the theoretical prediction [114, 142, 143]. Figure reproduced with permission from Ref. [144]

mesoscopic superposition states involving 29 photons have been realized [148], and superpositions of coherent states involving 111 photons have been reported for a superconducting qubit coupled to a waveguide cavity resonator [149]. Superconducting qubits such as quantronium, transmon, and fluxonium qubits [150] have achieved coherence times on the order of 100μs [151, 152]. A superposition of two hyperfine levels in a trapped $^9Be^+$ ion has shown coherence times of 10s [102], and superpositions of two vibrational states for a trapped $^{40}Ca^+$ ion have achieved motional center-of-mass-mode coherence times around 100ms . For an experimental realization of a decoherence-free subspace using trapped pairs of ions, coherence times of up to 34s have been observed [101, 102].

Such experiments, with their ability to realize highly nonclassical superposition states and exert exquisite control over the dominant decoherence processes, also play important roles in testing the universal validity of quantum mechanics [147, 153–159]. The goal here is to look for (or rule out) hypothetical physical mechanisms [10] that may induce a nonunitary evolution of the wave function for a closed system. The challenge in testing such collapse models is to distinguish any observation of a collapse from what may phenomenologically be explained in terms of decoherence [160, 161]. The pervasiveness of decoherence, and the fact that the effectiveness of the collapse mechanism increases with the size of the system (just as decoherence typically becomes more effective for larger systems), make the unambiguous detection of a novel collapse mechanism difficult. Many experiments have been proposed for probing the mass regimes relevant to collapse models. Examples include matter-wave interferometry [144, 147, 162], micromechanical oscillators [154, 156], interference of free nanoparticles [157, 163], molecular nanorotors [159], and even space-based experiments involving systems on the order of 10^{10} atoms [158]. The motivation for situating experiments in space comes from considerations of decoherence, since such an environment would offer very low background gas pressures (thus minimizing collisional decoherence), very low temperatures (minimizing thermal decoherence), and microgravity (minimizing decoherence due to gravitational time dilation [164]); the last property may even allow for tests of quantum gravity models [158]. None of these experiments have yet been realized, although matter-wave interferometry with molecular clusters between 10^6 and 10^8amu (the regime relevant to testing the collapse theories proposed in Refs. [165, 166]) may soon be within reach of experimental feasibility [144, 147].

It should be noted that not all reported experimental observations of a loss of coherence are due to a proper decoherence process as defined by an entanglement-mediated transfer of information to an environment. Frequently, such losses are caused by fluctuations (noise) acting locally on the system, manifesting itself in the form of dephasing for an ensemble of systems (see the brief discussion in Sect. 2 regarding the distinction between these types of coherence loss). For instance, many instances of an observed loss of coherence in superconducting quantum systems fall into this latter category, such as the dephasing of superpositions of supercurrents in SQUIDs [116], dephasing of superpositions of charge states in charge qubits [117], and dephasing of superpositions of macroscopically distinct phase states in phase qubits [118, 119]. The same is true for a number of experiments on ion traps. Here, dephasing between the internal qubit states of the ion has been observed due to sources such as the magnetic trapping field [125–128], laser fluctuations [120, 121], ac-Stark shifts [123], off-resonant excitations [122], and detuning errors [124]. Loss of coherence in ion traps has also been studied in terms of the dephasing of superpositions of motional states induced by a random noise process simulated by introducing fluctuations of the trap frequency [25, 84].

Above, we have alluded to the connection between experimental advances and the development of realistic decoherence models. This connection may be nicely illustrated with the example of collisional decoherence. The importance of collisional decoherence in accounting for preferred macroscopic states was already recognized by Zeh in his 1970 paper [2]. There, he discussed the example of the two spatially distinct configurations of a sugar molecule (right-handed and left-handed) that "interact in different ways with their environment." A dominant form of such interactions are environmental scattering processes that resolve differences between the spatial configurations, leading to the dynamical superselection of "handedness" as the stable state. This, Zeh suggested, explained why superpositions of the right-handed and left-handed configurations are not observed, even though such superpositions also form energy eigenstates, which are usually the preferred states for microscopic systems (see also Refs. [23, 167–169]).

One of the first models of collisional decoherence was studied in the classic paper by Joos and Zeh [170]. This work was an extension of research on the quantum Zeno effect Joos had done as a doctoral student under Zeh's supervision [171]. In his interview with Freitas [3], Zeh later recalled that he had instructed Joos to studiously steer clear of any reference to matters of interpretation, in order to avoid the kind of criticism and rejection from the academic community Zeh had experienced [4, 5] with his papers from the 1970s:

> I suggested that he might think about the role of the environment, but without ever talking about Everett. So he concentrated on consequences on the reduced density matrix. This was something nobody could deny.

Joos and Zeh's paper presented a master-equation treatment of the reduced density matrix and derived some classic order-of-magnitude estimates of decoherence rates. These estimates indicated the extreme efficiency with which superpositions of macroscopially distinct positions would be decohered. Even an environment as subtle, weak, and inescapable as the cosmic background radiation was found to be capable of having an appreciable effect: Joos and Zeh used the example of a dust grain of size 10^{-3}cm delocalized over a distance roughly equal to its size, and calculated a decoherence timescale on the order of one second.

In 1985, when Joos and Zeh's paper was published, such order-of-magnitude estimates were sufficient, as experiments that would actually resolve these collisional-decoherence dynamics in a gradual, controlled, and measurable fashion were not available. This changed around 2000 when the matter-wave interferometry experiments enabled an experimental test of models of collisional decoherence [112, 113, 140–142]. Now precise quantitative results were needed to compare with experimental values. Previous work by Gallis and Fleming [172] and Diósi [173] had already brought improvements and generalizations to Joos and Zeh's master equation, by relaxing the no-recoil assumption for the central particle, relaxing an assumption regarding the relative wavelengths of system and environmental particles, and addressing a mathematical issue that had resulted in decoherence rates that were too large by a factor of 2π (see also Ref. [174]). In 2003, Hornberger and Sipe [140] gave a rigorous derivation of the master equation for collisional decoherence. It used an approach quite different from Diósi's, and its predictions were found to agree well with the decrease in fringe visibility observed in the matter-wave experiments with fullerenes shown in Fig. 2 [140–142].

But this was not the final word on modeling. Hornberger and coworkers also gave a general, nonperturbative treatment based on the quantum linear Boltzmann equation [175–181], which provided a unified account of the dynamical interplay between phase-space decoherence and dissipation. Previous results were recovered as limiting cases, and it was shown that the pointer states dynamically selected by the environmental interaction were exponentially localized solitonic wave functions obeying the classical equations of motion [178, 180]. Subsequently, insights gleaned from molecular interferometry experiments have led to further refinements of these models. For example, near-field interferometry with massive molecules was found to be highly sensitive to molecular rotations [182, 183]. This observation helped motivate the generalization of previous models of collisional decoherence, in which the central particle was treated as a pointlike particle with no orientational degrees of freedom, to the derivation of master equations describing the spatio-orientational decoherence of rotating, anisotropic, nonspherical molecules immersed into an environment of gas particles [184–188]. These models were found to make predictions in good agreement with experimental data [187].

4 Concluding Remarks

An early version of what would become Zeh's 1970 paper [2] was rejected with the argument that "quantum theory does not apply to macroscopic objects" [4]. Even if such an application was not necessarily considered explicitly forbidden, a common view seems to have been that such an application would be either too difficult in practice (given the complexity of larger systems) or unnecessary (since the systems are known to behave classically, i.e., not to exhibit quantum effects). Indeed, for a long time the distinction between quantum and classical was often synonymous with the distinction between microsopic and macrosopic. In such light, Zeh's early work may be seen as motivated by his sense that this conflation was in error, leading him to his suggestion that everything, no matter its size, not only *could*, but *should* be described by a (realistically interpreted) quantum-mechanical wave function. This approach he referred to as a "universally valid quantum theory," which he defined in his 1970 paper as a "universe [described] by a wave function that obeys the Schrödinger equation" [2, p. 73].

To Schrödinger himself, the consideration of an amplifying chain of entanglement from microscopic to macroscopic scales had served merely as a *reductio ad absurdum*, meant to illustrate the "ridiculous" consequences of applying quantum mechanics in this way. Zeh not only viewed these consequences in a fresh and unprejudiced way, but, crucially, he also realized that the very quantum property of entanglement, "*the* characteristic trait of quantum mechanics, the one that enforces its entire departure from classical lines of thought" (Schrödinger's famous words [189, p. 555]), could provide a dynamical explanation of the emergence of classicality. Thus, in a twist of irony, the justification for why macroscopic systems may be treated classically came from Zeh's radically quantum-mechanical treatment of such systems. In this way, decoherence theory showed why the quantum–classical and microscopic–macroscopic boundaries, although fundamentally independent of each other, *appear* to coincide in practice. This became especially clear with Joos and Zeh's 1985 paper [170] and Zurek's estimate of a decoherence timescale [190],[6] both of which demonstrated that, in general, the larger the system, the stronger its decoherence, and, furthermore, that on mesoscopic and macroscopic scales such decoherence is virtually instantaneous compared to the timescales for dissipation and classical noise effects.

---

[6] Among these early studies, one should also mention the papers by Walls and Milburn [145] and by Caldeira and Leggett [191], which investigated the influence of damping on the coherence of superpositions. In particular, the master equation derived by Caldeira and Leggett [192] has been widely used in modeling decoherence and dissipation [172, 193, 194], and Camilleri [4] has suggested that Caldeira and Leggett's work "was of decisive importance not only for the mathematical treatment and conceptual development of environment-induced decoherence, but also for the interpretational context." It is important to bear in mind, however, that while damping (dissipation) is always accompanied by a loss of coherence, decoherence may be substantial even when there is no loss of energy from the system, as already demonstrated by one of the first models of decoherence [195].

The fundamental idea of decoherence theory—that ubiquitous, uncontrollable entanglement leads to local nonobservability of interference and dynamical superselection, as already spelled out in Zeh's seminal 1970 paper [2]—has proven uniquely fruitful, inspiring and making possible new experiments, as well as revolutionizing our theoretical understanding and experimental control of the quantum-to-classical transition. Given the current interest in the construction of devices for quantum computing, it is clear that decoherence will continue to play a central role in quantum science for the foreseeable future. If anything, its role will only amplify, as ever-larger multiqubit systems are realized and quantum phenomena involving coherence and entanglement are explored on increasingly macroscopic scales. The search for deviations from standard quantum mechanics, while not as prominent a research effort as quantum information science, will surely provide additional impetus for theoretical and experimental studies of decoherence.

We must be grateful to Zeh not only for his original insight, but also for the tenacity with which he pursued his ideas. He was a pioneer, and a bold and independent thinker. His voice will be missed.

References

1. K. Camilleri, M. Schlosshauer, Stud. Hist. Phil. Mod. Phys. **49**, 73 (2015)
2. H.D. Zeh, Found. Phys. **1**, 69 (1970)
3. Interview with H. Dieter Zeh, conducted at Waldhilsbach, Germany, by Fábio Freitas, July 25–26, 2008. Audio recording and transcript deposited at the American Institute of Physics
4. K. Camilleri, Stud. Hist. Phil. Mod. Phys. **40**, pages 290 (2009)
5. O. Freire Jr., Stud. Hist. Phil. Mod. Phys.**40**, 280 (2009). https://doi.org/10.1016/j.shpsb.2009.09.002
6. H.D. Zeh, *Quantum Decoherence*, ed. by B. Duplantier, J.-M. Raimond, V. Rivasseau (Birkhäuser, Basel, 2006), pp. 151–175
7. H.D. Zeh, Found. Phys. **3**, 109 (1973)
8. H. D. Zeh, Z. Naturf. **71a**, 195 (2016)
9. H.D. Zeh, Found. Phys. Lett. **13**, 221 (2000)
10. A. Bassi, G.C. Ghirardi, Phys. Rep. **379**, 257 (2003)
11. H.D. Zeh, Found. Phys. Lett. **12**, 197 (1999)
12. H. Vaihinger, *The Philosophy of "As if": A System of the Theoretical, Practical and Religious Fictions of Mankind*, 2nd ed. (Routledge & Kegan Paul, London, 1935). (translated by C. K. Ogden)
13. M. Hemmo, I. Pitowsky, Stud. Hist. Phil. Mod. Phys. **38**, 333 (2007). https://doi.org/10.1016/j.shpsb.2006.04.005
14. S. Saunders, J. Barrett, A. Kent, D. Wallace (eds.), *Many Worlds? Everett, Quantum Theory and Reality* (Oxford University Press, Oxford, 2010)
15. H.D. Zeh, in *New Developments on Fundamental Problems Quantum Physics (Oviedo II)*. ed. by M. Ferrero, A. van der Merwe (Kluwer, Dordrecht, 1997), pp. 441–452
16. E. Joos, *Decoherence: Theoretical, Experimental, and Conceptual Problems*, ed. by P. Blanchard, D. Giulini, E. Joos, C. Kiefer, I.-O. Stamatescu (Springer, Berlin, 2000), pp. 1–17
17. W. Heisenberg, *Niels Bohr and the Development of Physics: Essays Dedicated to Niels Bohr on the Occasion of his Seventieth Birthday*, ed. by W. Pauli, L. Rosenfeld, V. Weisskopf (McGraw Hill, New York, 1955), pp. 12–29

18. M. Brune, E. Hagley, J. Dreyer, X. Maître, A. Maali, C. Wunderlich, J.M. Raimond, S. Haroche, Phys. Rev. Lett. **77**, 4887 (1996)
19. X. Maître, E. Hagley, J. Dreyer, A. Maali, C.W.M. Brune, J.M. Raimond, S. Haroche, J. Mod. Opt. **44**, 2023 (1997)
20. J.M. Raimond, M. Brune, S. Haroche, Phys. Rev. Lett. **79**, 1964 (1997)
21. D. Howard, Phil. Sci. **71**, 669 (2004)
22. H.D. Zeh, *Science and Ultimate Reality: Quantum Theory, Cosmology and Complexity*, ed. by J.D. Barrow, P.C.W. Davies, C.L. Harper Jr. (Cambridge University Press, Cambridge, 2004), pp. 103–120
23. H.D. Zeh, in *Decoherence: Theoretical, Experimental, and Conceptual Problems*. ed. by P. Blanchard, D. Giulini, E. Joos, C. Kiefer, I. Stamatescu. Lecture Notes in Physics No. 538. (Springer, Berlin, 2000), pp. 19–42
24. H.D. Zeh, *Decoherence and the Appearance of a Classical World in Quantum Theory*, ed. by E. Joos, H.D. Zeh, C. Kiefer, D. Giulini, J. Kupsch, I.-O. Stamatescu, 2nd ed., Chap. 2 (Springer, New York, 2003)
25. C.J. Myatt, B.E. King, Q.A. Turchette, C.A. Sackett, D. Kielpinski, W.M. Itano, C. Monroe, D.J. Wineland, Nature **403**, 269 (2000)
26. W.H. Zurek, Rev. Mod. Phys. **75**, 715 (2003). https://doi.org/10.1103/RevModPhys.75.715
27. H.D. Zeh, *The Physical Basis of the Direction of Time, edition*, 5th edn. (Springer, Berlin, 2007)
28. K. Żelse Żfiyczkowski, P. Horodecki, M. Horodecki, R. Horodecki, Phys. Rev. A **65**, 012101 (2001). https://doi.org/10.1103/PhysRevA.65.012101
29. J. Lee, I. Kim, D. Ahn, H. McAneney, M.S. Kim, Phys. Rev. A **70**, 024301 (2004). https://doi.org/10.1103/PhysRevA.70.024301
30. J.T. Barreiro, P. Schindler, O. Gühne, T. Monz, M. Chwalla, C.F. Roos, M. Hennrich, R. Blatt, Nat. Phys. **6**, 943 (2010)
31. D. Braun, Phys. Rev. Lett. **89**, 277901 (2002). https://doi.org/10.1103/PhysRevLett.89.277901
32. F. Benatti, R. Floreanini, M. Piani, Phys. Rev. Lett. **91**, 070402 (2003). https://doi.org/10.1103/PhysRevLett.91.070402
33. M.S. Kim, J. Lee, D. Ahn, P.L. Knight, Phys. Rev. A **65**, 040101 (2002). https://doi.org/10.1103/PhysRevA.65.040101
34. L. Jakóbczyk,https://doi.org/10.1088/0305-4470/35/30/313 J. Phys. A: Math. Theor. **35**, 6383 (2002)
35. G. Bacciagaluppi, *The Stanford Encyclopedia of Philosophy*, ed. by E.N. Zalta (2012). http://plato.stanford.edu/archives/win2012/entries/qm-decoherence
36. M. Schlosshauer, Rev. Mod. Phys. **76**, 1267 (2004)
37. W.H. Zurek, Philos. Trans. R. Soc. Lond. Ser. A **356**, 1793 (1998)
38. J.N. Butterfield, *Quantum Mechanics: Scientific Perspectives on Divine Action*, ed. by R.J. Russell, P. Clayton, K. Wegter-McNelly, J. Polkinghorne (Vatican Observatory and The Center for Theology and the Natural Sciences, Vatican City State, 2001), pp. 111–140. (also available as an eprint from the Pittsburgh Philosophy of Science Archive at http://philsci-archive.pitt.edu/archive/00000203)
39. D. Wallace, Stud. Hist. Philos. Mod. Phys. **33**, 637 (2002)
40. D. Wallace, Stud. Hist. Philos. Mod. Phys. **34**, 87 (2003)
41. D. Wallace, *Many Worlds? Everett, Quantum Theory and Reality*, ed. by S. Saunders, J. Barrett, A. Kent, D. Wallace (Oxford University Press, Oxford, 2010), pp. 53–72
42. W.H. Zurek, Prog. Theor. Phys. **89**, 281 (1993)
43. W.H. Zurek, Phys. Rev. A **71**, 052105 (2005)
44. W.H. Zurek, Nat. Phys. **5**, 181 (2009)
45. W.H. Zurek, Phys. Today **67**, 44 (2014)
46. M. Zwolak, C.J. Riedel, W.H. Zurek, Sci. Rep. **6**, 25277 (2016)
47. W.H. Zurek, Phys. Rev. Lett. **90**, 120404 (2003)

48. W.H. Zurek, *Science and Ultimate Reality*, ed. by J.D. Barrow, P.C.W. Davies, C.H. Harper (Cambridge University Press, Cambridge, England, 2004), pp. 121–137
49. W.H. Zurek, Phil. Trans. R. Soc. A **376**, 20180107 (2018)
50. R. Clifton, Br. J. Philos. Sci. **47**, 371 (1996)
51. G. Bacciagaluppi, M. Hemmo, Stud. Hist. Philos. Mod. Phys. **27**, 239 (1996)
52. G. Bene, (2001). http://arxiv.org/abs/quant-ph/0104112quant-ph/0104112
53. G. Bacciagaluppi, Found. Phys. **30**, 1431 (2000)
54. R.B. Griffiths, J. Stat. Phys. **36**, 219 (1984)
55. R. Omnès, *The Interpretation of Quantum Mechanics* (Princeton University Press, Princeton, 1994)
56. R.B. Griffiths, *Consistent Quantum Theory* (Cambridge University Press, Cambridge, 2002)
57. J.P. Paz, W.H. Zurek, Phys. Rev. D **48**, 2728 (1993)
58. A. Albrecht, Phys. Rev. D **46**, 5504 (1992)
59. A. Albrecht, Phys. Rev. D **48**, 3768 (1993)
60. J. Twamley, Phys. Rev. D **48**, 5730 (1993)
61. M. Gell-Mann, J.B. Hartle, *Quantum Classical Correspondence: The 4th Drexel Symposium on Quantum Nonintegrability*, ed. by D.H. Feng and B.L. Hu (International Press, Cambridge, Massachussetts, 1998), pp. 3–35
62. C.J. Riedel, W.H. Zurek, M. Zwolak, Phys. Rev. A **93**, 032126 (2016)
63. H. Ollivier, D. Poulin, W.H. Zurek, Phys. Rev. Lett. **93**, 220401 (2004)
64. H. Ollivier, D. Poulin, W.H. Zurek, Phys. Rev. A **72**, 042113 (2005)
65. R. Blume-Kohout, W.H. Zurek, Found. Phys. **35**, 1857 (2005)
66. R. Blume-Kohout, W.H. Zurek, Phys. Rev. A **73**, 062310 (2006)
67. C.J. Riedel, W.H. Zurek, Phys. Rev. Lett. **105**, 020404 (2010)
68. C.J. Riedel, W.H. Zurek, New J. Phys. **13**, 073038 (2011)
69. C.J. Riedel, W.H. Zurek, M. Zwolak, New J. Phys. **14**, 083010 (2012)
70. A. Streltsov, W.H. Zurek, Phys. Rev. Lett. **111**, 040401 (2013)
71. W.H. Zurek, Phys. Rev. A **87**, 052111 (2013)
72. W.H. Zurek, Phil. Trans. R. Soc. A **376**, 20170315 (2018)
73. C.A. Fuchs, N.D. Mermin, R. Schack, Am. J. Phys. **82**, 749 (2014)
74. J.P. Paz, W.H. Zurek, *Coherent Atomic Matter Waves, Les Houches Session LXXII*, vol. 72, ed. by R. Kaiser, C. Westbrook, F. David. Les Houches Summer School Series (Springer, Berlin, 2001), pp. 533–614
75. H.-P. Breuer, F. Petruccione, *The Theory of Open Quantum Systems* (Oxford University Press, Oxford, 2002)
76. E. Joos, H.D. Zeh, C. Kiefer, D. Giulini, J. Kupsch, I.-O. Stamatescu, *Decoherence and the Appearance of a Classical World in Quantum Theory, edition*, 2nd edn. (Springer, New York, 2003)
77. M. Schlosshauer, *Decoherence and the Quantum-to-Classical Transition* (Springer, Berlin/Heidelberg, 2007)
78. K. Hornberger, in *Entanglement and Decoherence: Foundations and Modern Trends*, ed. by A. Buchleitner, C. Viviescas, M. Tiersch. Lecture Notes in Physics, vol. 768 (Springer, Berlin, 2009), pp. 221–276
79. M. Schlosshauer, Phys. Rep. **831**, 1 (2019). https://doi.org/10.1016/j.physrep.2019.10.001
80. D.A. Lidar, K.B. Whaley, *Irreversible Quantum Dynamics*, vol. 622, edi. by F. Benatti, R. Floreanini. Springer Lecture Notes in Physics (Springer, Berlin, 2003), pp. 83–120. (also available as eprint quant-ph/0301032)
81. D.A. Lidar, Adv. Chem. Phys. **154**, 295 (2014)
82. D.A.R. Dalvit, J. Dziarmaga, W.H. Zurek, Phys. Rev. A **62**, 013607 (2000)
83. J.F. Poyatos, J.I. Cirac, P. Zoller, Phys. Rev. Lett. **77**, 4728 (1996)
84. Q.A. Turchette, C.J. Myatt, B.E. King, C.A. Sackett, D. Kielpinski, W.M. Itano, C. Monroe, D.J. Wineland, Phys. Rev. A **62**, 053807 (2000)
85. A.R.R. Carvalho, P. Milman, R.L. de Matos Filho, L. Davidovich, Phys. Rev. Lett. **86**, 4988 (2001)

86. L. Viola, S. Lloyd, Phys. Rev. A **58**, 2733 (1998)
87. L. Viola, E. Knill, S. Lloyd, Phys. Rev. Lett. **82**, 2417 (1999)
88. P. Zanardi, Phys. Lett. A **258**, 77 (1999)
89. L. Viola, E. Knill, S. Lloyd, Phys. Rev. Lett. **85**, 3520 (2000)
90. L.-A. Wu, D.A. Lidar, Phys. Rev. Lett. **88**, 207902 (2002)
91. L.-A. Wu, M.S. Byrd, D.A. Lidar, Phys. Rev. Lett. **89**, 127901 (2002)
92. A.M. Steane, Phys. Rev. Lett. **77**, 793 (1996)
93. P.W. Shor, Phys. Rev. A **52**, R2493 (1995)
94. A.M. Steane, *Decoherence and Its Implications in Quantum Computation and Information Transfer*, ed. by P. Turchi, A. Gonis (IOS Press, Amsterdam, 2001), pp. 284–298. (also available as eprint quant-ph/0304016)
95. F. Gaitan, *Quantum Error Correction and Fault Tolerant Quantum Computing* (CRC Press, Boca Raton, 2008)
96. D.A. Lidar, T.A. Brun (eds.), *Quantum Error Correction* (Cambridge University Press, Cambridge, 2013)
97. P.G. Kwiat, A.J. Berglund, J.B. Altepeter, A.G. White, Science **290**, 498 (2000)
98. J.B. Altepeter, P.G. Hadley, S.M. Wendelken, A.J. Berglund, P.G. Kwiat, Phys. Rev. Lett. **92**, 147901 (2004)
99. D. Kielpinski, V. Meyer, M.A. Rowe, C.A. Sackett, W.M. Itano, C. Monroe, D.J. Wineland, Science **291**, 1013 (2001)
100. C.F. Roos, G.P.T. Lancaster, M. Riebe, H. Häffner, W. Hänsel, S. Gulde, C. Becher, J. Eschner, F. Schmidt-Kaler, R. Blatt, Phys. Rev. Lett. **92**, 220402 (2004). https://doi.org/10.1103/PhysRevLett.92.220402
101. H. Häffner, F. Schmidt-Kaler, W. Hänsel, C.F. Roos, T. Körber, M. Chwalla, M. Riebe, J. Benhelm, U.D. Rapol, C. Becher, R. Blatt, Appl. Phys. B **81**, 151 (2005)
102. C. Langer, R. Ozeri, J.D. Jost, J. Chiaverini, B. DeMarco, A. Ben-Kish, R.B. Blakestad, J. Britton, D.B. Hume, W.M. Itano, D. Leibfried, R. Reichle, T. Rosenband, T. Schaetz, P.O. Schmidt, D.J. Wineland, Phys. Rev. Lett. **95**, 060502 (2005). https://doi.org/10.1103/PhysRevLett.95.060502
103. L. Viola, E.M. Fortunato, M.A. Pravia, E. Knill, R. Laflamme, D.G. Cory, Science **293**, 2059 (2001)
104. M. Mohseni, J.S. Lundeen, K.J. Resch, A.M. Steinberg, Phys. Rev. Lett. **91**, 187903 (2003)
105. Q. Zhang, J. Yin, T.-Y. Chen, S. Lu, J. Zhang, X.-Q. Li, T. Yang, X.-B. Wang, J.-W. Pan, Phys. Rev. A **73**, 020301(R) (2006)
106. D.A. Pushin, M.G. Huber, M. Arif, D.G. Cory, Phys. Rev. Lett. **107**, 150401 (2011)
107. J.M. Raimond, M. Brune, S. Haroche, Rev. Mod. Phys. **73**, 565 (2001)
108. R. Kaiser, C. Westbrook, F. David (eds.), *Coherent Atomic Matter Waves, Les Houches Session LXXII, Les Houches Summer School Series* (Springer, Berlin, 2001)
109. S. Haroche, J.-M. Raimond, *Exploring the Quantum: Atoms, Cavities, and Photons* (Oxford University Press, Oxford, 2006)
110. S. Deléglise, I. Dotsenko, C. Sayrin, J. Bernu, M. Brune, J.-M. Raimond, S. Haroche, Nature **455**, 510 (2008)
111. M. Arndt, O. Nairz, J. Vos-Andreae, C. Keller, G. van der Zouw, A. Zeilinger, Nature **401**, 680 (1999)
112. K. Hornberger, S. Uttenthaler, B. Brezger, L. Hackermüller, M. Arndt, A. Zeilinger, Phys. Rev. Lett. **90**, 160401 (2003)
113. L. Hackermüller, K. Hornberger, B. Brezger, A. Zeilinger, M. Arndt, Appl. Phys. B **77**, 781 (2003)
114. L. Hackermüller, K. Hornberger, B. Brezger, A. Zeilinger, M. Arndt, Nature **427**, 711 (2004)
115. Y.Y. Fein, P. Geyer, P. Zwick, F. Kiałka, S. Pedalino, M. Mayor, S. Gerlich, M. Arndt, Nat. Phys. **15**, 1242 (2019). https://doi.org/10.1038/s41567-019-0663-9
116. I. Chiorescu, Y. Nakamura, C.J.P.M. Harmans, J.E. Mooij, Science **21**, 1869 (2003)
117. D. Vion, A. Aassime, A. Cottet, P. Joyez, H. Pothier, C. Urbina, D. Esteve, M.H. Devoret, Science **296**, 886 (2002)

118. Y. Yu, S. Han, X. Chu, S.-I. Chu, Z. Wang, Science **296**, 889 (2002)
119. J.M. Martinis, S. Nam, J. Aumentado, C. Urbina, Phys. Rev. Lett. **89**, 117901 (2002)
120. S. Schneider, G.J. Milburn, Phys. Rev. A **57**, 3748 (1998)
121. C. Miquel, J.P. Paz, W.H. Zurek, Phys. Rev. Lett. **78**, 3971 (1997)
122. A. Steane, C.F. Roos, D. Stevens, A. Mundt, D. Leibfried, F. Schmidt-Kaler, R. Blatt, Phys. Rev. A **62**, 042305 (2000). https://doi.org/10.1103/PhysRevA.62.042305
123. H. Häffner, S. Gulde, M. Riebe, G. Lancaster, C. Becher, J. Eschner, F. Schmidt-Kaler, R. Blatt, Phys. Rev. Lett. **90**, 143602 (2003). https://doi.org/10.1103/PhysRevLett.90.143602
124. D. Leibfried, B. DeMarco, V. Meyer, D. Lucas, M. Barrett, J. Britton, W.M. Itano, B. Jelenković, C. Langer, T. Rosenband, D.J. Wineland, Nature **422**, 412 (2003)
125. F. Schmidt-Kaler, S. Gulde, M. Riebe, T. Deuschle, A. Kreuter, G. Lancaster, C. Becher, J. Eschner, H. Häffner, R. Blatt, J. Phys. B **36**, 623 (2003)
126. S. Brouard, J. Plata, Phys. Rev. A **70**, 013413 (2004)
127. T. Grotz, L. Heaney, W.T. Strunz, Phys. Rev. A **74**, 022102 (2006)
128. H. Häffner, C.F. Roos, R. Blatt, Phys. Rep. **469**, 155 (2008). https://doi.org/10.1016/j.physrep.2008.09.003
129. A.V. Kuhlmann, J. Houel, A. Ludwig, L. Greuter, D. Reuter, A.D. Wieck, M. Poggio, R.J. Warburton, Nat. Phys. **9**, 570 (2013)
130. C. Arnold, V. Loo, A. Lemaître, I. Sagnes, O. Krebs, P. Voisin, P. Senellart, L. Lanco, Phys. Rev. X **4**, 021004 (2014). https://doi.org/10.1103/PhysRevX.4.021004
131. J. Fischer, M. Trif, W.A. Coish, D. Loss, Solid State Commun. **149**, 1443 (2009)
132. B. Urbaszek, X. Marie, T. Amand, O. Krebs, P. Voisin, P. Maletinsky, A. Högele, A. Imamoglu, Rev. Mod. Phys. **85**, 79 (2013)
133. A. Delteil, W.-b. Gao, P. Fallahi, J. Miguel-Sanchez, A. Imamo ğelse ğfilu, Phys. Rev. Lett. **112**, 116802 (2014). https://doi.org/10.1103/PhysRevLett.112.116802
134. P. Tighineanu, C.L. Dreeßen, C. Flindt, P. Lodahl, A.S. Sørensen, Phys. Rev. Lett. **120**, 257401 (2018)
135. K.Y. Fong, W.H.P. Pernice, H.X. Tang, Phys. Rev. B **85**, 161410 (2012). https://doi.org/10.1103/PhysRevB.85.161410
136. Y. Zhang, J. Moser, J. Güttinger, A. Bachtold, M.I. Dykman, Phys. Rev. Lett. **113**, 255502 (2014). https://doi.org/10.1103/PhysRevLett.113.255502
137. T. Miao, S. Yeom, P. Wang, B. Standley, M. Bockrath, Nano Lett. **14**, 2982 (2014)
138. J. Moser, A. Eichler, J. Güttinger, M.I. Dykman, A. Bachtold, Nat. Nanotechnol. **9**, 1007 (2014)
139. O. Maillet, F. Vavrek, A.D. Fefferman, O. Bourgeois, E. Collin, New J. Phys. **18**, 073022 (2016)
140. K. Hornberger, J.E. Sipe, Phys. Rev. A **68**, 012105 (2003)
141. K. Hornberger, J.E. Sipe, M. Arndt, Phys. Rev. A **70**, 053608 (2004)
142. K. Hornberger, L. Hackermüller, M. Arndt, Phys. Rev. A **71**, 023601 (2005)
143. K. Hornberger, Phys. Rev. A **73**, 052102 (2006)
144. K. Hornberger, S. Gerlich, S. Nimmrichter, P. Haslinger, M. Arndt, Rev. Mod. Phys. **84**, 157 (2012)
145. D.F. Walls, G.J. Milburn, Phys. Rev. A **31**, 2403 (1985)
146. M. Brune, S. Haroche, J.M. Raimond, L. Davidovich, N. Zagury, Phys. Rev. A **45**, 5193 (1992)
147. M. Arndt, K. Hornberger, Nature Phys. **10**, 271 (2014)
148. A. Auffeves, P. Maioli, T. Meunier, S. Gleyzes, G. Nogues, M. Brune, J.M. Raimond, S. Haroche, Phys. Rev. Lett. **91**, 230405 (2003)
149. B. Vlastakis, G. Kirchmair, Z. Leghtas, S.E. Nigg, L. Frunzio, S.M. Girvin, M. Mirrahimi, M.H. Devoret, R.J. Schoelkopf, Science **342**, 607 (2013). https://doi.org/10.1126/science.1243289
150. M.H. Devoret, R.J. Schoelkopf, Science **339**, 1169 (2013)
151. C. Rigetti, J.M. Gambetta, S. Poletto, B.L.T. Plourde, J.M. Chow, A.D. Córcoles, J.A. Smolin, S.T. Merkel, J.R. Rozen, G.A. Keefe, M.B. Rothwell, M.B. Ketchen, M. Steffen, Phys. Rev. B **86**, 100506 (2012). https://doi.org/10.1103/PhysRevB.86.100506

152. A.P. Sears, A. Petrenko, G. Catelani, L. Sun, H. Paik, G. Kirchmair, L. Frunzio, L.I. Glazman, S.M. Girvin, R.J. Schoelkopf, Phys. Rev. B **86**, 180504 (2012). https://doi.org/10.1103/PhysRevB.86.180504
153. A.J. Leggett, J. Phys. Condens. Matter **14**, R415 (2002)
154. W. Marshall, C. Simon, R. Penrose, D. Bouwmeester, Phys. Rev. Lett **91**, 130401 (2003)
155. A. Bassi, E. Ippoliti, S.L. Adler, Phys. Rev. Lett. **94**, 030401 (2005)
156. I. Pikovski, M.R. Vanner, M. Aspelmeyer, M.S. Kim, Č Brukner, Nat. Phys. **8**, 393 (2012)
157. C. Wan, M. Scala, G.W. Morley, A.A. Rahman, H. Ulbricht, J. Bateman, P.F. Barker, S. Bose, M.S. Kim, Phys. Rev. Lett. **117**, 143003 (2016)
158. R. Kaltenbaek, M. Aspelmeyer, P.F. Barker, A. Bassi, J. Bateman, K. Bongs, S. Bose, C. Braxmaier, Č. Brukner, B. Christophe, M. Chwalla, P.-F. Cohadon, A.M. Cruise, C. Curceanu, K. Dholakia, L. Diósi, K. Döringshoff, W. Ertmer, J. Gieseler, N. Gürlebeck, G. Hechenblaikner, A. Heidmann, S. Herrmann, S. Hossenfelder, U. Johann, N. Kiesel, M. Kim, C. Lämmerzahl, A. Lambrecht, M. Mazilu, G.J. Milburn, H. Müller, L. Novotny, M. Paternostro, A. Peters, I. Pikovski, A. Pilan Zanoni, E.M. Rasel, S. Reynaud, C.J. Riedel, M. Rodrigues, L. Rondin, A. Roura, W.P. Schleich, J. Schmiedmayer, T. Schuldt, K.C. Schwab, M. Tajmar, G.M. Tino, H. Ulbricht, R. Ursin, V. Vedral, EPJ Q. Technol. **3**, 5 (2016). https://doi.org/10.1140/epjqt/s40507-016-0043-7
159. B.A. Stickler, B. Papendell, S. Kuhn, B. Schrinski, J. Millen, M. Arndt, K. Hornberger, New J. Phys. **20**, 122001 (2018)
160. M. Tegmark, Found. Phys. Lett. **6**, 571 (1993)
161. S. Nimmrichter, K. Hornberger, Phys. Rev. Lett. **110**, 160403 (2013)
162. S. Nimmrichter, K. Hornberger, P. Haslinger, M. Arndt, Phys. Rev. A **83**, 043621 (2011)
163. O. Romero-Isart, A.C. Pflanzer, F. Blaser, R. Kaltenbaek, N. Kiesel, M. Aspelmeyer, J.I. Cirac, Phys. Rev. Lett. **107**, 020405 (2011)
164. I. Pikovski, M. Zych, F. Costa, Č. Brukner, Nat. Phys. 668 (2015)
165. S.L. Adler, J. Phys. A **40**, 2935 (2007)
166. A. Bassi, D.-A. Deckert, L. Ferialdi, EPL **92**, 50006 (2010)
167. R.A. Harris, L. Stodolsky, J. Chem. Phys. **74**, 2145 (1981)
168. J. Trost, K. Hornberger, Phys. Rev. Lett. **103**, 023202 (2009). https://doi.org/10.1103/PhysRevLett.103.023202
169. M. Bahrami, A. Shafiee, A. Bassi, Phys. Chem. Chem. Phys. **14**, 9214 (2012)
170. E. Joos, H.D. Zeh, Z. Phys. B Condens. Matter **59**, 223 (1985)
171. E. Joos, Phys. Rev. D **29**, 1626 (1984). https://doi.org/10.1103/PhysRevD.29.1626
172. M.R. Gallis, G.N. Fleming, Phys. Rev. A **42**, 38 (1990)
173. L. Diósi, Europhys. Lett. **30**, 63 (1995)
174. S.L. Adler, J. Phys. A Math. Gen. **39**, 14067 (2006)
175. K. Hornberger, Phys. Rev. Lett. **97**, 060601 (2006)
176. K. Hornberger, EPL **77**, 50007 (2007)
177. K. Hornberger, B. Vacchini, Phys. Rev. A **77**, 022112 (2008)
178. M. Busse, K. Hornberger, J. Phys. A Math. Theor. **42**, 362001 (2009)
179. B. Vacchini, K. Hornberger, Phys. Rep. **478**, 71 (2009)
180. M. Busse, K. Hornberger, J. Phys. A Math. Theor. **43**, 015303 (2010)
181. M. Busse, P. Pietrulewicz, H.-P. Breuer, K. Hornberger, Phys. Rev. E **82**, 026706 (2010)
182. M. Gring, S. Gerlich, S. Eibenberger, S. Nimmrichter, T. Berrada, M. Arndt, H. Ulbricht, K. Hornberger, M. Müri, M. Mayor, M. Böckmann, N.L. Doltsinis, Phys. Rev. A **81**, 031604(R) (2010)
183. B.A. Stickler, K. Hornberger, Phys. Rev. A **92**, 023619 (2015)
184. K. Walter, B.A. Stickler, K. Hornberger, Phys. Rev. A **93**, 063612 (2016)
185. B.A. Stickler, B. Papendell, K. Hornberger, Phys. Rev. A **94**, 033828 (2016)
186. B. Papendell, B.A. Stickler, K. Hornberger, New J. Phys. **19**, 122001 (2017)
187. B.A. Stickler, F.T. Ghahramani, K. Hornberger, Phys. Rev. Lett. **121**, 243402 (2018)
188. B.A. Stickler, B. Schrinski, K. Hornberger, Phys. Rev. Lett. **121**, 040401 (2018)
189. E. Schrödinger, Proc. Cambridge Philos. Soc. **31**, 555 (1935)

190. W.H. Zurek, *Frontiers of Nonequilibrium Statistical Mechanics*, ed. by G.T. Moore, M.O. Scully (Plenum Press, New York, 1986), pp. 145–149. (first published in 1984 as Los Alamos report LAUR 84-2750)
191. A.O. Caldeira, A.J. Leggett, Phys. Rev. A **31**, 1059 (1985)
192. A.O. Caldeira, A.J. Leggett, Physica A **121**, 587 (1983)
193. M.R. Gallis, Phys. Rev. A **45**, 47 (1992)
194. J.R. Anglin, J.P. Paz, W.H. Zurek, Phys. Rev. A **55**, 4041 (1997)
195. W.H. Zurek, Phys. Rev. D **26**, 1862 (1982). https://doi.org/10.1103/PhysRevD.26.1862

Experimental Decoherence in Molecule Interferometry

Markus Arndt, Stefan Gerlich, and Klaus Hornberger

1 Emergence of Classical Behavior in Quantum Experiments

This contribution is written in honor of Dieter Zeh, a pioneer in developing our understanding of the transition from quantum behavior to the appearance of classicality [1]. Starting with his work with Joos and Zeh [2], decoherence theory has evolved from a stance on the foundations of quantum physics to a practical guide of what must be done and what has to be averted if one seeks to observe quantum phenomena in increasingly complex and macroscopic systems. Indeed, current advances in quantum technologies such as matter-wave interferometry, quantum-assisted sensing and metrology, quantum simulation, and quantum computing owe their success to no small degree to a better understanding and control of decoherence phenomena.

In a nutshell, decoherence theory focuses on how a quantum system affects its environment. Unavoidable interactions correlate the quantum state of the system with its environment in a way that quantum information gets disseminated into the uncontrolled surroundings. Since this information is in principle available to an outside observer, the system behaves effectively as if it had been subjected to a measurement device with the outcome being discarded. The associated loss of coherence can be much faster than all other time scales in the experiment, and typically exceeds

M. Arndt (✉) · S. Gerlich
Faculty of Physics, University of Vienna, Boltzmanngasse 5, 1090 Vienna, Austria
e-mail: markus.arndt@univie.ac.at; markus.aspelmeyer@univie.ac.at
URL: https://www.quantumnano.at

S. Gerlich
e-mail: stefan.gerlich@univie.ac.at

K. Hornberger
Faculty of Physics, University of Duisburg-Essen, Lotharstraße 1-12, 47048 Duisburg, Germany
e-mail: klaus.hornberger@uni-due.de

C. Kiefer (ed.), *From Quantum to Classical*, Fundamental Theories of Physics 204,
https://doi.org/10.1007/978-3-030-88781-0_4

by many orders of magnitude other environmentally induced rates, such as friction or thermalization. Decoherence theory is by now a well-accepted field of quantum physics; introductions to the general concept and ideas can be found in [3–6].

Decoherence is often put forward as a solution to the measurement problem of quantum mechanics. However, the theory involves no collapse of a wave function. It presupposes unitary quantum dynamics and the superposition principle to hold not only for the quantum system but also for its environment and, ultimately the "universe". Decoherence theory therefore cannot explain why the pointer of a macroscopic measurement apparatus ends up at a specific position of the dial instead of any another, that would be equally accessible from the initial state. But it can answer why no such pointer is ever encountered in a superposition of positions, i.e. it can predict when and how the environment selects the basis of pointer states [7, 8]. Decoherence theory has found sympathy among those who think that quantum physics can be explained without invoking a measurement postulate [9], e.g. by adhering to a many-worlds interpretation. While no experiment so far has been able to discriminate between interpretations of quantum mechanics, work in the lab can clearly corroborate the validity of decoherence theory and emphasize its importance for our understanding of modern quantum technology.

From an experimental point of view, genuine decoherence is often hard to distinguish from effects of *phase averaging* that may lead to a similar loss of coherence in the observed quantum system. In phase averaging an interference phenomenon is washed out because the quantum evolution depends on noisy classical parameters that are uncontrolled in practice, but could be measured in principle. For instance, the fringes in matter-wave interferometry may get blurred due to vibrations or fluctuating external electromagnetic fields. The interference contrast could then be improved by monitoring the grating accelerations or field strengths and by correlating these signals with the detector recordings. This is to be contrasted with genuine decoherence, where the quantum nature of the environmental degrees of freedom and their entanglement with the quantum system rule out regaining coherence, because Hilbert space and state manipulation grow exponentially with the number of degrees of freedom. In the following we will therefore make a clear distinction between the phenomena of decoherence and phase averaging.

2 Experimental Decoherence in Macromolecule Interferometry

The *decoherence program* of Dieter Zeh and his coworkers has inspired a plethora of different experimental studies. While in the 90s of the last century one could still occasionally hear the question "Do you really believe in decoherence theory?", this framework has meanwhile become accepted and respected as what it is: the consequent application of unitary quantum mechanics to complex systems. Decoherence has also gained importance in the rapidly growing field of quantum information research, where it is the ultimate challenge in quantum computing [10] and quantum simulation [11]. The progressive coupling of Schrödinger-kittens to the environment has also been experimentally demonstrated on the level of a single field mode, by probing high-finesse microwave cavities with circular Rydberg atoms [12]. Many related studies have been performed in a large variety of systems.

Here, we focus on applications in macromolecule interferometry as performed at the University of Vienna. While the concepts are universal, we illustrate them in the specific context of the interferometers that we have built throughout the last two decades: from far-field diffraction (FFD) of hot molecules [13–15], via the first Talbot-Lau interferometer (TLI) with molecules [16], Kapitza-Dirac-Talbot-Lau interferometry (KDTLI) for highly polarizable molecules [17], Talbot-Lau lithography with fullerenes (TLL) [18], Optical Time-Domain Matter-Wave interferometry (OTIMA) with organic clusters [19] and antibiotic polypeptides [20] to the most recent addition in the series, the Long-Baseline Universal Matter-Wave interferometer (LUMI) [21]. The latter sets the current complexity record for matter-waves, with molecules exceeding the mass of 25 kDa. We refer to existing literature on how to prepare the sources, detectors, and interferometers [22–24] and also for novel concepts in nanoparticle interferometry [24, 25].

2.1 Coherence in the Molecular Beam

The formation of a molecular beam requires a volatilization mechanism. For many stable molecules up to $m \simeq 1000\,\text{Da}$ this is most simply realized by sublimating or evaporating them in a Knudsen cell. For large organic molecules or complex peptides this is done most efficiently using nanosecond or even femtosecond laser desorption [26]. In some cases, the molecules reach temperatures up to 1000 K. How can we understand that even at such high temperatures coherence can be prepared and maintained? The longitudinal velocity of a thermal beam follows a distribution $f(v)$, which in practical applications is often a Maxwell-Boltzmann distribution shifted by v_0 towards higher velocities, $f(v) \propto v^2 \exp\left[-m(v - v_0)^2)/2k_B T\right]$. This is due to collisions in the source.

Particle beams with a high temperature and mass give rise to a distribution of small de Broglie wavelengths, e.g. from 5 pm for thermal fullerenes [27] down to 50 fm for hot oligoporphyrin derivatives at $m = 25\,\text{kDa}$ and $v = 300\,\text{m/s}$ [21]. While these wavelengths are extraordinarily small in comparison to those encountered in atom interferometry, and even though they can be up to 10^5 smaller than the diameter of the molecule itself, these are no fundamental constraints to matter-wave coherence. This is because the effective transverse motional temperature is usually orders of magnitude lower than the longitudinal temperature found in an effusive source: tight collimation to about $\theta_{\text{coll}} \simeq 10\,\mu\text{rad}$ in far-field diffraction can select a transverse molecular temperature down to 10 nK, in exchange for a dramatically reduced beam flux. In near-field interferometry the collimation requirement is typically one hundred times less strict, associated with a transverse temperature on the order of $100\,\mu\text{K}$.

In molecular beams with a broad longitudinal velocity distribution the interference pattern is still an average of the fringes associated with fixed longitudinal velocities, as described by the correspondingly short longitudinal coherence length $L_c \propto \lambda_{dB} v/\Delta v$. Also the high internal temperature, which can often be assumed equal to the motional temperature in effusive cells, is a specific feature of many experiments in molecule interferometry.

2.2 Phase Averaging and Dispersion

Since coherence signifies the persistence of constant phase relations between different parts of the matter wave across space or time, one may be tempted to denote anything destroying this relation as decoherence. It is, however, often useful to be more precise and to distinguish phase averaging and decoherence. The former can be caused by a variety of fluctuating parameters of the experiment—changes in the source position, vibrations of the experimental setup, or random noise in electric or magnetic fields. In contrast to that, we reserve the denotation *decoherence* for processes that lead to quantum correlations between the interfering particle and an (internal or external) environment.

2.2.1 Phase Averaging by Apparatus Noise

Noise due to vibrations in the interferometer is at the same time the most important and the most trivial reason for loss of coherence. The phase difference associated with neighboring paths through a 3-grating interferometer, $\Delta\varphi = (x_1 - 2x_2 + x_3) \cdot 2\pi/d$, is determined by the positions x_i and the period d of all gratings. This shows that controlling all of these positions on the nanometer level may be required to maintain a stable interferometer phase. While this constitutes a substantial experimental challenge, grating vibrations can today be well controlled using passive or active isolation systems, even in demanding cases [21, 28].

Some of the best current atom interferometers aim at a relative precision in accelerometry of better than 10^{-10} to 10^{-12}. While it is difficult to eliminate all vibrations, precise phase readings can still be achieved by correlating complementary interferometer output ports. Even if each of them exhibits random phase noise complementary ports remain highly correlated. This way a differential readout can suppress common mode noise to an impressive level [29, 30].

2.2.2 Van der Waals Dispersion

Because of its conceptual and technological simplicity, far-field diffraction of the fullerene C_{60} [13] was the first experiment to demonstrate the matter-wave nature of hot large molecules. The molecular volume, structure, and internal excitation contributes already several interesting aspects to our discussion on phase averaging and decoherence. Naively speaking, *de Broglie* interference should be characterized only by the value of the de Broglie wavelength $\lambda_{\text{dB}} = h/mv$, determined by Planck's quantum of action h, the particle's mass m and velocity v. However, in practice the interaction with the gratings depends on several molecular properties.

Fullerenes are composed of 60 covalently bound carbon atoms, arranged in the shape of a miniaturized soccer ball of 0.7 nm inter-nuclear diameter. Because of their size and electron delocalization across the bonds, the static electric polarizability of fullerenes is a sizeable $\alpha_{\text{stat}} = 4\pi\varepsilon_0 \times 89\,\text{Å}^3$ [31, 32]. Even in absence of a static molecular dipole moment its quantum fluctuations will induce image dipoles in the grating, which are always oriented to attract the particle to the grating wall. This Casimir-Polder potential V_{CP} [33] is often distinguished from the simpler van der Waals approximation $V_{\text{vdW}}(r) = -C_3/r^3$, valid at close distances r, and the asymptotic far-distance behavior $V_{\text{CP}}(r) = -C_4/r^4$. The constants C_3 and C_4 depend on the particle polarizability and the dielectric properties of the surface. The Casimir-Polder interaction is conservative and does not introduce any decoherence or phase averaging, even though it is determined by the the internal particle structure.

Due to the presence of this dispersion force the matter-wave will in general acquire a sizeable phase shift $\Delta\varphi = \int dz V_{\text{CP}}/\hbar v$ when passing the grating structure. It depends strongly on both the particle's velocity and its distance to the grating. In far-field diffraction of atoms or molecules, this effect may already substantially modify the diffracted intensity distribution, favouring higher diffraction orders over lower ones [15, 27, 34, 35]. In near-field interferometry, the strong dispersion can become influential [16] and even detrimental: already for fullerenes travelling at 200 m/s across a Talbot-Lau interferometer with a grating period of $d = 266\,\text{nm}$, the acceptable velocity band can shrink to $\Delta v/v \simeq 0.5\%$. Even though all these velocity-dependent phase shifts are well understood, a finite longitudinal velocity distribution will quickly superimpose too many modified interference patterns, such that the fringe visibility gets lost.

The solution to this challenge is to either develop molecular sources with sufficiently good velocity selection and coherence length–a mayor aim for several years to come–or to replace the nanomechanical diffraction gratings by standing waves of light. The phase shift in optical gratings is still dispersive but the distance dependence in a sinusoidal structure is much more forgiving than the r^{-3}-dependence of the van der Waals potential. Theory and many experiments have proven that light gratings facilitate high contrast matter-wave interference, over large velocity bands and even for highly polarizable molecules [17].

2.2.3 Rotational Effects

In contrast to atoms, most molecules cannot be considered spherical. They are revolving as aspherical rotors, implying that any orientation-dependence of the molecule-wall interaction may have an impact on the interference pattern. Indeed, large polarizability anisotropies can be found in molecules such as carotenes, phthalocyanines or porphyrins. Since massive molecules require low velocities to stay within acceptable bounds for their de Broglie wavelengths and since the polarizability typically increases with molecular mass, the impact of anisotropies in large molecules is further increased due to the greater grating interaction time. In addition, all molecules without perfect inversion symmetry, i.e. the vast majority of molecules, exhibit a permanent electric dipole moment.

Most molecule-wall interactions therefore depend on the orientation state of the particle. The latter can be usually characterized only by a rotational temperature, since neither the initial orientation nor the particle's angular momentum can be controlled in any currently existing or foreseeable molecular beam source. Even tiny local charge imbalances inside the grating slit will then diffract the molecule differently for different orientations with respect to the slit [36]. This expectation is consistent with the observation that polar molecules have never led to sizable fringe contrast in either far-field diffraction at nanogratings [37] or near-field interferometry at larger gold gratings [38]. However, as suggested by theory and confirmed by experiment, high-contrast interference of polar molecules is still possible upon diffraction at optical phase gratings [39] or photo-depletion gratings [20].

2.2.4 The Role of Molecular Vibrations

Molecules are no static entities; apart from rotating they may vibrate or even undergo structural state changes in free flight. Typical time scales of these processes for molecules with about the size of C_{60} or beta-carotene are 100–1000 fs for vibrations, 100 ps for rotations and 1 ns for structural changes. It is remarkable that the electric dipole moment may fluctuate by as much as 300% in hot functionalized azobenzenes [40] or α-tocopherol [41], and that we still observe high-contrast matter-wave interference with these molecules, even in the presence of external electromagnetic fields and field gradients.

This can be explained by noting that on the fast time scale of the dipole fluctuations the external field varies adiabatically slow. In consequence, if an electric field $\mathbf{E}$ is applied inside the interferometer all that happens is that the interference pattern gets shifted. The associated force $\chi_{\text{tot}}(\mathbf{E}\nabla)\mathbf{E}$ is determined by a time-averaged scalar quantity, the electric susceptibility $\chi_{\text{tot}} = \alpha_{\text{stat}} + \langle \mathbf{d} \cdot \mathbf{d} \rangle / 3kT$. It involves not only the static polarizability α_{stat} but also the temporal average of the thermally fluctuating dipole moment $\mathbf{d}$, with T the vibrational temperature of the molecule. Figuratively speaking, one may thus say that the vibrational effect on the interference fringes allows one taking a look into the molecule and its thermally averaged nanosecond dynamics. This holds even though nature forbids us to know which path the molecule took during its flight through the interferometer.

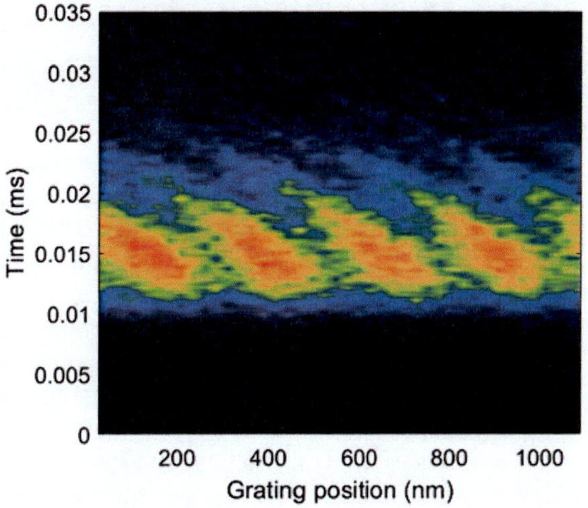

Fig. 1 Matter-wave fringes in fullerene interferometry for different flight times of C_{60} molecules through the LUMI interferometer (with grating period 266 nm). The Coriolis force entails a velocity dependence of the phase of the interference fringes. The phase difference between fast and slow molecules can easily exceed 2π (Reproduced from Ref. [32])

2.2.5 Inertial Forces

If a matter-wave interferometer consisting of three equal gratings is accelerated it accumulates a phase $\varphi = k_{\text{eff}} a T^2$. Here k_{eff} is the effective reciprocal lattice vector of the diffraction gratings, a the transverse acceleration and T the transit time between two subsequent gratings. Gravity would cause an acceleration of $a \simeq 9.81\,\text{m/s}^2$, while a rotation of the interferometer around the slit direction at angular frequency Ω will cause a Coriolis acceleration $a = 2v\Omega$ for particles of longitudinal velocity v.

To date, no molecule interferometer has been optimized towards sensitivity to these inertial forces and it would be difficult to compete with atom interferometry because of the prevalent phase-space density and signal-to-noise ratio in current molecular beam sources. However, all advanced molecular de Broglie experiments are significantly influenced by both gravity and Earth's rotation. While an overall phase and fringe shift would not matter in many experiments, a finite velocity spread Δv entails a finite phase distribution $\Delta\varphi = 2k_{\text{eff}} a T^2(\Delta v/v)$. To observe high-contrast fringes in the interference pattern we require $\Delta\varphi \ll 2\pi$. In consequence, molecule and high-mass matter-wave interferometers need to either compensate both gravity and the Coriolis force by an appropriate choice of the interferometer alignment [32], or the velocity spreads must be narrowed down sufficiently by beam cooling or velocity selection.

Figure 1 shows the effect of the Coriolis force on fullerene interference fringes in the LUMI interferometer. Averaging over different velocities here corresponds to a vertical sum over each column of that image, which would reduce the contrast substantially.

2.3 *Decoherence Due to Internal States*

2.3.1 The Fallacy of the Internal Clock

Atoms turn into optical or radio-frequency clocks when we prepare and probe superpositions of their electronic or hyperfine states. In analogy, complex molecules may be viewed as rotational or vibrational clocks. The analogy to a clock becomes intuitive when we consider rotating polar molecules, where the dipole moment acts as the hand of the clock.

Consider for instance the antibiotic ciprofloxacin, $C_{17}H_{18}FN_3O_3$, which carries many different clocks. It is composed of 41 atoms with a total molecular weight of 331.3 Da. It has 123 different degrees of freedom, shared among three directions of translation, highly excited rotations around three randomly oriented axes and 117 vibrational modes. In its thermal ground state, the molecule has an electric dipole moment of 2.7 Debye, which may be regarded as a hand of a clock, rotating at a frequency of the order of 10^{10} Hz when heated to about 500 K. In addition every vibrational normal mode is populated and 'counts time' on its own.

When the molecule is diffracted at a thin standing light wave, first-order far-field diffraction requires two neighboring paths through the grating towards the detector to differ by a single de Broglie wavelength. Naively, one might wonder whether this path-length difference is associated with different arrival times that can be read off from the clock, which would reveal which-path information and cause decoherence. However, interference and detection of each molecule takes place at a well-defined time, no matter whether it had existed as a superposition of different wavelets inside the interferometer. Since external and internal degrees of freedom are decoupled, the same molecule arriving at the same detector at the same time must be in the same internal state and the internal clock remains in phase. The prize to pay is coherence: the longitudinal coherence of the molecular beam must be sufficiently large for the high velocity components in the longer arm to catch-up with low velocity components in the shorter arm of the interferometer. Like for point particles without internal clocks, the longitudinal coherence length L_c discussed in Sect. 2.1 must exceed $n\lambda_{\text{dB}}$ for nth-order far-field diffraction to be observable [42].

2.3.2 Molecular Internal Clocks in the Gravitational Field

The issue of the internal clock has been recently revisited in the context of the gravitational time dilation predicted by general relativity: What happens if we prepare an atom in a quantum superposition of two internal clock states and propagate this ticking clock in a superposition along two vertically separated arms of an interferometer? Given that the red shift in the gravitational field of the Earth amounts to $\Delta\nu/\nu \simeq 10^{-16}$ per meter height difference, how long can that superposition be maintained before the proper times in the upper and the lower arm differ sufficiently to prevent any interference due to the which-path information provided by the clock?

As analyzed in Ref. [43] the gravitationally induced dephasing between internal and external degrees of freedom should become observable in near-future rare-earth atom interferometers featuring path separations throughout seconds and on the meter scale. However, unlike genuine decoherence, this loss of coherence is expected to be fully reversible: if the spatial superposition is maintained for sufficiently long, the internal clock states will oscillate between being in and out of phase, periodically admitting or preventing matter-wave interference.

One may extend this idea to high-mass molecules as typically used in KDTLI or LUMI interferometry. These molecules rotate and vibrate in typically 6000 modes, most of them ticking at different frequencies. The above-mentioned interference revivals should then be suppressed or strongly delayed, thus providing a universal path to classical behavior whenever the rotational or vibrational clock is red-shifted in the gravitational field [44]. However, the frequencies of vibrational modes are about three orders of magnitude smaller than the transition frequencies in optical clocks and existing beam splitters allow splitting the molecular beam over microns rather than on the meter scale. To observe this type of red-shift decoherence in large molecules one still needs to bridge more than eight orders of magnitude in the product of beam separation and coherence time. Increasing the mass is less favorable here, as the required coherence time scales with the particle mass as $1/\sqrt{M}$.

2.4 Decoherence Due to the Environment

Decoherence due to momentum transfer with environmental degrees of freedom is by far the most direct and most relevant mechanism in macromolecular quantum optics. Already 14 years before the first fullerene diffraction was realized in Vienna [13], decoherence of high-mass superpositions was studied by Joos and Zeh [2]. They considered several mechanisms of which collisional and thermal decoherence were seen as the most dominant processes.

In the decoherence literature one often finds comparisons of different decoherence rates for a delocalized quantum particle, that are obtained by linearizing its interaction with the environment. These rates grow rapidly above all bounds as the delocalization length is increased. For the usual scattering-type interactions this is clearly unphysical, an artifact of the linearization, since the decoherence rate cannot exceed the total scattering rate. In this context, the historical note may be in order that in the early days of decoherence studies some theory colleagues declared that fullerene diffraction will never be observed—based on a linearized description of gas scattering. This only illustrates that all theory must be put into context and that one must be aware of the underlying assumptions.

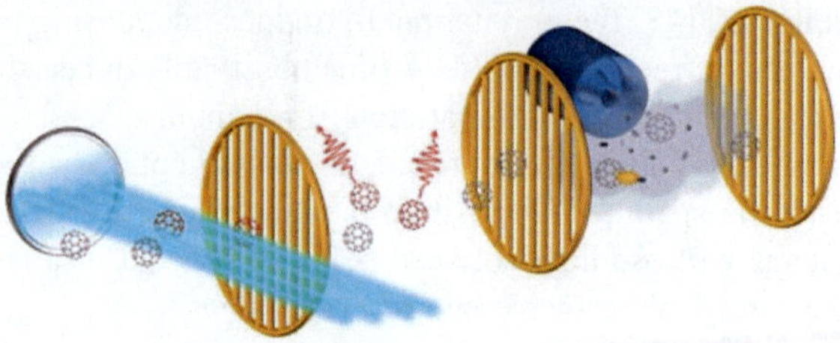

Fig. 2 Molecules traversing a Talbot-Lau interferometer may lose coherence by emission of thermal radiation or collisions with residual gas in the vacuum chamber. Both processes have been studied in fullerene interferometry [45, 46]. The internal temperature can be adjusted by laser heating before the interferometer and calibrated by the enhanced electron emission close to the heating stage. The number of collisions can be set by additional gas admitted to the chamber, and controlled to a few percent with an ionizing pressure gauge

2.4.1 Collisional Decoherence

A first systematic study of collisional decoherence was realized [47] and theoretically analyzed [45] in Talbot-Lau interferometry of fullerenes in Vienna [16] (see Fig. 2). An array of 475 nm wide slits with a period of 991 nm had been etched in to three gold gratings of 500 nm thickness. They were positioned in a distance of 440 mm from each other to form a near-field matter-wave interferometer with noticeable but still tolerable influence of Casimir-Polder forces in the molecule-grating interaction. High-contrast interference of about 30% fringe visibility could be routinely observed and this quantity was monitored as a function of the background pressure for different collision gases (decohering agents).

Linearly raising the base pressure in the chamber from 10^{-7} mbar to 10^{-6} mbar reduced the contrast exponentially down to the background noise, see Fig. 3. Also the mean count rate decreased somewhat, in agreement the Lambert-Beer law, but the transmission remained sufficiently high to allow for reliable measurements. While head-on collisions would kick the molecules to beyond the detector range, small angle collisions were of particular relevance. Even a deflection of the fullerene by 1 μrad could reduce the interference fringe contrast, since it would shift the fringe between a maximum and a minimum. The cross section for such small-angle collisions, determined by the long-range Casimir-Polder interaction between two molecules, can exceed the geometrical cross section by almost two orders of magnitude, as was the case for interactions between the interfering fullerenes and nitrogen molecules [45] (Fig. 4).

2.4.2 Emission of Thermal Radiation by the Molecule

Molecules and nanoparticles differ from atoms, neutrons, or electrons in that they can be regarded as small thermal bodies. Given the number N of atoms, covalently bound

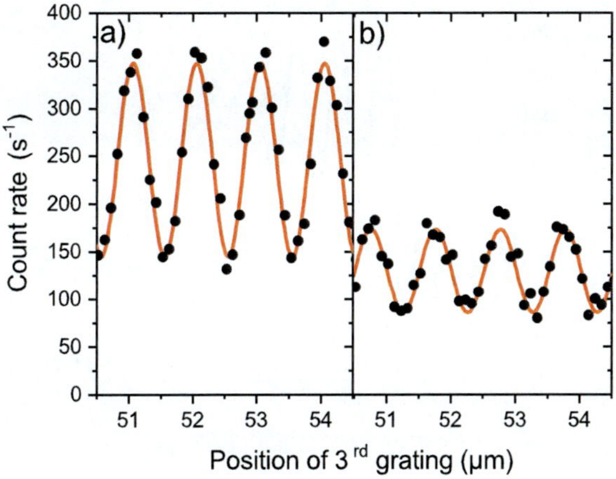

Fig. 3 **a** Typical interferogram of C_{70} molecules in a TLI at a residual gas pressure of 3×10^{-8}mbar. The observed visibility of 42% is in good agreement with theory. **b** Already a slight increase of the background gas pressure to 5×10^{-7}mbar due to the addition of argon gas leads to a significant drop of the interference contrast to 34%, and to a loss of counts as a certain fraction of the molecules are scattered outside the detection area (Reproduced from Ref. [48])

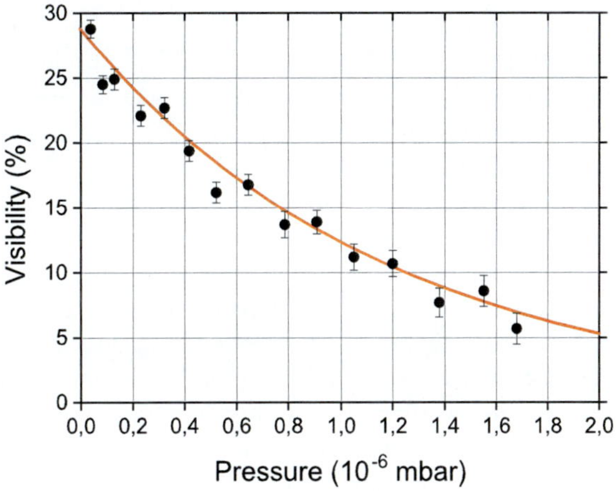

Fig. 4 Reduction of the interference visibility of C_{70} molecules in a TLI due to collisions with the background gas (argon) as a function of gas pressure. The data is in excellent agreement with the expected exponential loss of contrast (solid line) (Reproduced from Ref. [48]) in a different scaling

into a single entity, they provide a reservoir of $3N - 6$ modes to store vibrational energy.

C_{60} fullerenes are therefore loaded with about 6–7 eV of vibrational energy in 174 modes when they leave an effusive source at 900 K, and it is possible to heat them temporarily to microcanonical temperatures beyond 3000 K. Three competing processes can then be observed, which justify the view that a single fullerene may already be regarded as a lump of classical matter: hot fullerenes emit radiation in a black-body-like spectrum [49], they evaporate carbon dimers [50], and they undergo thermionic emission [51], that is they emit electrons like a tungsten wire in a light

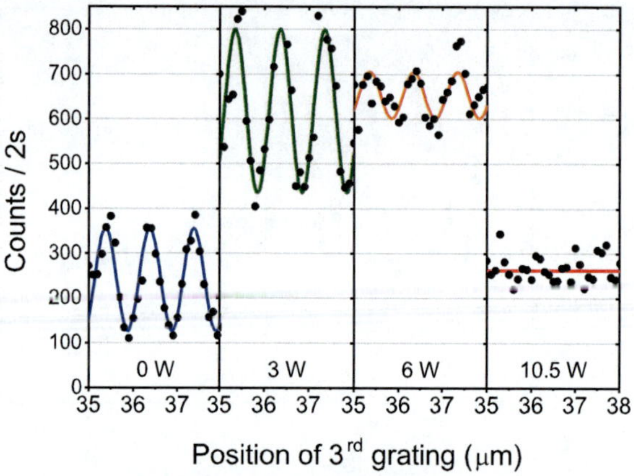

Fig. 5 Interferograms of C_{70} molecules in a TLI at different laser power settings in the heating stage. The count rate initially increases due to the increased thermal ionization probability at higher internal temperatures before it drops again at even higher laser power settings as a result of ionization and fragmentation in the heating stage. The interference contrast on the other hand drops continuously from initially 47 to 0% at $P = 10.5\,\mathrm{W}$ due to the rising emission probability of position resolving photons within the interferometer (Reproduced from Ref. [46])

bulb. Either of these processes can result in substantial decoherence if it occurs while the molecules are delocalized in a matter-wave experiment.

Decoherence due to thermal emission of radiation was explored in an experiment on Talbot-Lau interference with C_{70} fullerenes [46]. Multi-pass laser absorption was employed to heat the molecules internally to temperatures beyond 3000 K, see Fig. 5.

The maximum path separation in the interferometer exceeded 1 μm so that even a single thermal photon with a wavelength of 2 μm emitted close to the second grating would reveal sufficient which-path information to fully decohere the interference pattern. A full theoretical description needs to account for the spectral distribution of the thermal radiation, the temperature decrease due to radiative cooling [46, 52], and the emission position, since the resolving power of a thermal photon is highest close to the middle grating and lowest close to the outer ones. Very good agreement was found between the predictions of decoherence theory and the experimental observation, see Fig. 6. It is striking that the internal temperature needs to exceed 1500 K before sizeable decoherence could be observed, growing gradually and with the expected functional dependence when the temperature was increased. On a historical note, it is worth realizing that a source temperature of 900 K was already required to obtain sufficient vapor pressure to perform the experiment. A small (and initially unforeseen) factor of two in temperature made the difference between the successful first realization of a molecule interferometer and the first observation of thermal decoherence.

Recent macromolecule experiments succeeded in demonstrating interference with up to 2000 atoms for a coherence time of about 10 ms [21]. While the emission of infrared photons is to be expected in such a setting, the grating period was a mere

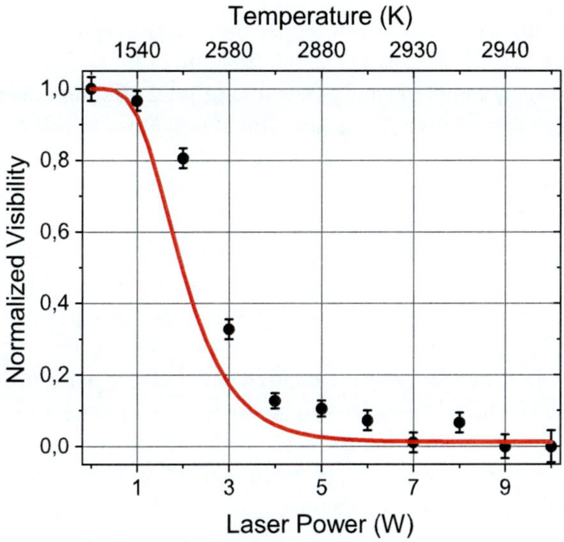

Fig. 6 Loss of visibility due to thermal decoherence plotted versus heating power (lower scale) and mean internal temperature of the C_{70} molecules when they enter the Talbot-Lau interferometer (upper scale) at a mean velocity of $v_m = 100\,\text{m/s}$ (Reproduced from Ref. [46])

266 nm, about ten times shorter than the most probable photon emitted at 1000 K. Since the photon wavelength is large compared to the path separation, momentum diffusion due to many weak photons is then the expected form of decoherence. This effect was still too weak to become influential but should be considered for even more complex particles.

3 Perspectives in Matter-Wave Decoherence

3.1 Experimental Perspectives

Interference experiments for matter waves are amongst the most sensitive devices for measuring forces, geometric phases, and accelerations. Atom interferometers are about to reach path separations of half a meter [53], coherence times of half a minute [54], and path lengths of hundreds of meters in free fall towers. Molecule interferometers, for which such length and time scales are still challenging, are venturing into the realm of biochemistry, targeting the delocalization of proteins, DNA and precursors of life, such as viroids and viruses. The quest of pushing matter-wave interferometry to ever larger objects was promoted by John Clauser, who suggested delocalizing even *little rocks and life viruses* [55]. The little rocks are nowadays called nanoparticles, and viruses may not be considered genuinely alive, but an inspiration like this is at the heart of our own research program [13, 56, 57].

In a development complementary to studying freely falling matter wave beams, an increasing number of research groups aim at preparing *trapped* nanoparticles in specific quantum states of their center of mass. This requires cooling these objects into the deep quantum regime [58] and will allow novel interferometric quantum experiments [24, 59–62].

Substantial progress in feedback cooling [63, 64], cavity cooling [65–67], resolved side-band cooling [68], and coherent scattering cooling [69, 70] brings many of these ideas closer to realization. The 1D ground state in a trap has been recently reached [71].

Beyond the center of mass, also the rotational motion of a nanoparticle may exhibit matter-wave interference. It gives rise to distinctive quantum recurrence phenomena [25, 72]. Cooling and controlling the rotation dynamics of a levitated nanoparticle is therefore an important aim [73]. In recent experiments our group demonstrated precision optical control of silicon nanorotors [74, 75], suggesting that cooling to the rotational quantum regime will soon be feasible [76–78]. Matter wave interference in the rotational degrees of freedom will then be subject to orientational decoherence effects [79, 80].

3.2 Probing New Physics

Aside from the many applications in sensing and metrology, there is another motivation for demonstrating matter wave interference with ever more massive and complex objects: The highly macroscopic quantum states [81, 82] involved allow probing unexplored territories where new physical effects might be found. While it is hard to say what to expect from the unknown, many suggestions and speculative ideas have been put forward that might be testable with matter-wave interferometry.

For instance, it has been suggested that fluctuations of space-time on the Planck scale could have an observable effect [83–85], even though matter-wave experiments currently employ de Broglie wavelengths of $<30\,\text{fm}$ [21]. The diffusion of these matter waves due to space-time fluctuations might become detectable in the same way as agitated water molecules affect the motion of pollen in Brownian motion–provided one can boost the mass of the interfering objects by another six orders of magnitude. This sounds like a big step, and it is. But macromolecule interferometry operates already with masses almost eight orders of magnitudes greater than the electron, which stood at the beginning of matter-wave experiments.

It has also been suggested that decoherence due to gravitational waves might be observable [86, 87]. The random noise of gravitational waves would cause contractions of the arm length in a matter-wave interferometer and thus stochastically blur its fringes. A promising route will open up if one succeeds in creating large-momentum beam splitters for ultra-fast hydrogen atoms, for instance by diffracting 80 eV hydrogen atoms coherently at the hexagonal 243 pm lattice of single-layer graphene [88].

As described in Sect. 2.3.2, the interface between general relativity and quantum physics can also be probed by delocalized atoms with a single fast-ticking internal clock or by molecules with slower but abundant vibrational clocks [43, 44].

Decoherence could also be induced by particles beyond the standard model [89]. In particular, numerous astronomical and cosmological observations hint at the existence of dark matter with an average density of 0.4 amu/cm^3 in our galaxy. It might consist of weakly interacting, massive particles (WIMPs) at low density or of ultra-light dark particles in quantum degenerate dark-matter waves (or of anything in between). In case of high-mass WIMPs any of the rare collisions with the matter-wave beam would kick the nanoparticle beyond the detector area, so that dark matter leads to a loss of counts rather than decoherence. However, for the intermediate mass regime between 1 and 10^6 eV/c^2 collisions would be strong enough to cause decoherence while keeping the nanoparticles in the experiment. An observed additional loss of coherence would then be an indicator for dark matter–provided all natural decoherence processes are kept under control. It has been suggested that this effect may become visible in matter-wave experiments with massive particles if one can profit from a coherent enhancement of the scattering cross section for dark-matter de Broglie wavelengths greater than the size of the colliding partner [90, 91].

Matter-wave interferometry is particularly well-suited to probe theories that postulate an objective, spontaneous collapse process for the quantum wave function [92]. Such models replace the Schrödinger equation with a nonlinear stochastic equation, causing macroscopically delocalized wave functions to collapse randomly and thus effecting a transition from quantum to classical behavior. The most prominent one is the model of continuous spontaneous localization (CSL) devised by Ghirardi et al. [93] and Pearle [94]. It was introduced *ad hoc*, but it can also be seen as a representative of a generic class of 'classicalizing' modifications of quantum mechanics derived from natural consistency requirements [81]. Importantly, CSL also implies a weak gradual heating of macroscopic objects.

There are two parameters in the CSL hypothesis, a localization length r_c and a rate parameter λ. While there is no *a priori* constraint for them, the model only serves its purpose of 'solving' the measurement problem by reinstating objective realism on the macroscale, if the length r_c is in the microscopic domain and the rate λ is sufficiently large. Only then will macroscopic objects behave classically. On the other hand, experimental observations can provide upper bounds for λ given r_c. Currently, the best bounds on the CSL parameters are from the observed lack of heating in classical setups [95–97], but the predicted heating can be reduced by tweaking the model.

The most stringent constraints due to quantum experiments are set by atom interferometry for r_c on the meter scale [98] and by macromolecule interferometry for $r_c \simeq 100$ nm [99]. Since the collapse rate λ scales with the square of the interfering particle mass it is favorable to use large molecules. On the other hand, a special type of interference using large squeezed atomic Bose-Einstein condensates might allow ruling out the entire relevant parameter space of CSL, provided the number of detected atoms can be resolved sufficiently well [100].

As the mass of the interfering particle is increased, one may wonder what happens when the superposed object becomes sufficiently massive to turn into a sizeable source of gravity. It has also been argued [101] that a massive object in superposition of position states will deform space-time locally and thus create a superposition of space-times. The latter might collapse on a time scale given by the energy-time uncertainty relation involving the gravitational self-energy. This can be cast into a dynamical decoherence-type equation [102] where the collapse rate scales with the square of the mass, speaking again in favor of high mass interferometry.

In a similar spirit, the idea of gravitational self-interaction may motivate the so-called Schrödinger-Newton equation, leading to a rather unusual nonlinear evolution of the center-of-mass wavefunction, with a partial shrinking to a soliton-like solution and a partial spread to infinite distances [103]. Such a behavior would be expected for particle masses of 10^{11} Da only after several 10^4 s of free evolution time, but the compatibility with special relativity is not clear [104].

Finally, recent studies have also suggested that one might probe the quantumness of gravity by placing two superposed micro-spheres in close vicinity: if their positions become entangled by gravity, gravity would need to be quantized. Such ideas will become testable if one succeeds in placing particles with masses in the range of 10^{14} Da in superposition states with a delocalization range as wide as 100 μm [105, 106]. Preparing and probing such a macroscopic superposition state is a highly ambitious goal and many decoherence studies will still be required to push matter-wave physics to this range.

The work by Dieter Zeh and many of the pioneers of decoherence theory has paved the path for a better understanding of the limitations of modern quantum technologies and it has triggered a plethora of new experiments to improve our understanding of the foundations of physics. His name will always stay connected to decoherence research.

Acknowledgements M.A. thanks Anton Zeilinger and Lucia Hackermüller, Björn Brezger and Stefan Uttenthaler for intriguing first years in molecule interferometry that culminated in our joint experiments on collisional and thermal decoherence as included and referenced here. We thank the QNP team in Vienna for contributing the experimental basis to this review and Filip Kialka for comments on the manuscript. We acknowledge funding by the Austrian Science Funds in the project and P32543 N, as well as by the Gordon and Betty Moore foundation in project 10771.

References

1. H.D. Zeh, Found. Phys. **1**, 69 (1970)
2. E. Joos, H.D. Zeh, Z. Phys. B. **59**, 223 (1985)
3. E. Joos, H.D. Zeh, C. Kiefer, D. Giulini, J. Kupsch, I.-O. Stamatescu, *Decoherence and the Appearance of a Classical World in Quantum Theory*, 2nd edn. (Springer, Berlin, 2003)
4. W.H. Zurek, Rev. Mod. Phys. **75**, 715 (2003)
5. K. Hornberger, *Introduction to Decoherence Theory*. Lecture Notes in Physics (Springer, Berlin, 2008)
6. M. Schlosshauer, Phys. Rep. **831**, 1 (2019)
7. W.H. Zurek, Phys. Rev. D **24**, 1516 (1981)
8. W.H. Zurek, Phys. Rev. D **26**, 1862 (1982)
9. W.H. Zurek, Philos. Trans. R. Soc. A **376**, 20180107 (2018)
10. M.A. Nielsen, I.L. Chuang, *Quantum Computation and Quantum Information* (Cambridge University Press, Cambridge, 2010)
11. F. Tacchino, A. Chiesa, S. Carretta, D. Gerace, Adv. Quant. Technol. **3**, 1900052 (2019)
12. M. Brune, E. Hagley, J. Dreyer, X. Maître, A. Maali, C. Wunderlich, J.M. Raimond, S. Haroche, Phys. Rev. Lett. **77**, 4887 (1996)
13. M. Arndt, O. Nairz, J. Voss-Andreae, C. Keller, G. van der Zouw, A. Zeilinger, Nature **401**, 680 (1999)
14. T. Juffmann, A. Milic, M. Müllneritsch, P. Asenbaum, A. Tsukernik, J. Tüxen, M. Mayor, O. Cheshnovsky, M. Arndt, Nat. Nanotechn. **7**, 297 (2012)
15. C. Brand, M. Sclafani, C. Knobloch, Y. Lilach, T. Juffmann, J. Kotakoski, C. Mangler, A. Winter, A. Turchanin, J. Meyer et al., Nat. Nanotechnol. **10**, 845 (2015)
16. B. Brezger, L. Hackermüller, S. Uttenthaler, J. Petschinka, M. Arndt, A. Zeilinger, Phys. Rev. Lett. **88**, 100404 (2002)
17. S. Gerlich, L. Hackermüller, K. Hornberger, A. Stibor, H. Ulbricht, M. Gring, F. Goldfarb, T. Savas, M. Müri, M. Mayor et al., Nat. Phys. **3**, 711 (2007)
18. T. Juffmann, S. Truppe, P. Geyer, A.G. Major, S. Deachapunya, H. Ulbricht, M. Arndt, Phys. Rev. Lett. **103** (2009)
19. P. Haslinger, N. Dörre, P. Geyer, J. Rodewald, S. Nimmrichter, M. Arndt, Nat. Phys. **9**, 144 (2013)
20. A. Shayeghi, P. Rieser, G. Richter, U. Sezer, J.H. Rodewald, P. Geyer, T.J. Martinez, M. Arndt, Nat. Communs. **11**, 1447 (2020)
21. Y.Y. Fein, P. Geyer, P. Zwick, F. Kiałka, S. Pedalino, M. Mayor, S. Gerlich, M. Arndt, Nat. Phys. **15**, 1242 (2019)
22. K. Hornberger, S. Gerlich, P. Haslinger, S. Nimmrichter, M. Arndt, Rev. Mod. Phys. **84**, 157 (2012)
23. T. Juffmann, H. Ulbricht, M. Arndt, Rep. Progr. Phys. **76**, 086402 (2013)
24. M. Arndt, K. Hornberger, Nat. Phys. **10**, 271 (2014)
25. B.A. Stickler, B. Papendell, S. Kuhn, B. Schrinski, J. Millen, M. Arndt, K. Hornberger, New J. Phys. **20**, 122001 (2018)
26. J. Schätti, P. Rieser, U. Sezer, G. Richter, P. Geyer, G.G. Rondina, D. Häussinger, M. Mayor, A. Shayeghi, V. Köhler et al., Commun. Chem. **1**, 93 (2018)
27. O. Nairz, M. Arndt, A. Zeilinger, Am. J. Phys. **71**, 319 (2003)
28. A. Peters, K. Yeow-Chung, S. Chu, Nature **400**, 849 (1999)
29. R.H. Parker, C. Yu, W. Zhong, B. Estey, H. Müller, Science **360**, 191 (2018)
30. P. Asenbaum, C. Overstreet, M. Kim, J. Curti, M.A. Kasevich, Phys. Rev. Lett. **125**, 191101 (2020)
31. M. Berninger, A. Stefanov, S. Deachapunya, M. Arndt, Phys. Rev. A **76**, 013607 (2007)
32. Y.Y. Fein, F. Kiałka, P. Geyer, S. Gerlich, M. Arndt, New J. Phys. **22**, 033013 (2020)
33. H.B.G. Casimir, D. Polder, Phys. Rev. **73**, 360 (1948)
34. R.E. Grisenti, W. Schöllkopf, J.P. Toennies, G.C. Hegerfeldt, T. Köhler, M. Stoll, Phys. Rev. Lett. **85**, 2284 (2000)

35. V.P.A. Lonij, C.E. Klauss, W.F. Holmgren, A.D. Cronin, Phys. Rev. Lett. **105**, 233202 (2010)
36. B.A. Stickler, K. Hornberger, Phys. Rev. A **92**, 023619 (2015)
37. C. Knobloch, B.A. Stickler, C. Brand, M. Sclafani, Y. Lilach, T. Juffmann, O. Cheshnovsky, K. Hornberger, M. Arndt, Fortschr. Phys. **65**, 1600025 (2017)
38. L. Hackermüller, S. Uttenthaler, K. Hornberger, E. Reiger, B. Brezger, A. Zeilinger, M. Arndt, Phys. Rev. Lett. **91**, 090408 (2003)
39. K. Hornberger, S. Gerlich, H. Ulbricht, L. Hackermüller, S. Nimmrichter, I. Goldt, O. Boltalina, M. Arndt, New J. Phys. **11**, 043032 (2009)
40. M. Gring, S. Gerlich, S. Eibenberger, S. Nimmrichter, T. Berrada, M. Arndt, H. Ulbricht, K. Hornberger, M. Müri, M. Mayor et al., Phys. Rev. A **81**, 031604(R) (2010)
41. L. Mairhofer, S. Eibenberger, J.P. Cotter, M. Romirer, A. Shayeghi, M. Arndt, Angew. Chem. Int. Ed. **56**, 10947 (2017)
42. M. Arndt, L. Hackermüller, E. Reiger, Braz. J. Phys. **35**, 216 (2005)
43. M. Zych, F. Costa, I. Pikovski, C. Brukner, Nat. Commun. **2**, 505 (2011)
44. I. Pikovski, M. Zych, F. Costa, A. Brukner, Nat. Phys. **11**, 668 (2015)
45. K. Hornberger, J.E. Sipe, Phys. Rev. A **68**, 12105 (2003)
46. L. Hackermüller, K. Hornberger, B. Brezger, A. Zeilinger, M. Arndt, Nature **427**, 711 (2004)
47. K. Hornberger, S. Uttenthaler, B. Brezger, L. Hackermüller, M. Arndt, A. Zeilinger, Phys. Rev. Lett. **90**, 160401 (2003)
48. L. Hackermüller, K. Hornberger, B. Brezger, A. Zeilinger, M. Arndt, Appl. Phys. B **77**, 781 (2003)
49. K. Hansen, O. Echt, Phys. Rev. Lett. **78**, 2337 (1997)
50. M.S. Dresselhaus, G. Dresselhaus, P.C. Eklund, *Science of Fullerenes and Carbon Nanotubes*, 2nd edn. (Academic Press, San Diego, 1998)
51. E. Campbell, G. Ulmer, I. Hertel, Phys. Rev. Lett. **67**, 1986 (1991)
52. K. Hornberger, L. Hackermüller, M. Arndt, Phys. Rev. A **71**, 023601 (2005)
53. P. Asenbaum, C. Overstreet, T. Kovachy, D.D. Brown, J.M. Hogan, M.A. Kasevich, Phys. Rev. Lett. **118**, 183602 (2017)
54. V. Xu, M. Jaffe, C.D. Panda, S.L. Kristensen, L.W. Clark, H. Müller, Science **366**, 745 (2019)
55. J. Clauser, *De Broglie-wave interference of small rocks and live viruses* (Kluwer Academic, 1997), pp. 1–11
56. M. Arndt, K. Hornberger, A. Zeilinger, Phys. World **3**, 35 (2005)
57. T. Juffmann, S. Nimmrichter, M. Arndt, H. Gleiter, K. Hornberger, Found. Phys. **42**, 98 (2010)
58. D.E. Chang, C.A. Regal, S.B. Papp, D.J. Wilson, J. Ye, O. Painter, H.J. Kimble, P. Zoller, Proc. Natl. Acad. Sci. USA **107**, 1005 (2010)
59. O. Romero-Isart, Phys. Rev. A **84**, 052121 (2011)
60. J. Bateman, S. Nimmrichter, K. Hornberger, H. Ulbricht, Nat. Communs. **5**, 4788 (2014)
61. C. Wan, M. Scala, G.W. Morley, A.A. Rahman, H. Ulbricht, J. Bateman, P.F. Barker, S. Bose, M.S. Kim, Phys. Rev. Lett. **117**, 143003 (2016)
62. H. Rudolph, K. Hornberger, B.A. Stickler, Phys. Rev. A **101**, 011804 (2020)
63. T. Li, S. Kheifets, M.G. Raizen, Nat. Phys. **7**, 527 (2011)
64. J. Gieseler, B. Deutsch, R. Quidant, L. Novotny, Phys. Rev. Lett. **109**, 103603 (2012)
65. P. Asenbaum, S. Kuhn, S. Nimmrichter, U. Sezer, M. Arndt, Nat. Communs. **4**, 2743 (2013)
66. N. Kiesel, F. Blaser, U. Delic, D. Grass, R. Kaltenbaek, M. Aspelmeyer, Proc. Natl. Acad. Sci. USA **110**, 14180 (2013)
67. J. Millen, P. Fonseca, T. Mavrogordatos, T. Monteiro, P. Barker, Phys. Rev. Lett. **114** 123602 (2015)
68. N. Meyer, A. de los Rios Sommer, P. Mestres, J. Gieseler, V. Jain, L. Novotny, R. Quidant Phys. Rev. Lett. **123**, 153601 (2019)
69. U. Delic, M. Reisenbauer, D. Grass, N. Kiesel, V. Vuletic, M. Aspelmeyer, Phys. Rev. Lett. **122**, 123602 (2019)
70. D. Windey, C. Gonzalez-Ballestero, P. Maurer, L. Novotny, O. Romero-Isart, R. Reimann, Phys. Rev. Lett. **122**, 123601 (2019)

71. U. Delić, M. Reisenbauer, K. Dare, D. Grass, V. Vuletić, N. Kiesel, M. Aspelmeyer, Science **367**, 892 (2020)
72. Y. Ma, K.E. Khosla, B.A. Stickler, M. Kim, Phys. Rev. Lett. **125**, 053604 (2020)
73. T.M. Hoang, Y. Ma, J. Ahn, J. Bang, F. Robicheaux, Z.-Q. Yin, T. Li, Phys. Rev. Lett. **117**, 123604 (2016)
74. S. Kuhn, A. Kosloff, B.A. Stickler, F. Patolsky, K. Hornberger, M. Arndt, J. Millen, Optica **4**, 356 (2017)
75. S. Kuhn, B.A. Stickler, A. Kosloff, F. Patolsky, K. Hornberger, M. Arndt, J. Millen, Nat. Communs. **8**, 1670 (2017)
76. B.A. Stickler, S. Nimmrichter, L. Martinetz, S. Kuhn, M. Arndt, K. Hornberger, Phys. Rev. A **94**, 033818 (2016)
77. S. Kuhn, G. Wachter, F.-F. Wieser, J. Millen, M. Schneider, J. Schalko, U. Schmid, M. Trupke, M. Arndt, Appl. Phys. Lett. **111** (2017)
78. J. Schäfer, H. Rudolph, K. Hornberger, B.A. Stickler, Phys. Rev. Lett. **126**, 163603 (2021)
79. B.A. Stickler, B. Papendell, K. Hornberger, Phys. Rev. A **94** (2016)
80. B. Papendell, B.A. Stickler, K. Hornberger, New J. Phys. **19** (2017)
81. S. Nimmrichter, K. Hornberger, Phys. Rev. Lett. **110**, 160403 (2013)
82. B. Schrinski, S. Nimmrichter, K. Hornberger, Phys. Rev. Res. **2**, 033034 (2020)
83. I.C. Percival, W.T. Strunz, Proc. Royal Soc.: Math. Phys. Eng. **435**, 431 (1997)
84. J. Wang, M. Yang, J. Jellinek, G. Wang, Phys. Rev. A **74** (2006)
85. P.M. Bonifacio, C.H.T. Wang, J.T. Mendonça, R. Bingham, Class. Quant. Grav. **26**, 145013 (2009)
86. B. Lamine, M.T. Jaekel, S. Reynaud, Eur. Phys. J. D **20**, 165 (2002)
87. B. Lamine, R. Hervé, A. Lambrecht, S. Reynaud, Phys. Rev. Lett. **96** 050405 (2006)
88. C. Brand, M. Debiossac, T. Susi, F. Aguillon, J. Kotakoski, P. Roncin, M. Arndt, New J. Phys. **21**, 033004 (2019)
89. C.J. Riedel, I. Yavin, Phys. Rev. D **96**, 023007 (2017)
90. C. Riedel, Phys. Rev. D **88** (2013)
91. C.J. Riedel, Phys. Rev. A **92** (2015)
92. A. Bassi, K. Lochan, S. Satin, T.P. Singh, H. Ulbricht, Rev. Mod. Phys. **85**, 471 (2013)
93. G.C. Ghirardi, A. Rimini, T. Weber, Phys. Rev. D **34**, 470 (1986)
94. P. Pearle, Phys. Rev. A **39**, 2277 (1989)
95. K. Piscicchia, A. Bassi, C. Curceanu, R.D. Grande, S. Donadi, B.C. Hiesmayr, A. Pichler, Entropy **19**, 319 (2017)
96. M. Carlesso, M. Paternostro, H. Ulbricht, A. Vinante, A. Bassi, New J. Phys. **20**, 083022 (2018)
97. A. Vinante, M. Carlesso, A. Bassi, A. Chiasera, S. Varas, P. Falferi, B. Margesin, R. Mezzena, H. Ulbricht, Phys. Rev. Lett. **125**, 100404 (2020)
98. T. Kovachy, P. Asenbaum, C. Overstreet, C.A. Donnelly, S.M. Dickerson, A. Sugarbaker, J.M. Hogan, M.A. Kasevich, Nature **528**, 530 (2015)
99. S. Nimmrichter, K. Hornberger, P. Haslinger, M. Arndt, Phys. Rev. A **83**, 043621 (2011)
100. B. Schrinski, K. Hornberger, S. Nimmrichter, arXiv preprint arXiv:2008.13580 (2020)
101. R. Penrose, Gen. Rel. Grav. **28**, 581 (1996)
102. L. Diosi, Phys. Lett. **105 A**, 199 (1984)
103. A. Großardt, Thesis *Die Schrödinger-Newton-Gleichung als Modell selbstgravitierender Quantensysteme* (2013)
104. D. Giulini, C. Kiefer, *Gravitationswellen* (Springer Spektrum, 2018)
105. C. Marletto, V. Vedral, Nature **547** (2017)
106. S. Bose, A. Mazumdar, G.W. Morley, H. Ulbricht, M. Toros, M. Paternostro, A.A. Geraci, P.F. Barker, M.S. Kim, G. Milburn, Phys. Rev. Lett. **119**, 240401 (2017)

When Zeh Meets Feynman: How to Avoid the Appearance of a Classical World in Gravity Experiments

Markus Aspelmeyer

1 Introduction

There is renewed and growing interest in producing hard experimental evidence that gravity requires a quantum description. The origins of this quest can be traced back to at least the Chapel Hill Conference in 1957 on "The role of gravitation in physics". At this conference, to make the point "*that we're in trouble if we believe in quantum mechanics but don't quantize gravitational theory*" [1], Feynman proposed a gedankenexperiment with macroscopic superpositions of massive objects that are coupled via gravity. Since the interaction involves superpositions of the gravitational field, a correct description of the experiment would also have to involve a quantum description of gravity. More than 50 years later, Claus Kiefer and I were part of a working group on "Historical Roots of Quantum Gravity Research" at the Max Planck Institute for the History of Sciences, where the newly published transcript of the Chapel Hill conference was heavily discussed during several sessions in Berlin [2]. Claus brought the transcript to the attention of Dieter Zeh, who responded with a fantastic comment on Feynman's contributions [3]. I am following up on this debate by addressing a question that was asked by another Chapel Hill participant, Louis Witten: "*What prevents this from becoming a practical experiment?*".

Feynman's initial response to Witten's question was clearly in the spirit of Zeh's decoherence: "*You might argue this way: Somewhere in your apparatus this idea of [probability] amplitude has been lost. You don't need it any more, so you drop it. The wave packet would be reduced (or something). Even though you don't know where*

This paper is dedicated to Dieter Zeh and his remarkable work.

M. Aspelmeyer (✉)
Faculty of Physics, University of Vienna, Boltzmanngasse 5, 1090 Vienna, Austria

Institute for Quantum Optics and Quantum Information (IQOQI) Vienna,
Austrian Academy of Sciences, Boltzmanngasse 3, 1090 Vienna, Austria
e-mail: markus.aspelmeyer@univie.ac.at

C. Kiefer (ed.), *From Quantum to Classical*, Fundamental Theories of Physics 204,
https://doi.org/10.1007/978-3-030-88781-0_5

it's reduced, it's reduced. And then you can't do an experiment which distinguishes interfering alternatives from just plain odds (like with dice)." In other words: the attempt to amplify quantum coherence to the level of macroscopic superpositions would likely result in loss of information to the environment. It is fair to say that at that time there was no prospect for actual experiments that involve gravitational source masses in the quantum regime. Amazingly, this has changed. Looking at today's landscape of quantum experiments involving objects of increasing mass and complexity in spatial superpositions [4, 5] and the fast progress in quantum controlling the motion of solid state objects [6] such experiments seem to be possible in principle—provided that we manage to prevent decoherence from happening too fast.

2 Feynman's Gedanken Experiment

The original idea put forward by Feynman is simple: "*One should think about designing an experiment which uses a gravitational link and at the same time shows quantum interference*" [1]. Concretely, a Stern-Gerlach magnet produces a spatial superposition of a gravitational source mass carrying a net spin of 1/2, resulting in $|\Psi_0\rangle_s = |x_L\rangle_s + |x_R\rangle_s$, with $x_{L,R}$ being two positions (say left, right) of the center of mass along the x-axis that are macroscopically distinct ($\langle x_L|x_R\rangle_s \ll 1$). The mass then couples via gravity to a distant test mass $|\Phi_0\rangle_t = |x_0\rangle_t$ at location x_0. If quantum mechanics holds, the test mass experiences two possible accelerations $a_{L,R} = Gm_s/(|x_{L,R} - x_0|^2)$ (G: Newton's constant; m_s: source mass) that result in entanglement between the masses once the separation in test mass displacement is sufficiently large, i.e. $|\Psi_0\rangle \otimes |\Phi_0\rangle \rightarrow |x_L\rangle_s |x_L\rangle_t + |x_R\rangle_s |x_R\rangle_t$ for $\langle x_L|x_R\rangle_t \ll 1$.

From a non-relativistic quantum physics point of view the generation of entanglement is a trivial consequence of gravitational coupling between the two masses, where each branch of the superposition is treated separately. On the level of an effective description of instantaneous interactions, entanglement is generated by the classical Newtonian potential. We know, however, from independent astrophysical observations (including the recent detection of gravitational waves) that gravity requires a field description. If one takes into account that the interaction is mediated locally via the gravitational field, entanglement between the masses can only be created when the mediator itself is genuinely non-classical (see e.g. [7], or [8] for a more general no-go theorem). This is the case discussed by Feynman.[1] It is also instructive to analyze the situation from the geometric viewpoint of general relativity. In the low-energy limit of solar system dynamics or table-top experiments gravitational interaction is understood as geodesic motion in a non-Minkowski space-time metric g_{00} that is

[1] Feynman uses a somewhat more drastic wording: "*But aside from that possibility* [that quantum mechanics fails], *if you believe in quantum mechanics up to any level then you have to believe in gravitational quantization in order to describe this experiment.*". Here he explicitly refers to the ability to apply the superposition principle, i.e. probability amplitudes, to different gravitational field configurations, and *not* the necessity to invoke quanta of the field (gravitons).

sourced by the mass density configuration $\varrho(\underline{r})$. The Newtonian potential ϕ is then simply given by the flat-space perturbation $h_{00} = 1 - g_{00}$ as $\phi = -\frac{1}{2}h_{00} = -G\frac{M}{r}$. A test mass freely falling in a fixed metric will not be able to become entangled in either position or momentum. Entanglement can only occur if the space-time metric itself is in a quantum superposition [9].

So by saying that the generation of entanglement in Feynman's gedankenexperiment requires a quantum description of the gravitational field, what one actually means is that the observation of entanglement is inconsistent with the assumption of a purely classical source mass configuration—and hence with a well-defined space-time metric. At the same time, the outcome will likely be fully consistent with the quantum theoretical predictions based on an effective quantum field theory in the linearized regime [7, 10, 11]. This is reminiscent of quantum optics experiments in the early 1970s that were designed to exclude semiclassical theories of radiation [12, 13], and that culminated in the seminal Bell experiments [14–16] to rule out the broad class of local-realistic models as an underlying description for quantum phenomena [17]. Violating a Cauchy-Schwartz inequality in a photon correlation experiment does not imply quantization of the electromagnetic field; it rather excludes the possibility of being describable by a genuinely classical theory, while at the same time being consistent with the quantum field theoretic predictions of electrodynamics [13]. Similarly, gravity experiments whose outcome cannot be explained by a purely classical source mass configuration would rule out the possibility of semiclassical gravity [18, 19]. In such a theory, the quantum mechanical energy-momentum operator $\hat{T}_{\mu\nu}$ is replaced by its expectation value $\langle\hat{T}_{\mu\nu}\rangle$ as a source for the gravitational field [20, 21]. As a consequence, the space-time metric is always well defined—since even a mass in a superposition would contribute an actual energy density at each of its possible locations in the superposition—and can hence never reproduce phenomena due to quantum interference.[2] Formally, the resulting dynamics is captured by the non-linear Schrödinger-Newton equation [25], whose predictions for massive quantum objects provide an independent testbed for semiclassical gravity [26–28].

In the following we will focus on the actual challenge when trying to implement such experiments: to maintain full control over quantum coherence at the level of the gravitational source (and test) masses. This can only be achieved by sufficiently isolating the experiment from environmental influences such that all decoherence rates are small compared to the time scales of the wanted gravitational coupling. Otherwise, environmental decoherence [29] will result in a mass distribution that cannot, even in principle, be distinguished from an incoherent mixture and therefore will always be consistent with the predictions of semiclassical gravity (as is for example the case in the first experimental attempt of Page and Geilker [18]).

[2] Note that this is also one underlying assumption in gravitational collapse hypotheses as formulated by Penrose [22] and others [23, 24]. This implies that any observation of gravitationally induced entanglement would rule out the validity of these models by principle.

3 Witten's Challenge: Becoming a Practical Experiment

We restrict the discussion to the generation of quantum entanglement via gravitational interaction, as it is at the heart of most previous debates [1, 18, 19]. Two concrete possibilities have been recently proposed in the literature: (i) Entanglement via gravitational phase evolution of superposition states [30, 31], which is a direct adaptation of Feynman's thought experiment, and (ii) entanglement via gravitational interaction of two quantum harmonic oscillators [32–34]. Since both cases deal with table-top experimental settings, it is sufficient to use the linearized, low-energy regime of general relativity, i.e. Newtonian dynamics. For both approaches we consider two masses m that are kept at a fixed distance d along the z-axis and that are free to move along one orthogonal direction (x-axis). The particles are prepared in independent initial states $|\Psi_0\rangle_1$, $|\Phi_0\rangle_2$ and are assumed to interact via gravity only.

In the first scenario [30, 31] we start out with each of the masses being prepared in a spatial superposition state of size $\Delta x = |x_L - x_R|$, i.e. $|\Psi_0\rangle_1 = |x_L\rangle_1 + |x_R\rangle_1$ and $|\Phi_0\rangle_2 = |x_L\rangle_2 + |x_R\rangle_2$. Again, $x_{L,R}$ are two macroscopically distinct positions of the center of mass of each particle along the x-axis, i.e. $\langle x_L|x_R\rangle \ll 1$. In the presence of a gravitational potential ϕ each particle wavefunction picks up a dynamical phase $\varphi = \frac{1}{\hbar}\int m\phi\, dt$. This was used in the seminal "COW" experiment by Colella, Overhauser and Werner to generate quantum interference of neutrons induced by Earth's gravitational field [35]. Here, in contrast, the source for the gravitational field is one of the particles. The potential experienced by the other particle can take the values $\phi_{0,1} = Gm/|r_{0,1}|$ (r: center-of-mass distance) with $|r_0| = d$ and $|r_1| = \sqrt{d^2 + \Delta x^2}$. Since the source mass is in a superposition state, the resulting conditional "COW" phase shifts $\varphi_{0,1}(t) = \frac{1}{\hbar}m\phi_{0,1}t$ can generate entanglement between the previously independent particles.[3] Specifically, the evolution of the 2-particle state is given by $|\Psi_0\rangle_1 \otimes |\Phi_0\rangle_2 \rightarrow |x_L\rangle_1 |\tilde{x}_L\rangle_2 + |x_R\rangle_1 |\tilde{x}_R\rangle_2$, with $|\tilde{x}_L\rangle_2 = |x_L\rangle_2 + e^{i\Delta\varphi}|x_R\rangle_2$, $|\tilde{x}_R\rangle_2 = e^{i\Delta\varphi}|x_L\rangle_2 + |x_R\rangle_2$ and $\langle\tilde{x}_L|\tilde{x}_R\rangle_2 \propto |1 + \cos(2\Delta\varphi)|$, where $\Delta\varphi = \varphi_1 - \varphi_0$. As a result, entanglement is generated for $\Delta\varphi > 0$ and at a rate $\Gamma_{\text{ent}} = \frac{d}{dt}\Delta\varphi \approx (G/\hbar)(m^2\Delta x^2/d^3)$, consistent with [30, 31].

In the second scenario, two harmonically bound masses are prepared in pure quantum states with wavepacket extensions $\Delta x = \eta\sigma_0$ (η: expansion parameter; $\sigma_0 = \sqrt{\hbar/(m\omega_0)}$: quantum ground state position uncertainty; ω_0: trap frequency). In most cases, this will likely be realized by squeezed states of motion quantified by $\eta = e^s$, where s is the corresponding squeezing parameter. Newtonian gravity facilitates a $1/r$ coupling between the masses via the Hamiltonian $\hat{H}_G = -Gm^2/|\hat{r}|$, with the center of mass distance $|\hat{r}| = \sqrt{d^2 + |\hat{x}_1 - \hat{x}_2|^2}$. Taylor expansion yields the relevant entangling interaction term $\hat{H}_{\text{int}} \approx (Gm^2/d^3)\hat{x}_1\hat{x}_2 = \hbar\Gamma_{\text{ent}}\hat{x}_1\hat{x}_2/\Delta x^2$, with $\Gamma_{\text{ent}} \approx (G/\hbar)(m^2\Delta x^2/d^3)$ being the rate at which entanglement is created and consistent with [32–34].

Somewhat surprisingly, even though both scenarios are experimentally very different they generate entanglement at the same rate Γ_{ent} (in the ideal case of pure

[3] For $\Delta\phi = \pi$ this is known as an entangling CSIGN gate in quantum information [36].

states). The relevant parameters are the mass m, the center-of-mass distance d and the size of the delocalization Δx of each of the masses. For scenario 1, Δx is the separation between two center-of-mass states of a spatial superposition, for scenario 2 it is the spatial extent of a Gaussian wavepacket. To provide an example: if we assume that we can keep two masses at a distance $d \approx 1\,\mu\text{m}$ and can prepare a delocalization $\Delta x \approx 100\,\text{nm}$, it would require at least $N \approx 10^{11}$ silicon atoms to achieve entanglement rates on the order of seconds. Given that the confinement of these atoms needs to be on a scale much smaller than d^3 this will require solid-state densities for each mass. In the following we will therefore restrict the discussion to quantum states of motion of solids (rather than atomic gases or molecules) and to the idealized case of spherical masses. In order to avoid unphysical constraints (center-of-mass distance smaller than object sizes) we parametrize the center-of-mass distance via $d = 2R\alpha$, with particle radius R and $\alpha > 1$ (e.g. $\alpha = 2$ corresponds to a surface-to-surface distance between the masses of $2R$). Using $m \approx 4R^3\rho$ (ρ: material density) this results in $\Gamma_{\text{ent}} \approx (G/\hbar)(2\rho^2/\alpha^3)R^3\Delta x^2$. Coming back to Witten's question: it is obvious that environmental decoherence will be able to prevent this from becoming a practical experiment. So what are the sources of decoherence and how can we overcome them?

4 Zeh's Decoherence: Avoiding the Appearance of a Classical World

Let us start again with scenario 1, the "CSIGN" approach to gravitational entanglement. As of today it is unclear how well one can realize the initial state $|\Psi_0\rangle = |x_L\rangle + |x_R\rangle$ that is required for this protocol, namely a superposition of macroscopically distinct states of the center of mass position of a massive object. The trend in existing experiments indicates a clear trade-off between separation Δx and mass m. For example, while experiments with individual neutrons or atoms achieve separations over tens of centimeters [35, 37], experiments with macromolecules comprising up to 2000 atoms achieve separations over hundreds of nanometers [38], and experiments with solid-state resonators made of more than 10^{12} atoms separations only on the level of femtometers [39].[4] There are proposals in the literature to overthrow this trend, both for dielectric [40, 41] and superconducting [42, 43] solids, however thus far no experiment has entered these extreme regimes of macroscopic quantum superpositions. For now we assume that the preparation of such states is possible and ask for the minimum requirements on the initial state to implement the entanglement protocol. A first strict bound is provided by the achievable vacuum level in the experiment. Cryogenic XUV experiments operating at environment temperatures $T_e \approx 1\,\text{K}$ can achieve gas densities that correspond to pressure levels below 10^{-17} mbar [44]. Typical de-Broglie wavelengths of hydrogen molecules, the dominating rest gas

[4] To state a fun fact: in all of the existing macroscopic quantum experiments the trade-off between mass and superposition size is of the same order of magnitude with $\Delta x \cdot m \approx 10^{-28}$ kg · m.

species in this regime, are on the order of $\lambda_{th} = 2\pi\hbar/(\sqrt{2\pi m_{H2} k_B T_e}) \approx 1\,\text{nm}$, which is likely much smaller than the superposition size Δx. This means that scattering of a single molecule is sufficient to localize the superposition state. As a consequence, decoherence from rest gas scattering occurs at a rate [29, 45] $\Gamma_g = (\lambda_{th}/\hbar)(16\pi/3)pR^2 \approx 2 \times 10^{26} pR^2$ (p: gas pressure, R: particle radius). To generate entanglement in the presence of rest gas requires $\Gamma_{\text{ent}} > \Gamma_g$, which reduces to the condition $R\Delta x^2 > 10^{26}(\hbar/G)(\alpha^3/\rho^2)p \approx 10^{-12}(1/\rho^2)$ for $p = 10^{-17}$ mbar $= 10^{-15}$ Pa and $\alpha = 2$. Ideally, one chooses large particles at large delocalization, since the mass coupling ($\propto R^3$) grows faster than the decoherence due to scattering ($\propto R^2$). Practically, as is mentioned above, there will be a trade-off between those two numbers. Looking at current experiments, silica nanoparticles ($\rho \approx 2 \times 10^3$ $\text{kg} \cdot \text{m}^{-3}$) with $R \approx 75\,\text{nm}$ have now been prepared in their quantum ground state of motion [49–51]. They could provide a promising starting point for this experiment, albeit at a required superposition size of $\Delta x > 2\,\mu\text{m}$, which seems challenging. At the other end of the spectrum, a Planck-mass lead particle ($R \approx 70\,\mu\text{m}$, $\rho \approx 10^4$ $\text{kg} \cdot \text{m}^{-3}$) would require $\Delta x > 12\,\text{nm}$, but it is unclear as of yet if quantum control can be extended to this mass regime.

Another omnipresent source of decoherence is scattering, absorption and emission of blackbody radiation, which provide a temperature limit to both the environment temperature T_e (for scattering and absorption) and the internal particle temperature T_i (for emission). The thermal wavelength of the photons $\lambda_{th}^{e,i} = \pi^{2/3}\hbar c/(k_B T_{e,i}) \approx$ 5 mm for $T_{e,i} \approx 1\,\text{K}$ is significantly larger than Δx and hence decoherence occurs at a rate $\Gamma_{bb} = \Lambda_{bb}\Delta x^2$. Scattering of blackbody photons is described by [29, 45] $\Lambda_{bb}^{sc} = (1/\lambda_{th}^{e})^9(8!8\zeta(9)\pi^5 cR^6/9)\text{Re}\{(\epsilon-1)/(\epsilon+2)\}^2 \approx 3 \times 10^{36} R^6 T_e^9$ (ζ: Riemann Zeta function, ϵ: dielectric constant), and emission and absorption by $\Lambda_{bb}^{e,i} = (1/\lambda_{th}^{e,i})^6(16\pi^9 cR^3/189)\text{Im}\{(\epsilon-1)/(\epsilon+2)\} \approx 7 \times 10^{25} R^3 T_{e,i}^6$. Here we have assumed the worst case scenario of large dispersion and absorption, for which $(\epsilon-1)/(\epsilon+2) \approx 1$. Using again the examples from above, for a 75 nm silica particle $\Gamma_{\text{ent}} > \Gamma_{bb}$ results in bounds $T_{i,e} < \left[(G/\hbar)(2\rho^2/\alpha^3)(1.4 \times 10^{-26})\right]^{1/6} \approx$ 4 K from blackbody emission and absorption, and $T_e < \left[(G/\hbar)(2\rho^2/\alpha^3)(1/R^3)\right.$ $\left.(3 \times 10^{-37})\right]^{1/9} \approx 40\,\text{K}$ from blackbody scattering. For a Planck mass lead particle, one obtains $T_{i,e} < 7\,\text{K}$ from emission and absorption, and $T_e < 6\,\text{K}$ from scattering. The limits on internal temperature are independent on the object size and ultimately set by the material absorption properties. In the case of levitated superconductors or magnets, small internal temperatures should be easily achievable as state preparation will likely occur at cryogenic temperatures in a magnetic trap [43], while optically levitated particles will need to implement either active cooling methods or switch to "dark" traps such as electrostatic traps [46–48]. On the other hand, the available environment temperature will set an ultimate bound on the object size. For the extreme case of a 10cm glass object, such as the mirrors used in gravitational wave detectors, environment temperatures below 300mK would be required to avoid localization via blackbody scattering on the time scale relevant for entanglement generation.

The second scenario, gravitationally coupled oscillators, is facing similar constraints. Gas scattering and blackbody radiation remain as significant sources of decoherence. The bounds on the delocalization Δx are now bounds on the minimal extension (or amount of squeezing) of the initial oscillator wavepacket. That this will be experimentally demanding can be seen by sticking to our previous example: achieving a wavepacket size $\Delta x > 2\,\mu\text{m}$ for a $R = 75\,\text{nm}$ silica particle as used in recent experiments ($\omega_0/2\pi \approx 10^5\,\text{Hz}$) would require to extend the ground-state wavepacket $\sigma \approx 3\,\text{pm}$ by a factor of at least 6×10^5. Compared to the CSIGN approach, additional requirements need to be met. Trapped oscillators will dissipate energy to the environment at a rate γ (e.g. via photon recoil, trapped flux, etc.), which results in thermal decoherence at a rate $\Gamma_{th} = \gamma(k_B T_e/\hbar\omega_0)$. Fluctuations in trap position and frequency will add heating rates [34] $\Gamma_x = \pi\omega_0^2 S_x(\omega_0)/(4\sigma_0^2)$ and $\Gamma_\omega = \pi\omega^2 S_x(2\omega_0)/16$, with $S_{x,\omega}$ being the power spectral densities of position and frequency fluctuations at the relevant frequencies in the system, respectively. Current quantum experiments with massive levitated particles [49–51] are bounding these values to $\sqrt{S_x(\omega_0)} < 10^{-16}\,\text{m}\cdot\text{Hz}^{-0.5}$ and $\sqrt{S_\omega(2\omega_0)} < 10^{-4}\,\text{Hz}^{-0.5}$ in the frequency regime around $\omega_0/2\pi = 10^5\,\text{Hz}$.[5] At last the coupling needs to overcome the thermal excitations in the oscillator system. Concretely, in the absence of dissipation, the logarithmic negativity E_N that detects entanglement in the system for $E_N > 0$ is given by $E_N \approx \frac{1}{\ln 2}(4g_s/\omega - 4\bar{n})$ [52], resulting in entanglement for $\Gamma_{\text{ent}} > \bar{n}\omega_0$, or $R^3\Delta x^2\rho^2 > (2/\pi) \times 10^{-23}\bar{n}\omega_0$. This condition can be further relaxed for example by taking into account time-dependent interactions, which can significantly alter the system dynamics [53]. Coming back to examples: optically trapped nanoparticles ($R = 75\,\text{nm}, \rho = 2 \times 10^3\,\text{kg}\cdot\text{m}^{-3}, \omega_0/2\pi = 1 \times 10^5\,\text{Hz}$) in the quantum regime have been realized with a state purity $\bar{n} \approx 0.5$ ($\bar{n}$: harmonic oscillator thermal occupation) and with photon-recoil limited dissipation as low as $\gamma \approx 1 \times 10^{-3}\,\text{Hz}$. For these parameters, entanglement can be generated for $\Delta x > 3\,\text{cm}$, when all other sources of decoherence can be neglected (which is the case for $p < 10^{-6}\,\text{Pa}$, $T_e < 40\,\text{K}$, $T_i < 4\,\text{K}$, $\sqrt{S_x(\omega_0)} < 10^{-15}\,\text{m}\cdot\text{Hz}^{-1/2}$, $\sqrt{S_\omega(2\omega_0)} < 10^{-3}\,\text{Hz}^{-1/2}$).[6] The required large wavepacket size is a direct consequence of the energy penalty for generating entanglement between harmonic oscillators with non-unit purity. Relaxing these initial conditions to experimentally realistic values will likely require much better cooling in combination with non-stationary entanglement protocols [53] or effective free-fall configurations [34]. Planck-mass lead spheres would likely be suspended at lower frequencies, say $\omega_0/2\pi = 1 \times 10^3\,\text{Hz}$. For an initial state purity of $\bar{n} \approx 0.5$ entanglement generation then requires $\Delta x > 20\,\text{nm}$ (along with $\gamma < 10^{-5}\,\text{Hz}$, $p < 10^{-15}\,\text{Pa}$, $T_e < 6\,\text{K}$, $T_i < 8\,\text{K}$, $\sqrt{S_x(\omega_0)} < 10^{-23}\,\text{m}\cdot\text{Hz}^{-1/2}$, $\sqrt{S_\omega(2\omega_0)} < 10^{-2}\,\text{Hz}^{-1/2}$). Even though this appears to be a significantly smaller displacement when compared to

[5] In the absence of any other low-frequency noise sources, position noise $\sqrt{S_x(\omega)}$ is expected to scale with $1/\omega^2$. Extrapolation to below 100 Hz will likely require a dedicated laboratory protected against other vibrational noise sources or LIGO-scale vibration isolation.

[6] Note that these numbers only define the experimental requirements once the initial conditions have been implemented. Preparing the initial states can require even more stringent environment conditions, see for example the wavepacket expansion protocol analyzed in [34]. Also note that for $\Delta x > d$ higher-order terms in the Taylor expansion of the interaction need to be taken into account.

the nanosphere example, the relative wavepacket expansion $\eta = \Delta x/\sigma_0$ (σ_0: ground state size) is beyond current experimental reach in both cases. For example, the requirements $\eta > 10^9$ for nanoparticles and $\eta > 10^7$ for Planck-mass spheres are in stark contrast to the thus far experimentally achieved values of $\eta \approx 1.2$ [54–56].

Finally, entanglement could also be generated between suspended mechanical oscillators. A recent experiment has demonstrated gravitational coupling between two millimetre-sized gold spheres using a torsional pendulum operating in the mHz-regime [57]. Assuming one can extend these parameters ($R \approx 10^{-3}$ m, $\rho \approx 2 \times 10^4\,\mathrm{kg \cdot m^{-3}}$, $\omega_0/2\pi \approx 10^{-2}$ Hz) to the quantum regime ($\bar{n} = 0.5$) it would only require $\Delta x > 600$ fm to generate entanglement. However, since the respective entanglement rates are extremely small, this comes at the cost of unrealistic pressure requirements ($P < 10^{-22}$ Pa). Using $p < 10^{-15}$ Pa from above results in a consistent requirement set $\Delta x > 2$ nm, or a wavepacket expansion $\eta > 3 \times 10^5$, in combination with $\gamma < 3 \times 10^{-8}$ Hz ($Q > 2 \times 10^6$), $p < 10^{-15}$ Pa, $T_e < 2.5$ K, $T_i < 7$ K, $\sqrt{S_x(\omega_0)} < 10^{-16}\,\mathrm{m \cdot Hz^{-1/2}}$, $\sqrt{S_\omega(2\omega_0)} < 10^4\,\mathrm{Hz^{-1/2}}$. Similarly, if two kilogram-scale mirrors from current gravitational wave detector experiments ($R = 10^{-1}$ m, $\rho = 2 \times 10^3\,\mathrm{kg \cdot m^{-3}}$, $\omega_0/2\pi \approx 10^2$ Hz) could be operated at conditions $\bar{n} < 0.5$, $p < 10^{-15}$ Pa, $T_e < 0.4$ K, $T_i < 4$ K and $\gamma < 80$ Hz ($Q > 10$), the generation of gravitationally induced entanglement would require a wavepacket size $\Delta x > 2$ nm, or an initial wavepacket expansion of $\eta > 10^{10}$.

Clearly, these considerations show that we are still far away from realistic experiments and that we need better ideas (or more patience) for going the next step.

5 Conclusion

The topics discussed at the Chapel Hill conference included two of the most outstanding questions in gravitational physics: *Do gravitational waves exist?*, and *Do we require a quantum description of gravity?*. At that time, both questions were of theoretical nature and experimental approaches did not exist beyond the power of thought experiments. In Feynman's words: "*There exists, however, one serious difficulty, and that is the lack of experiments. Furthermore, we are not going to get any experiments, so we have to take a viewpoint of how to deal with problems where no experiments are available.*" [1].

Today, more than 60 years later, we have an experimental answer to the first question. Laser interferometric gravitational wave detectors around the world unambiguously show almost on a daily basis that gravitational energy is radiated [58] and gravitational wave astronomy is now opening a completely new field in physics. In a parallel development, quantum experiments involving massive macroscopic objects have reached a level of maturity that we can confidently begin to tackle the second question. One possibility is to follow Feynman's initial idea and entangle objects via the gravitational field.[7] We have seen that the known mechanisms of decoherence impose strict and demanding limits on the experimental boundary conditions for this approach. And even though current experiments operate far away from these regimes, decoherence can in principle be tackled until we bump into a fundamental (or technological) show stopper. Thus far, this has not been the case. It is, however, also fair to say that optimistic claims about the technical feasibility of these experiments are -at least from today's view- highly exaggerated. It is certainly worthwhile to also investigate other phenomena involving coherent source mass distributions that provide us with observations that are inconsistent with (semi-)classical gravity [19]. Most likely, the outcome of such experiments will be consistent with the predictions of an effective quantum field theory. Beyond the possibility that this is not the case, the actual benefit of this undertaking lies in ruling out semiclassical models of gravity and, probably more importantly, provide empirical evidence that gravity in fact requires a quantum description. An even more thrilling perspective is to gain access to the field degrees of freedom of gravity, either directly via optical clocks [59] or indirectly by observing decay of entanglement between the masses through additional entanglement with the field. This represents an even more demanding experimental challenge, but who knows what the next 60 years will bring.

References

1. C.M. DeWitt, D. Rickles (eds.), *The Role of Gravitation in Physics. Report from the 1957 Chapel Hill Conference*. Max Planck Research Library for the History and Development of Knowledge (2011)
2. C. Kiefer, Conceptual problems in quantum gravity and quantum cosmology. ISRN Math. Phys. **2013**, 1–17 (2013)
3. H.D. Zeh, Feynman's interpretation of quantum theory. Eur. Phys. J. H **36**, 63–74 (2011)
4. K. Hornberger, S. Gerlich, P. Haslinger, S. Nimmrichter, M. Arndt, Colloquium: quantum interference of clusters and molecules. Rev. Mod. Phys. **84**, 157–173 (2012)
5. M. Arndt, K. Hornberger, Testing the limits of quantum mechanical superpositions. Nat. Phys. **10**, 271–277 (2014)
6. M. Aspelmeyer, T.J. Kippenberg, F. Marquardt, Cavity optomechanics. Rev. Mod. Phys. **86**, 1391 (2014)
7. A. Belenchia, R.M. Wald, F. Giacomini, E. Castro-Ruiz, Č Brukner, M. Aspelmeyer, Quantum superposition of massive objects and the quantization of gravity. Phys. Rev. D **98**, 126009 (2018)
8. T.D. Galley, F. Giacomini, J.H. Selby, A no-go theorem on the nature of the gravitational field beyond quantum theory. Preprint at https://arxiv.org/abs/2012.01441 (2020)
9. M. Christodoulou, C. Rovelli, On the possibility of laboratory evidence for quantum superposition of geometries. Phys. Lett. B **792**, 64–68 (2019)

[7] I refer to this as the "0"-level of excitement on the Dvali scale. Next levels in increasing order include the demonstration of the effect for a propagating field (level "1") and the inclusion of field degrees of freedom, e.g. loss of entanglement visibility via entanglement with the field (level "2"). This classification followed from discussions with Gia Dvali at the workshop "Primordial black holes, de Sitter space and quantum tests of gravity" hosted by the Hamburg Academy of Sciences in February 2019.

10. J.F. Donoghue, General relativity as an effective field theory: the leading quantum corrections. Phys. Rev. D **50**, 3874–3888 (1994)
11. M.P. Blencowe, Effective field theory approach to gravitationally induced decoherence. Phys. Rev. Lett. **111**, 021302 (2013)
12. J. Clauser, Experimental limitations to the validity of semiclassical radiation theories. Phys. Rev. A **6**, 49–54 (1972)
13. J. Clauser, Experimental distinction between the quantum and classical field-theoretic predictions for the photoelectric effect. Phys. Rev. D **9**, 853 (1974)
14. S.J. Freedman, J.F. Clauser, Experimental test of local hidden-variable theories. Phys. Rev. Lett. **28**, 938–941 (1972)
15. A. Aspect, J. Dalibard, G. Roger, Experimental test of Bell's inequalities using time- varying analyzers. Phys. Rev. Lett. **49**(25), 1804–1807 (1982)
16. G. Weihs, T. Jennewein, C. Simon, H. Weinfurter, A. Zeilinger, Violation of Bell's inequality under strict einstein locality conditions. Phys. Rev. Lett. **81**, 5039–5043 (1998)
17. J.S. Bell, On the Einstein Podolsky Rosen paradox. Physics **1**, 195–200 (1964)
18. D. Page, C. Geilker, Indirect evidence for quantum gravity. Phys. Rev. Lett. **47**, 979–982 (1981)
19. W.G. Unruh, Steps towards a quantum theory of gravity, in *Quantum Theory of Gravity: Essays in Honor of the 60th Birthday of Bryce S. DeWitt*, ed. by S.M. Christensen (1984), pp. 234–242
20. C. Møller, The energy momentum complex in general relativity and problems, in *Les théories relativistes de la gravitation*, ed. by A. Lichnerowicz, M.-A. Tonnelat, Paris (CNRS, 1962), pp. 15–29
21. L. Rosenfeld, On quantization of fields. Nucl. Phys. **40**, 353–356 (1963)
22. R. Penrose, On the gravitization of quantum mechanics 1: quantum state reduction. Found. Phys. **44**, 557–575 (2014)
23. L. Diósi, Gravitation and quantum-mechanical localization of macro-objects. Phys. Lett. A **105**, 199–202 (1984)
24. S. Colin, T. Durt, R. Willox, Can quantum systems succumb to their own (gravitational) attraction? Classical Quant. Grav. **31**, 245003 (2014)
25. M. Bahrami, A. Großardt, S. Donadi, A. Bassi, The Schrödinger-Newton equation and its foundations. New J. Phys. **16**, 115007 (2014)
26. D. Giulini, A. Großardt, Gravitationally induced inhibitions of dispersion according to the Schrödinger-Newton equation. Classical Quant. Grav. **28**, 195026 (2011)
27. H. Yang, H. Miao, D.S. Lee, B. Helou, Y. Chen, Macroscopic quantum mechanics in a classical spacetime. Phys. Rev. Lett. **110**, 170401 (2013)
28. S. Colin, T. Durt, R. Willox, Crucial tests of macrorealist and semiclassical gravity models with freely falling mesoscopic nanospheres. Phys. Rev. A **93**, 062102 (2016)
29. E. Joos, H.D. Zeh, The emergence of classical properties through interaction with the environment. Zeitschrift für Physik B Condensed Matter **59**, 223–243 (1985)
30. S. Bose, A. Mazumdar, G.W. Morley, H. Ulbricht, M. Toroš, M. Paternostro, A.A. Geraci, P.F. Barker, M.S. Kim, G. Milburn, Spin entanglement witness for quantum gravity. Phys. Rev. Lett. **119**, 240401 (2017)
31. C. Marletto, V. Vedral, Gravitationally induced entanglement between two massive particles is sufficient evidence of quantum effects in gravity. Phys. Rev. Lett. **119**, 240402 (2017)
32. A. Al Balushi, W. Cong, R.B. Mann, Optomechanical quantum Cavendish experiment. Phys. Rev. A **98**, 043811 (2018)
33. T. Krisnanda, G.Y. Tham, M. Paternostro, T. Paterek, Observable quantum entanglement due to gravity. NPJ Quant. Inform. **6**, 12 (2020)
34. T. Weiss, M. Roda-Llordes, E. Torrontegui, M. Aspelmeyer, O. Romero-Isart, Large quantum delocalization of a levitated nanoparticle using optimal control: applications for force sensing and entangling via weak forces. Phys. Rev. Lett. **127**, 023601 (2021)
35. R. Colella, A. Overhauser, S. Werner, Observation of gravitationally induced quantum interference. Phys. Rev. Lett. **34**, 1472–1474 (1975)
36. M.A. Nielsen, I.L. Chuang, *Quantum Computation and Quantum Information* (Cambridge University Press, Cambridge, 2010)

37. T. Kovachy, P. Asenbaum, C. Overstreet, C.A. Donnelly, S.M. Dickerson, A. Sugarbaker, J.M. Hogan, M.A. Kasevich, Quantum superposition at the half-metre scale. Nature **528**, 530–533 (2015)
38. Y.Y. Fein, P. Geyer, P. Zwick, F. Kiałka, S. Pedalino, M. Mayor, S. Gerlich, M. Arndt, Quantum superposition of molecules beyond 25 kDa. Nat. Phys. **15**, 1242–1245 (2019)
39. A.D. O'Connell, M. Hofheinz, M. Ansmann, R.C. Bialczak, M. Lenander, E. Lucero, M. Neeley, D. Sank, H. Wang, M. Weides, J. Wenner, J.M. Martinis, A.N. Cleland, Quantum ground state and single-phonon control of a mechanical resonator. Nature **464**, 697–703 (2010)
40. O. Romero-Isart, A.C. Pflanzer, F. Blaser, R. Kaltenbaek, N. Kiesel, M. Aspelmeyer, J.I. Cirac, Large quantum superpositions and interference of massive nanometer-sized objects. Phys. Rev. Lett. **107**, 020405 (2011)
41. J. Bateman, S. Nimmrichter, K. Hornberger, H. Ulbricht, Near-field interferometry of a free-falling nanoparticle from a point-like source. Nat. Commun. **5**, 4788 (2014)
42. O. Romero-Isart, L. Clemente, C. Navau, A. Sanchez, J. Cirac, Quantum magnetomechanics with levitating superconducting microspheres. Phys. Rev. Lett. **109**, 147205 (2012)
43. H. Pino, J. Prat-Camps, K. Sinha, B.P. Venkatesh, O. Romero-Isart, On-chip quantum interference of a superconducting microsphere. Quant. Sci. Technol. **3**, 025001 (2018)
44. G. Gabrielse, Comparing the antiproton and proton, and opening the way to cold antihydrogen. Adv. Atomic Molecul. Opt. Phys. **45**(C), 1–39 (2001)
45. O. Romero-Isart, Quantum superposition of massive objects and collapse models. Phys. Rev. A **84**, 052121 (2011)
46. G.P. Conangla, A.W. Schell, R.A. Rica, R. Quidant, Motion control and optical interrogation of a levitating single nitrogen vacancy in vacuum. Nano Lett. **18**, 3956–3961 (2018)
47. T. Delord, P. Huillery, L. Nicolas, G. Hétet, Spin-cooling of the motion of a trapped diamond. Nature **580**, 56–59 (2020)
48. L. Dania, D.S. Bykov, M. Knoll, P. Mestres, T.E. Northup, Optical and electrical feedback cooling of a silica nanoparticle levitated in a Paul trap. Phys. Rev. Res. **3**, 013018 (2021)
49. U. Delić, M. Reisenbauer, K. Dare, D. Grass, V. Vuletić, N. Kiesel, M. Aspelmeyer, Cooling of a levitated nanoparticle to the motional quantum ground state. Science **367**, 892–895 (2020)
50. L. Magrini, P. Rosenzweig, C. Bach, A. Deutschmann-Olek, S.G. Hofer, S. Hong, N. Kiesel, A. Kugi, M. Aspelmeyer, Real-time optimal quantum control of mechanical motion at room temperature. Nature **595**, 373–377 (2021)
51. F. Tebbenjohanns, M.L. Mattana, M. Rossi, M. Frimmer, L. Novotny, Quantum control of a nanoparticle optically levitated in cryogenic free space. Nature **595**, 378–382 (2021)
52. M. Ludwig, K. Hammerer, F. Marquardt, Entanglement of mechanical oscillators coupled to a nonequilibrium environment. Phys. Rev. A **82**, 012333 (2010)
53. F. Galve, L.A. Pachón, D. Zueco, Bringing entanglement to the high temperature limit. Phys. Rev. Lett. **105**, 180501 (2010)
54. E.E. Wollman, C.U. Lei, A.J. Weinstein, J. Suh, A. Kronwald, F. Marquardt, A.A. Clerk, K.C. Schwab, Quantum squeezing of motion in a mechanical resonator. Science **349**, 952–955 (2015)
55. F. Lecocq, J.B. Clark, R.W. Simmonds, J. Aumentado, J.D. Teufel, Quantum nondemolition measurement of a nonclassical state of a massive object. Phys. Rev. X **5**, 041037 (2015)
56. J.-M. Pirkkalainen, E. Damskägg, M. Brandt, F. Massel, M.A. Sillanpää, Squeezing of quantum noise of motion in a micromechanical resonator. Phys. Rev. Lett. **115**, 243601 (2015)
57. T. Westphal, H. Hepach, J. Pfaff, M. Aspelmeyer, Measurement of gravitational coupling between millimetre-sized masses. Nature **591**, 225–228 (2021)
58. B.P. Abbott et al., Observation of gravitational waves from a binary black hole merger. Phys. Rev. Lett. **116**, 061102 (2016)
59. E. Castro Ruiz, F. Giacomini, Č. Brukner, Entanglement of quantum clocks through gravity, in *Proceedings of the National Academy of Sciences*, vol. 114 (2017), pp. E2303–E2309

Preferred Basis, Decoherence and the Quantum State of the Universe

Andrei O. Barvinsky and Alexander Yu. Kamenshchik

1 Introduction

As is well-known quantum mechanics and more generally quantum theory (including quantum statistical mechanics and quantum field theory) have had great achievements in the description of the microworld. Its arrival not only has brought a lot of new technologies but also has changed our vision of the world. Since the days of its creation the problems of interpretation of quantum mechanics have been attracting attention of a person working in this field as well as of a broader public, including philosophers, psychologists, biologists and even of the people of arts and literature (see e.g. [44]). The main feature of quantum mechanics, which distinguishes it from classical Newton mechanics is the fact, that even if one has a complete knowledge of a state of a system under consideration and would like to make a certain experiment, more than one alternative result of such an experiment is possible. The knowledge of the state of the system can permit us only to calculate the probabilities of different outcomes of the experiment, as was first understood by Max Born [27]. However, a natural question arises how can we see only one outcome of an experiment and what happens with all other alternatives? One can present this question in a slightly different form: how can we reconcile ourselves with the fact that the physics of the microworld is described by quantum mechanics while in the macroworld, which we perceive in our everyday experience we encounter the laws of classical physics?

A. O. Barvinsky (✉)
Theory Department, Lebedev Physics Institute, Leninsky Prospect 53, Moscow 119991, Russia
e-mail: barvin@td.lpi.ru; barvin@lpi.ru

A. Yu. Kamenshchik
Dipartimento di Fisica e Astronomia, Università di Bologna and INFN, via Irnerio 46, 40126 Bologna, Italy
e-mail: kamenshchik@bo.infn.it; Alexander.Kamenshchik@bo.infn.it

Landau Institute for Theoretical Physics, Russian Academy of Sciences, Kosygin street 2, Moscow 119334, Russia

C. Kiefer (ed.), *From Quantum to Classical*, Fundamental Theories of Physics 204,
https://doi.org/10.1007/978-3-030-88781-0_6

At the dawn of quantum mechanics the very existence of different alternatives and non-classicality of microworld was a source of some kind of dismay. The first attempt to cope with this situation was undertaken in the framework of the so called Copenhagen interpretation of quantum mechanics, which represents a collection of views and ideas of some of the founders of quantum mechanics. In particular, Niels Bohr has suggested the notion of complementarity between different notions and approaches, which seemed to resolve apparent contradictions of the quantum theory. Besides, he insisted on the existence of the so called classical realm, where all the results of experiments and observations were registered. Thus, the classical physics was considered not only as a limiting case of the quantum physics, but also as a prerequisite of its very existence [26]. Finally, it was von Neumann, who formulated the mathematically rigorous idea of the reduction of the wave packet and in such a way had given a constructive picture of events, occurring in the process of quantum measurement [70]. According to von Neumann, in quantum mechanics there coexisted two processes. One of them is a unitary deterministic evolution of the wave function, describing the quantum system, according to the Schrödinger equation. The second process takes place during quantum measurement and is called the reduction of the wave function, when one of the possible outcomes is realized, while others disappear into thin air. In the process of quantum measurement three players participate: an object, a measuring device and an observer, and the presence of the latter two corresponds in a way to the Bohr's classical realm. Thus, in this picture, everything which one calculates and predicts in quantum theory finds its explanation. Generally, one can say that according to the Copenhagen interpretation the classical and quantum mechanics were present in our physical world on equal footing. The Copenhagen interpretation has played a very important role in the development of quantum mechanics and its applications. It has made the physicists' imagination more free and have made them more accustomed to the idea that the deterministic ideal of classical mechanics is not an absolute goal of the physical theory. However, some other founders of quantum theory such as Planck, Einstein, Schrödinger and de Broglie were not happy with the Copenhagen interpretation and thought that some rebirth of the classical ideal was necessary. The most consistent attempt of such a rebirth was undertaken by Bohm and is known as de Broglie-Bohm interpretation [25].

On the other hand, the presence of two dynamical processes in quantum theory looked logically unsatisfactory and in 1957 Hugh Everett has published a short version of his Ph.D. thesis [36] under the title "Relative-state formulation of quantum mechanics". It contained a simple idea. We do not need the postulate of the reduction of the wave packet and hence, only one fundamental process exists in quantum theory—unitary evolution governed by the Schrödinger equation. All the outcomes of the experiment co-exist and the objective result of the measurement under consideration is the establishment of correlations between the measured and measuring subsystems, which are treated on equal footing. Thus, there is no need in a special classical realm too.

The Everett interpretation of quantum mechanics seemed to be quite logical and economical, however, this economy was achieved by means of the acception of parallel existence of different outcomes of a quantum measurement and this was a critical point. Indeed, behind the mask of the relative states of two or more subsystems loomed a disturbing image of the co-existence of parallel worlds and of the splitting reality. Probably this fact explains a rather troubled history of the Everett interpretation recognition by the scientific community [28]. In 1970 DeWitt published the paper in Physics Today [31] and then edited a book [32], which have given a new birth to the Everett interpretation of quantum mechanics under the name "many-worlds" interpretation.

For many years the many-worlds interpretation was treated as something rather exotic. However, now the situation is changing due to two main developments in quantum physics: progress in the study of quantum cosmology (see e.g. [52]) and the birth of quantum information (see e.g. [57]). In quantum cosmology one treats the universe as a unique quantum object. Thus, there is no place for a classical external observer and other agents, who could be responsible for the presumed reduction of the wave function. Quantum algorithms, in their turn, use essentially the parallel quantum processes, which also marginalises the idea of a quantum realm.

Generally, one can say that the idea that quantum theory is more fundamental than classical one becomes more and more popular. However, an important question remains: why in many situations we see the classical behaviour of macroscopic objects? What are the relations between classical and quantum theories? Heinz–Dieter Zeh has made an important contributions in this field. In the beginning of the seventies he has published two seminal papers in Foundations of Physics. In one of these papers [71] he has open a new direction—the study of decoherence. In another paper [72] he considered the problem of the choice of the preferred basis in the many-worlds interpretation of quantum mechanics. From our point of view these two approaches complement each other. In this paper we present a review of several works whose content and results are closely related to the works of Zeh.

The structure of the paper is the following: in the second section we discuss the problem of the preferred basis in quantum mechanics; in the third section we describe the dynamics of the preferred basis, while the fourth section is devoted to the interrelations between the phenomenon of decoherence and the many-worlds approaches. In the fifth section we discuss applications of the decoherence approach to cosmology in connection with the problem of ultraviolet divergences in quantum gravity; the sixth section is devoted to the notion of quantum-classical duality; in the seventh section we consider the relations between the many-worlds interpretation and Anthropic Principle; the eighth section is devoted to the cosmological model in which the notion of the density matrix of the universe arises; the ninth section contains brief concluding remarks.

2 The Problem of the Preferred Basis in the Many-Worlds Interpretation of Quantum Mechanics

As we have already mentioned, in the framework of the many-worlds interpretation Schrödinger evolution is the only process. Thus, the principle of superposition is applicable to all the states including macroscopic ones and all the outcomes of any measurement-like processes are realized simultaneously but in different "parallel universes". The very essence of the many-worlds interpretation can be expressed by one simple formula. Let us consider the wave function of a system, containing two subsystems (say, an object and a device), whose wave functions are respectively $|\Phi\rangle$ and $\Psi\rangle$ and let the process of interaction between these two subsystems be described by a unitary operator $\hat{U}$. The result of the action of this operator can be represented as

$$\hat{U}\,|\Phi\rangle_0\Psi\rangle_i = |\Phi\rangle_i\Psi\rangle_i. \tag{1}$$

Here the state $|\Psi\rangle_i$ is a quantum state of the object corresponding to a definite outcome of the experiment, while $|\Phi\rangle_0$ is an initial state of the measuring device. Now, let the initial state of the object be described by a superposition of quantum states:

$$|\Psi\rangle = \sum_i c_i|\Psi\rangle_i. \tag{2}$$

Then the superposition principle immediately leads to

$$\hat{U}|\Phi\rangle_0\Psi\rangle = \hat{U}|\Phi\rangle_0\sum_i c_i|\Psi\rangle_i = \sum_i c_i|\Phi\rangle_i\Psi\rangle_i. \tag{3}$$

Here $|\Phi\rangle_i$ describes the state of the measuring device, which has found the quantum object in the state $|\Psi\rangle_i$. The superposition (3) contains more than one term, while one sees only one outcome of measurement. The reduction of the wave packet postulate solves this puzzle by introducing another process eliminating in a non-deterministic way all the terms in the right-hand side of Eq. (3) but one. All the terms of the superposition are realized but in different universes.

Here, an inevitable question arises: decomposing the wave function of the universe one should choose a certain basis. The result of the decomposition essentially depends on it. Thus, the so called problem of the choice of the preferred basis arises. This problem was considered by different authors (see e.g. [1, 24, 30, 33, 56]). From our point of view the most convincing approach to the solution of this problem was proposed by Zeh [72] and further developed in our works [3, 5, 8, 9]. The essence of the problem can be formulated considering the same example of a quantum system consisting of two subsystems. Let us emphasize that now we would like to undertake a consideration of a general case without particular reference to measuring devices and quantum objects (for a moment we consider this division of a system into subsystems as granted). The only essential characteristics of the branching process is the defactorization of the wave function. This means that if at the initial moment the wave function of the system under consideration was represented by a direct product of the wave functions of the subsystems,

$$|\Psi\rangle = |\phi\rangle|\chi\rangle, \tag{4}$$

then after an interaction between the subsystems it becomes

$$\sum_i c_i|\phi\rangle_i|\chi\rangle_i, \tag{5}$$

where more than one coefficient c_i is different from zero. Apparently the decomposition (5) can be done in various manners. As soon as each term is associated with a separate universe, the unique prescription for the construction of such a superposition should be fixed. We believe that the correct choice of the preferred basis is the so called Schmidt [63] or bi-orthogonal basis. This basis is formed by eigenvectors of both the density matrices of the subsystems of the quantum system under consideration. These density matrices are defined as

$$\hat{\rho}_I = \mathrm{Tr}_{II}|\Psi\rangle\langle\Psi|, \tag{6}$$

$$\hat{\rho}_{II} = \mathrm{Tr}_I|\Psi\rangle\langle\Psi|. \tag{7}$$

Remarkably, the eigenvalues of the density matrices coincide and hence the number of non-zero eigenvalues is the same, in spite of the fact that the corresponding Hilbert spaces can be very different.

$$\hat{\rho}_I|\phi_n\rangle = \lambda_n|\phi_n\rangle, \tag{8}$$

$$\hat{\rho}_{II}|\chi_n\rangle = \lambda_n|\chi_n\rangle, \tag{9}$$

Consequently, the wave function is decomposed as

$$|\Psi\rangle = \sum_n \sqrt{\lambda_n}|\phi_n\rangle|\chi_n\rangle. \tag{10}$$

The bi-orthogonal basis was first used at the dawn of quantum mechanics by Schrödinger [64, 65] for the study of correlations between quantum systems. Recently, this basis has been actively used for measuring the degree of entanglement, in particular in relation to quantum computing [35].

We believe that the bi-orthogonal basis being defined by the fixing of the decomposition of the system into subsystems has a fundamental character and determines the worlds which result from the defactorization process. However, the subdivision of the system into subsystems which implies the branching of the worlds should satisfy some reasonable criteria which we are not ready to formalize at the moment (see, however our paper [9] for the analysis of some relatively simple cases). One can say, that the decomposition into the subsystems should be such that the corresponding preferred basis would be rather stable. For example, when one treats a quantum mechanical experiment of the Stern-Gerlach type, it is natural to consider the measuring device and the atom as subsystems. In the case when we consider a system with some kind of internal symmetry, like in quantum chromodynamics, the division of the system into subsystems which belong to singlet representations of the internal symmetry group looks also reasonable from the point of view of stability of the bi-orthogonal preferred basis of the many-worlds interpretation.

3 Dynamics of the Preferred Basis

Let us consider a system, consisting of two subsystems. In the preferred basis its wave function has the form

$$\Psi\rangle = \sum_n c_n|n\rangle_I|n\rangle_{II}. \tag{11}$$

The whole wave function $|\Psi(t)\rangle$ satisfies the Schrödinger equation

$$i\frac{\partial|\Psi(t)\rangle}{\partial t} = H|\Psi(t)\rangle, \tag{12}$$

with

$$H = H_I + H_{II} + V, \tag{13}$$

where V is the interaction Hamiltonian between the subsystems. One can show [5, 8] that the evolution of the preferred basis vectors is unitary and is governed by some effective Hamiltonians:

$$i\frac{\partial}{\partial t}|m\rangle_I = \mathcal{H}_I|m\rangle_I, \tag{14}$$

$$\mathcal{H}_I = i\sum_m\left(\frac{\partial|m\rangle_I}{\partial t}{}_I\langle m|\right). \tag{15}$$

(The equations for the second subsystem are quite analogous). Non-diagonal elements of $\mathcal{H}_I$ are given by

$$(\mathcal{H}_{\mathcal{I}})_{mn} = (H_I)_{mn} - \frac{{}_I\langle m|\mathrm{Tr}_{II}[V,|\Psi\rangle\langle\Psi|]|n\rangle_I}{p_m - p_n}, \tag{16}$$

and, similarly, for $\mathcal{H}_{II}$. The diagonal elements can be fixed as

$$(\mathcal{H}_I)_{mm} = (H_I)_{mm};\ (\mathcal{H}_{II})_{mm} = (H_{II})_{mm}, \tag{17}$$

whence it follows that

$$i\frac{\partial c_n}{\partial t} = {}_I\langle n|_{II}\langle n|V|\Psi\rangle. \tag{18}$$

Equations (16) and (17) completely determine the effective Hamiltonians which turn out to be complicated functionals not only of the Hamiltonian H, but also of the quantum state $|\Psi\rangle$ of the system. A similar consideration can be found in the paper by Kübler and Zeh [53].

The most unexpected conclusion from the unitary dynamics of the proposed basis is the following. Since the observer (identified, for example, with the subsystem I) observes and measures only one relative state of the second subsystem II in his many-worlds branch, he finds that this state undergoes a unitary evolution of the above type. This is in spite of the impure nature of this open subsystem II described by a non-factorizable density matrix.

The second conclusion is that the observer studying the dynamics of his relative state measures the effective Hamiltonian $\mathcal{H}_{II}$ and not the fundamental Hamiltonian H of the total system. This apparently means that research into nature at the most fundamental levels requires additional efforts in reconstructing the fundamental dynamical laws on the grounds of the observable reality. Other aspects of the relations between observable and fundamental Hamiltonians which we call "quantum-classical duality" were considered in [47] and will be briefly described in Sect. 6.

4 Preferred Basis and Decoherence

In his seminal paper [71] Heinz-Dieter Zeh has opened a new direction in quantum theory—decoherence. Let us briefly remind what it is. Let us consider for example the Stern-Gerlach experiment, where the initial state of the atom and of the device is described by the vector

$$|\Psi\rangle = |\Phi\rangle(c_1|\psi_\uparrow\rangle + c_2|\psi_\downarrow\rangle). \tag{19}$$

As a result of the measurement, we have

$$|\Psi\rangle_{\text{final}} = c_1|\Phi_\uparrow\rangle|\psi_\uparrow\rangle + c_2|\Phi_\downarrow\rangle|\psi_\downarrow\rangle. \tag{20}$$

Let us remember that in every measurement process a third system participates—it is the environment. Hence, the initial state of the general system is

$$|\Psi\rangle = |\chi\rangle|\Phi\rangle(c_1|\psi_\uparrow\rangle + c_2|\psi_\downarrow\rangle), \tag{21}$$

while its final state is

$$|\Psi\rangle_{\text{final}} = c_1|\chi_\uparrow\rangle|\Phi_\uparrow\rangle|\psi_\uparrow\rangle + c_2|\chi_\downarrow\rangle|\Phi_\downarrow\rangle|\psi_\downarrow\rangle. \tag{22}$$

Now, if we calculate the reduced density matrix tracing out the environmental degrees of freedom, we obtain

$$\rho_{\text{reduced}} = \text{Tr}_{\{\chi\}}|\Psi\rangle_{\text{final final}}\langle\Psi| = |c_1|^2|\Phi_\uparrow\rangle\langle\Phi_\uparrow|\psi_\uparrow\rangle\langle\psi_\uparrow| + |c_2|^2|\Phi_\downarrow\rangle\langle\Phi_\downarrow|\psi_\downarrow\rangle\langle\psi_\downarrow|, \tag{23}$$

where we have used the fact that

$$\langle\chi_\uparrow|\chi_\downarrow\rangle = 0. \tag{24}$$

The expression (23) represents a classical statistical mixture, which substitutes the quantum state due to tracing out the environmental degrees of freedom. This is the essence of the decoherence approach. It explains how the classical world arises from the quantum one and has proved its efficiency for description of a very wide range of phenomena in theoretical and experimental physics (see [45]). Thus, the idea of application of the decoherence approach to quantum cosmology was quite natural. This application was first suggested in the paper by Zeh [73] and further developed in papers by Kiefer [48, 49]. In these papers were considered both the minisuperspace variables and infinitely many perturbations. In the next section we shall discuss the problem of ultraviolet divergences arising in cosmological models with infinite number of degrees of freedom.

Before concluding this section we would like to make several remarks about the comparison between the many-worlds approach to the interpretation of quantum mechanics and the decoherence approach. As we have already told, what lies in the foundation of the decoherence approach is understanding the fact that in any process of quantum measurement there are not two, but three participants: namely, not only the quantum object and measuring device, but also the rest of the universe—the so called environment. After the measurement, we can construct the reduced density matrix, describing the object and device, tracing out unobservable degrees of freedom of the environment. It appears that in many cases this reduced density matrix becomes quickly practically diagonal in a certain "good" basis, whose states are sometimes called "pointer states" [74, 75] and behaves more or less classically. In such a way, the quantum state of the object and of the measuring device becomes a classical statistical mixture. However, from our point of view the decoherence approach to the problem of quantum measurement and to the problem of classical-quantum relations is less fundamental than the many-worlds approach.

First, there is an essential difference between statistical principles in classical and quantum physics. In classical physics the probability is "the measure of our ignorance" of the initial conditions or of the details of interaction while in quantum physics we cannot get rid of the probability even in principle, where there is no analogue to the "Laplace demon", who can calculate everything. Thus, the transition to a classical statistical mixture does not resolve the problem of choice between different alternatives.

Second, the decoherence properties of reduced density matrix depend crucially on the choice of the basis. Thus, the classicality is introduced into the theory already at the level of the choice of the basis. In the bi-orthogonal preferred basis approach, described above, the basis is defined by the chosen decomposition of the system under consideration into subsystems. After that, one can study the dynamics of different elements of the basis and to see if they behave classically [8]. It appears, that sometimes classicality exists as a stable phenomenon, sometimes as a temporary phenomenon and sometimes it does not exist at all. Thus, the many-worlds interpretation, insisting on the primary role of the quantum theory with respect to the classical one, describes a wider class of phenomena. Nevertheless, for a large class of situations, the predictions of both approaches are close. It happens when the bi-orthogonal basis is close to the pointer basis.

5 Decoherence and Ultraviolet Divergences in Quantum Cosmology

The first question which arises when one tries to explain the classicalization of a quantum Universe using the decoherence approach, is connected with the definition of an environment. Indeed, in contrast to the usual description of quantum-mechanical experiment in a laboratory, there is no external environment, because the object of quantum cosmology is the whole Universe. Thus, we should treat some part of degrees of freedom as essential and observables, while the others could be treated as an environment with subsequent tracing them out in transition to a reduced density matrix. It is natural to believe that inhomogeneous degrees of freedom play the role of environment while macroscopic variables such as a cosmological radius or initial value of the inflaton scalar field are treated as observables. It is easy to guess that there is an infinite number of environmental degrees of freedom and hence, calculating a reduced density matrix encounters the problem of ultraviolet divergences, which was analyzed in papers [11–13].

To tackle this problem we should consider the wave function of the universe in the one-loop approximation, describing simultaneously with homogeneous also inhomogeneous degrees of freedom [6, 7, 10]. It is convenient to write both the no-boundary [42] and tunneling [69] cosmological wave functions in the form:

$$\Psi(t|\varphi, f) = \frac{1}{\sqrt{v_\varphi^*(t)}} \exp\Big(\mp I(\varphi)/2 + iS(t, \varphi)\Big) \times \prod_n \psi_n(t, \varphi|f_n), \tag{25}$$

$$\psi_n(t, \varphi|f_n) = \frac{1}{\sqrt{v_n^*(t)}} \exp\left(-\frac{1}{2}\Omega_n(t) f_n^2\right), \tag{26}$$

$$\Omega_n(t) = -a^k(t)\frac{\dot{v}_n^*(t)}{v_n^*(t)}. \tag{27}$$

Here, the sign minus or plus in front of Euclidean action $I(\varphi)$ in the exponential of (25) corresponds to the no-boundary and to the tunneling wave functions of the

Universe, respectively, f_n describe amplitudes of inhomogeneous modes, while v_n correspond to solutions of linearized second-order differential equations. The exponential k in the expression for the frequency function Ω_n, depends on the spin s of the field under consideration and on its parametrization. For the "standard" parametrization $k = 3 - 2s$. Inclusion of inhomogeneous modes into the wave function of the Universe was first considered in [40, 54]. One can show [6] that the diagonal of the reduced density matrix corresponding to the wave function (25)

$$\rho(t|\varphi) \equiv \rho(t|\varphi, \varphi) = \int \prod_n df_n |\,\Psi(t|\varphi, f)\,|^2 \tag{28}$$

is

$$\rho(t|\varphi) = \frac{\sqrt{\Delta_\varphi}}{|v_\varphi(t)|} \exp\Big(\mp I(\varphi) - \Gamma_{1-\text{loop}}(\varphi)\Big), \tag{29}$$

where

$$\Delta_\varphi \equiv i a^k (v_\varphi^* \dot{v}_\varphi - \dot{v}_\varphi^* v_\varphi) \tag{30}$$

is the Wronskian of the φ-mode functions and $\Gamma_{1-\text{loop}}$ is the one-loop effective action calculated on the DeSitter instanton of the radius $1/H(\varphi)$, where $H(\varphi)$ is the effective Hubble constant. When $H(\varphi) \to \infty$,

$$\Gamma_{1-\text{loop}} = Z \ln \frac{H(\varphi)}{\mu}, \tag{31}$$

where Z is the anomalous scaling of the theory, μ is a renormalization scale. It is easy to see that the condition of the normalizability of the wave function of the Universe is

$$Z > 1, \tag{32}$$

and this condition provides us with the selection criterium for particle physics models [6].

Information about the decoherence behaviour of the system is contained in the off-diagonal elements of the density matrix. In our case they read

$$\rho(t|\varphi, \varphi') = \left(\frac{\Delta_\varphi \Delta_{\varphi'}}{v_\varphi v_{\varphi'}^*}\right)^{\frac{1}{4}} \exp\left(-\frac{1}{2}\Gamma - \frac{1}{2}\Gamma' + i(S - S')\right) D(t|\varphi, \varphi'). \tag{33}$$

Here $D(t|\varphi, \varphi')$ is the so called decoherence factor defined by the formula

$$D(t|\varphi, \varphi') = \prod_n \left(\frac{4Re\Omega_n Re\Omega_n'^*}{(\Omega_n + \Omega_n'^*)^2}\right)^{\frac{1}{4}} \left(\frac{v_n v_n'^*}{v_n^* v_n'}\right)^{\frac{1}{4}}. \tag{34}$$

How to cope with ultraviolet divergences appearing in the sum of this type? This question was already discussed in [51, 58, 61]. In [11–13] we used the dimensional regularization [43]. As usual, the main effect of a dimensional regularization consists in changing the number of degrees of freedom involved in summation. For example, for a scalar field, the degeneracy number of harmonics in spacetime of dimensionality d changes from the well-known value (see e.g. [55])

$$\dim(n, 4) = n^2, \tag{35}$$

to

$$\dim(n, d) = \frac{(2n + d - 4)\Gamma(n + d - 3)}{\Gamma(n)\Gamma(d - 1)}. \tag{36}$$

Making analytical continuation and discarding the poles $1/(d - 4)$ one has finite values for $D(t|\varphi, \varphi')$. However, for scalar, photon and graviton fields one gets because of oversubtraction of UV-infinities a pathological behaviour:

$$|D(t|\varphi, \varphi')| \to \infty, \text{ at } |\varphi - \varphi'| \to \infty. \tag{37}$$

For example, for a massive scalar field

$$\ln |D(t|\varphi, \varphi')| \approx \frac{7}{64} m^3 \bar{a}(a - a')^2, \tag{38}$$

where

$$\begin{aligned} a &= \frac{1}{H(\varphi)} \cosh H(\varphi)t, \\ a' &= \frac{1}{H(\varphi')} \cosh H(\varphi')t, \\ \bar{a} &= \frac{a + a'}{2}. \end{aligned} \tag{39}$$

Such a form of a decoherence factor not only does not correspond to decoherence, but also renders the density matrix ill-defined, breaking the condition $\mathrm{Tr}(\rho^2) \leq 1$. However, there is a remedy: using the reparametrization of a bosonic scalar field

$$f \to \tilde{f} = a^{\mu} f, \; v_n \to \tilde{v}_n = a^{\mu} v_n, \tag{40}$$

one can get the new form of the frequency function

$$\Omega_n(t) = -i a^{3-2\mu}(t) \frac{\dot{\tilde{v}}_n^*(t)}{\tilde{v}_n^*(t)}. \tag{41}$$

In such a way one can suppress ultraviolet divergences. For the so-called conformal parametrization, $\mu = 1$, for the massive scalar field one has

$$\ln|\tilde{D}(t|\varphi, \varphi')| = -\frac{m^3 \pi \bar{a}(a - a')^2}{64}, \tag{42}$$

For the case of fermions this trick does not work [12]. Let us discuss it in more detail. The wave function of the Universe filled by fermions has the form [12, 29, 50]

$$\Psi(t, \varphi|x, y) = \Psi_0(t, \varphi) \prod_n \psi_n(t|x_n, y_n), \tag{43}$$

where x, y are Grassmann variables. Partial wave functions have the form

$$\psi_n(t|x_n, y_n) = v_n - \frac{i\dot{v}_n + \nu v_n}{m} x_n y_n, \tag{44}$$

where the functions v_n satisfy the second-order equation

$$\ddot{v}_n + (-i\dot{\nu} + m^2 + \nu^2) v_n = 0, \quad \nu = \frac{n + \frac{1}{2}}{a}. \tag{45}$$

As was shown in [50]

$$|D(a, \varphi|a', \varphi')| = \exp\left(-\frac{m^2(a - a')^2}{8} \sum_{n=1} \frac{n(n+1)}{\left(n + \frac{1}{2}\right)^2}\right). \tag{46}$$

One can try eliminating ultraviolet divergences here by dimensional regularization using the fact that for spinors in spacetime of dimensionahty d:

$$\dim(n, d) = \frac{\Gamma(n + 2^{(d-2)})\Gamma(n + 2^{(d-2)/2} - 1)}{[\Gamma(2^{(d-2)/2})]^2 \Gamma(n+1)\Gamma(n)}. \tag{47}$$

However, as the result of the renormalization procedure we have

$$|D(a, \varphi|a', \varphi')| = \exp\left(-\frac{m^2(a - a')^2}{8} I\right), \ I < 0, \tag{48}$$

and we encounter the same problem as in the case of bosons. Moreover, one cannot use the conformal reparametrization in this case because standard fermion variables are already presented in the conformal parametrization. However, there is another way to circumvent this problem. One can perform a non-local Bogoliubov transformation mixing Grassmann variables x and y. This transformation modifies the effective mass of fermions in Eq. (45) for their basis functions. Choosing it in a certain way one can suppress ultraviolet divergences. The reasonable idea is to fix this transformation by the requirement that decoherence is absent in static spacetime. Then:

$$|D(a, \varphi|a', \varphi')| = \exp\left(-\frac{\pi^2 m^2 (a - a')^2}{192} I\right), \quad I > 0, \tag{49}$$

and is finite. The main conclusion to be drawn from the above examples is the fact that consistency of the reduced density matrix might determine the very definition of the environment in quantum cosmology.

6 Classical—Quantum Duality

We have already expressed our opinion that quantum theory is more fundamental than the classical one and that classicality arises under some particular circumstances. In [8] we considered toy models with the Gaussian quantum states and simple interaction Hamiltonians between two subsystems. These Hamiltonians have the structure similar to $V = xp_y + yp_x$ and represent a generalization of the Hamiltonian xp_y introduced by von Neumann [70] to describe the process of quantum measurement. The study of these models confirmed our hypothesis about possible transient character of classicality.

Here we would like to dwell on other aspects of quantum—classical relations following a recent paper [47]. Speaking about the problem of time in quantum cosmology and generally, in quantum mechanics, one can remember that some analogue of the classical time can be introduced even in a system with one degree of freedom [62, 66]. The idea is very simple. Let us consider a particle with one spatial coordinate and a stable probability distribution for this coordinate. Naturally, the quantum state of such a particle is an eigenstate of its Hamiltonian. Then, one can suppose that behind this probability distribution there is a classical motion which we can observe stroboscopically. That means that we can detect its position many times and obtain a probability distribution for this position. Classically this measured probability is inversely proportional to the velocity of the particle. Indeed, the higher is the velocity of a particle in some region of the space the less is the time that it spends there. However in quantum mechanics this probability is given by the squared modulus of its wave function. Thus, we can write down the following equality [62, 66]

$$\psi^*(x)\psi(x) = \frac{1}{|v(x)|T}, \tag{50}$$

where T is a some kind of normalising time scale, for example, a half period of the motion of the particle. In this spirit, in the paper [62], the probability distributions for the energy eigenstates of the hydrogen atom with a large principal quantum number n were studied. It was shown that the distributions with the orbital quantum number l having the maximal possible value $l = n - 1$, being interpreted as in Eq. (50), describe the corresponding classical motion of the electron on the circular orbit. At the same time, the state with $l = 0$ cannot produce immediately a correct classical limit [62]. To arrive at such a limit, which represents a classical radial motion of a particle (i.e. on a degenerate ellipse) one should apply a coarse-graining procedure based on the Riemann–Lebesgue theorem. There is another interesting example: the harmonic oscillator with a large value of the quantum number n. In this case, making a coarse-graining of the probability density one can again reproduce a classical motion of the oscillator [60].

Usually, when one studies the question of the classical-quantum correspondence, one looks for situations where this correspondence is realised. However, it is reasonable to suppose that such situations are not always realised. Now we would like to attract attention to another phenomenon: a particular quantum-classical duality between the systems governed by different Hamiltonians [47]. We can consider a simple example. Let us suppose that we have a classical motion of the harmonic oscillator, governed by the law

$$x(t) = x_0 \sin \omega t. \tag{51}$$

The velocity is

$$\dot{x}(t) = \omega x_0 \cos \omega t. \tag{52}$$

Using Eq. (50), we can believe that behind this classical motion there is a stationary wave function

$$\psi(x) = \frac{1}{\sqrt{\pi}(x_0^2 - x^2)^{1/4}} e^{if(x)} \theta(x_0^2 - x^2), \tag{53}$$

where θ is the Heaviside theta-function and f is a real function. Now, applying the energy conservation law and the stationary Schrödinger equation we can find the corresponding potential for the quantum problem:

$$V(x) = \frac{m\omega^2 x_0^2}{2} + \frac{\hbar^2}{2m}\left(\frac{1}{2(x_0^2 - x^2)} + \frac{5x^2}{4(x_0^2 - x^2)^2} + if'' + if'\frac{x}{x_0^2 - x^2} - f'^2\right),$$
$$\text{if } x^2 < x_0^2. \tag{54}$$

Here, "prime" means the derivative with respect to x. To guarantee the reality of the potential and, hence, the hermiticity of the Hamiltonian, we must choose the phase function f such that

$$f' = C\sqrt{x_0^2 - x^2}, \tag{55}$$

where C is a real constant. Then the potential (54) is equal to

$$V(x) = \frac{m\omega^2 x_0^2}{2} + \frac{\hbar^2}{2m}\left(\frac{1}{2(x_0^2 - x^2)} + \frac{5x^2}{4(x_0^2 - x^2)^2} + C^2(x^2 - x_0^2)\right),$$
$$\text{if } x^2 < x_0^2. \tag{56}$$

Then for $x^2 > x_0^2$ we can treat the value of the potential as infinite since there the wave function is zero.

Naturally, the example constructed above is rather artificial. We have elaborated it to hint at the possibility of encountering a similar effect in cosmology. One can imagine a situation where behind the visible classical evolution of the universe looms a quantum system, whose Hamiltonian is quite different from the classical Hamiltonian governing this visible classical evolution.

7 Many-Worlds Interpretation, Probabilities and Anthropic Principle

As is well known, one of the main features of quantum mechanics is the probabilistic character of its predictions. During many years the Born rule, connecting the probabilities of the outcomes of a quantum measurement with the squared modules of the coefficients of expansion of the wave function of a system under consideration with respect to the eigenvectors of the operator, representing the measured quantity, was considered as a fundamental postulate. However, later the so called Finkelstein–Hartle–Graham theorem was proven [38, 39, 41]. The message of this theorem consists in the fact that the Born rule can be derived from other postulates of quantum mechanics if one considers a huge number of identical quantum systems and defines the probability as a relative frequency of a chosen outcome of an experiment with respect to the general number of trials. Finkelstein and Hartle considered this statement without connection with the many-worlds interpretation of quantum mechanics, while Graham worked in the framework of this interpretation. It seems to us that the Finkelstein–Hartle–Graham theorem looks especially harmonious in the framework of the many-worlds interpretation.

At the same time, the probability treatment of the predictions of quantum mechanics is quite natural when one speaks about multiple experiments or about multiple identical systems as it happens in the proof of the Finkelstein–Hartle–Graham theorem. What sense can it have when one considers the Universe as a whole? Can we say that one branch of the wave function of the Universe is more probable than another? Perhaps, it would be more consistent to interpret the wave function of the Universe as a fundamental object and to treat all its branches on equal footing? Here arises the idea to combine the many-worlds interpretation of quantum mechanics with the Anthropic Principle (see e.g. [2]). Firstly this idea was expressed in [3] and then further developed in [46]. At first glance, one can think that the many-worlds interpretation is "anti-anthropic" in its spirit because it deprives a human being of its privileged position in the Universe. At the same time, it can be combined quite harmoniously with the anthropic principle and cosmology [4]. Indeed, from the point of view of this interpretation the Universe is described by a unique wave function, which is adequate to a quantum reality where all possible versions of the evolution are realized (It is interesting that the fundamental paper by Hugh Everett representing the extended version of his PhD Thesis was called "The Theory of the Universal Wave Function" [37]). In some of the branches of the wave function of the universe not only the classical properties, but also the Life and Mind arise, while in other there is nothing similar. Thus, the world described by our branch of the wave function exists not because it had to arise because of some necessity, but because it was possible and all the possibilities are realized. It is not necessary to require that our world and other similar worlds are the most probable from the point of view of the measure on the Hilbert space, where the wave function of the Universe is defined. It could be just in the opposite way. To be exact, our well structured part of the unique quantum reality has a low probability while less organized branches are more probable. Such an approach matches well with the Anthropic Principle and common sense. Just like Life is localized in a rather small part of the usual space, it could be localized in a tiny part of the Hilbert space.

However, all of the above does not mean that everything is possible. Indeed, the requirements of consistency of theories can impose rather stringent restrictions on the concrete physical laws which govern dynamics in all possible branches of the wave function of the universe. In the next section we discuss a particular example of this.

8 Density Matrix of the Universe and the Cosmological Bootstrap

It is well known that pure states form only a part of possible quantum states of a physical system, and in all of the above examples impurity was a result of decoherence associated with the absence of access of the observer to a certain part of the system. However, what if absence of quantum coherence is fundamentally encoded in the quantum state of the Universe? Moreover, can the dilemma of pure versus mixed state of the Universe be solved at the dynamical level within some minimal assumptions on its initial quantum state? Such an approach was first put forward as Euclidean quantum gravity path integral prescription for the density matrix of the Universe in [14, 15] and then reinterpreted as microcanonical equipartition in real physical spacetime with Lorentzian signature [16, 17]. This model was further developed in [18–23].

The very idea that instead of a pure quantum state of the universe one can consider a density matrix in the form of the Euclidean quantum gravity path integral was pioneered by Page [59]. A corresponding mixed state of the universe arises naturally if there exists an instanton with two turning points (surfaces of vanishing external curvature). As shown in [14, 15] such an instanton naturally arises and can be dynam-

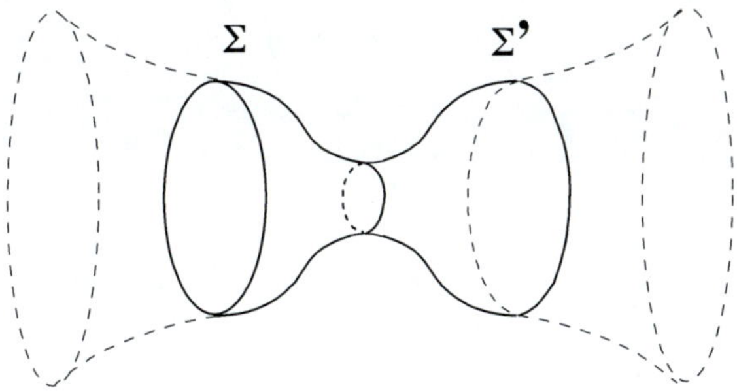

Fig. 1 Picture of instanton representing the density matrix. Dashed lines depict the Lorentzian Universe nucleating from the instanton at the minimal surfaces Σ and Σ′

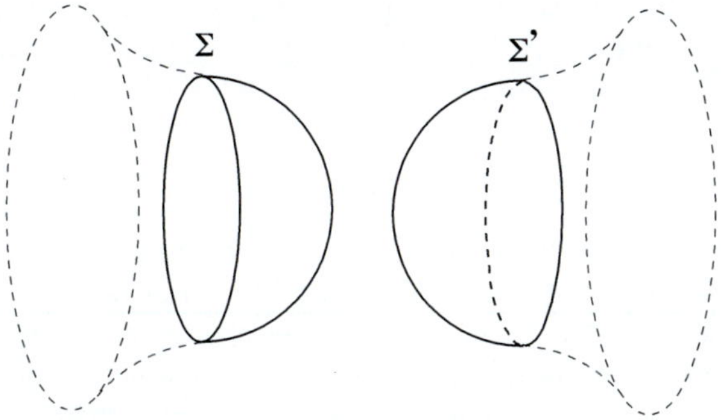

Fig. 2 Density matrix of the pure Hartle-Hawking state represented by the union of two no-boundary vacuum instantons

ically supported if one considers a closed Friedmann universe with the following two essential ingredients: effective cosmological constant and radiation corresponding to the set of conformally invariant quantum fields. The Euclidean Friedmann equation in this case is

$$\frac{\dot{a}^2}{a^2} = \frac{1}{a^2} - H^2 - \frac{C}{a^4}, \tag{57}$$

where $H^2 = \Lambda/3$ is an effective cosmological constant and the constant C characterizes the amount of radiation in the universe. The turning points for solutions of this equation are

$$a_\pm = \frac{1}{\sqrt{2}H}\sqrt{1 \pm (1 - 4CH^2)^{1/2}}, \;\; 4CH^2 \leq 1. \tag{58}$$

Figure 1 gives the picture of the instanton which underlies the density matrix of the universe, minimal and maximal values of the oscillating scale factor corresponding to these turning points. For the pure quantum state [42] the instanton bridge between Σ and Σ′ breaks down (see Fig. 2). However, the radiation stress tensor prevents these half instantons from decoupling—the minimal value a_- stays nonzero.

The relevant density matrix is the path integral over metric and matter field histories interpolating between their boundary values at Σ and Σ′,

$$\rho[\varphi, \varphi'] = e^{\Gamma} \int\limits_{g,\phi\big|_{\Sigma,\Sigma'} = (\varphi,\varphi')} D[g, \phi] \exp\big(-S_{\rm E}[g, \phi]\big). \tag{59}$$

Here $S_{\rm E}[g, \phi]$ is the Euclidean action of the model. We integrate over such field configurations g, ϕ, which at the boundaries Σ and Σ' coincides with the fields φ and φ' respectively (See the subscript of the functional integral). The partition function $e^{-\Gamma}$ for this density matrix follows from integrating out the field φ in the coincidence limit of its two-point kernel at $\varphi' = \varphi$. This corresponds to the identification of Σ' and Σ, the underlying Euclidean spacetime acquiring the "donut" topology $S^1 \times S^3$. The semiclassical saddle point of the path integral for $e^{-\Gamma}$ is just the instanton of the above type.

The metric of this instanton is conformally equivalent to the metric of the Einstein static universe:

$$ds^2 = d\eta^2 + d^2\Omega^{(3)}, \tag{60}$$

where η is the conformal time related to the cosmic time τ by the relation $d\eta = d\tau/a(\tau)$. This opens the possibility of exact calculations for conformally invariant quantum fields, because their effective action on this minisuperspace background is exhausted by the contribution of the conformal anomaly, relevant Casimir energy and free energy. Therefore, at the quantum level the Friedmann equation gets modified to

$$\frac{\dot a^2}{a^2} + B\left(\frac{1}{2}\frac{\dot a^4}{a^4} - \frac{\dot a^2}{a^4}\right) = \frac{1}{a^2} - H^2 - \frac{C}{a^4}, \tag{61}$$

where the amount of radiation constant C is given by the bootstrap equation[1]

$$m_P^2 C = m_P^2 \frac{B}{2} + \frac{dF(\eta)}{d\eta} \equiv \frac{B}{2} m_P^2 + \sum_\omega \frac{\omega}{e^{\omega\eta} - 1}. \tag{62}$$

Here $F(\eta)$ is the free energy which for the conformally coupled scalar field is given by the series of terms contributed by field-theoretical oscillators with frequencies $\omega/a(\tau)$ on a 3-spere of the radius $a(\tau)$

$$F(\eta) = \sum_\omega \ln\left(1 - e^{-\omega\eta}\right) = \sum_{n=1}^{\infty} n^2 \ln\left(1 - e^{-n\eta}\right). \tag{63}$$

Here η is the period of the cosmological instanton in units of the conformal time—effective inverse temperature of the gas of conformal particles. The constant $B = \beta/8\pi^2 M_P^2$ here describes the contribution associated with the conformal anomaly and Casimir energy of the model, where β is a dimensionless coefficient of the

[1] We use units with rescaled Planck mass m_P which is related to the gravitational coupling constant G and the reduced Planck mass M_P as $m_P^2 = 3\pi/4G = 6\pi^2 M_P^2$.

Gauss-Bonnet term of the stress tensor trace anomaly. Similar expressions hold for other conformally invariant fields of higher spins.

Let us emphasize that we have obtained a highly non-trivial system of equations. While the geometry of the instanton depends on the amount of radiation through the modified Friedmann equation, the amount of radiation, in turn, depends on the parameters of the instanton. We called this phenomenon "cosmological bootstrap".

The Friedmann equation can be rewritten as

$$\dot{a}^2 = \sqrt{\frac{(a^2 - B)^2}{B^2} + \frac{2H^2}{B}(a_+^2 - a^2)(a^2 - a_-^2)} - \frac{(a^2 - B)}{B} \tag{64}$$

and has the same two turning points $a_\pm$ as in the classical case provided

$$a_-^2 \geq B. \tag{65}$$

This requirement is equivalent to

$$C \geq B - B^2H^2, \quad BH^2 \leq \frac{1}{2}. \tag{66}$$

Together with $CH^2 \leq 1/4$ the admissible domain for instantons on a two-dimensional plane of C and H^2 reduces to the curvilinear wedge below the hyperbola and above the straight line to the left of the critical point (see Fig. 3):

$$C = \frac{B}{2}, \quad H^2 = \frac{1}{2B}.$$

More detailed analysis shows that cosmological instantons form one-parameter families classified by the number of oscillations of the scale factor during the instanton time period $k = 1, 2, \ldots$. Because of these oscillations they can be called garlands.

The suggested approach allows one to resolve the problem of the so-called infrared catastrophe for the no-boundary state of the Universe based on the Hartle-Hawking instanton. This problem is related to the fact that the Euclidean action on this instanton is negative and inversely proportional to the value of the effective cosmological constant. This means that the probability of the universe creation with an infinitely big size is infinitely high. The effect of conformal anomaly allows one to avoid this counter-intuitive conclusion.

Indeed, outside of the admissible domain for the instantons with two turning points, obtained above, one can also construct instantons with one turning point which smoothly close at $a_- = 0$ with $\dot{a}(\tau_-) = 1$. Such instantons correspond to the Hartle-Hawking pure quantum state. However, in this case the on-shell effective action, which reads for the set of solutions obtained above as

$$\Gamma_0 = F(\eta) - \eta\frac{dF(\eta)}{d\eta} + 4m_P^2\int_{a_-}^{a_+}\frac{da\dot{a}}{a}\left(B - a^2 - \frac{B\dot{a}^2}{3}\right), \tag{67}$$

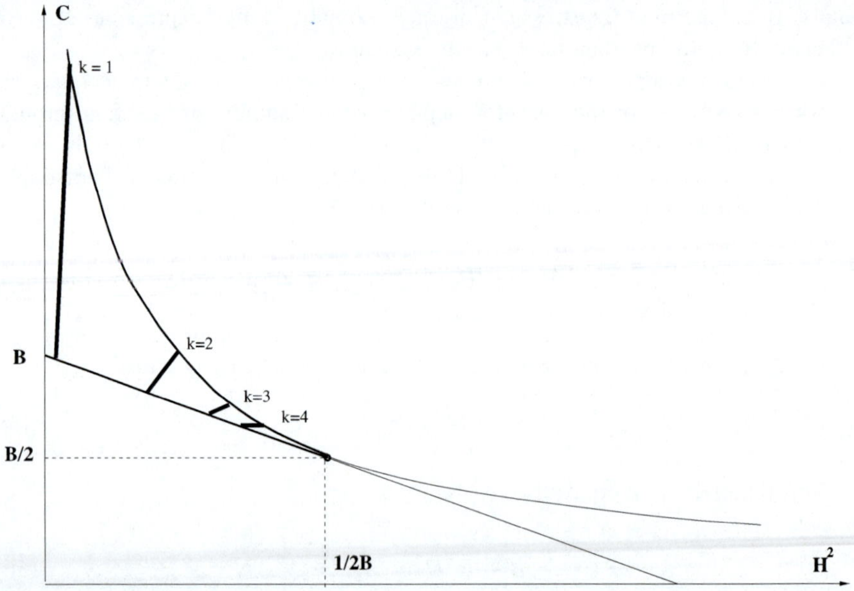

Fig. 3 The instanton domain in the (H^2, C)-plane is located between bold segments of the upper hyperbolic boundary and lower straight line boundary. The first one-parameter family of instantons is labeled by $k = 1$. Families of garlands are qualitatively shown for $k = 2, 3, 4$. $(1/2B, B/2)$ is the critical point of accumulation of the infinite sequence of garland families

diverges to plus infinity. Indeed, for $a_- = 0$ and $\dot{a}_- = 1$

$$\eta = \int_0^{a_+} \frac{da}{\dot{a}a} = \infty, \quad F(\infty) = F'(\infty) = 0, \tag{68}$$

and hence the effective Euclidean action diverges at the lower limit to $+\infty$. Thus,

$$\Gamma_0 = +\infty, \quad \exp(-\Gamma_0) = 0,$$

and this fact completely rules out all pure-state instantons, and only mixed quantum states of the universe with finite values of the effective Euclidean action Γ_0 turn out to be admissible. This is a dynamical mechanism of selection of mixed states in the cosmological ensemble described by the density matrix.

It turned out that under an appropriate definition of the microcanonical ensemble in the theory of spatially closed universes the picture of the above type can be derived from the canonically quantized gravity theory in physical spacetime with Lorentzian signature [16, 17]. The microcanonical density matrix of the Universe, defined as a projector on the subspace of solutions of the Wheeler-DeWitt equation and momentum constraints, can be represented by the same path integral (59) with the prescription that integration over the Euclidean metric lapse function (in the

Arnowitt-Deser-Misner (3+1)-decomposition) should run along the imaginary axis. This construction describes a kind of an ultimate equipartition in the physical phase space of the quantum constrained gravity theory. However, in terms of the observable spacetime geometry this equipartition is peaked about a set of cosmological instantons which, according to Fig. 3, are limited to a bounded range of values of the cosmological constant $\Lambda = H^2/3$. These instantons obtained above as fundamental in Euclidean quantum gravity framework, in fact, turn out to be the saddle points of the gravitational path integral in physical Lorentzian signature spacetime, located at the imaginary axis in the complex plane of the ADM lapse function.

The further development of this concept has shown that this model of initial conditions for a quantum state of the Universe suggests many interesting physical predictions. They include the energy scale of inflation $\Lambda \sim M_P^2/\beta$, which is inversely proportional to the conformal anomaly coefficient β [14, 15], potentially observable thermal imprint on primordial CMB spectrum [19, 20], new type of the hill-top inflation [21, 22] in the version of Higgs inflationary model with a large non-minimal inflaton-curvature coupling, etc. In particular, the anticipated hierarchy between the Planck scale and inflation scale suggests a very big value of β which could be accessible only in now rather popular theories with numerous higher-spin conformal particles [67]. Moreover, this model undergoes a test on applicability of semiclassical expansion because it stays below a well-known effective theory cutoff [34, 68], $M_P^2/\beta \ll M_P^2/N$, for large numbers N of gravitating quantum species [23].

To summarize this section, we conclude that relaxing a usually tacit assumption of purity for the quantum state of the Universe one can come to nontrivial predictions for basic cosmological parameters, which might lead to resolving such fundamental issues as the cosmological constant problem, the problem of cosmological landscape in large scale structure of the Universe and others.

9 Concluding Remarks

We have briefly reviewed here a series of results hopefully elucidating foundations of quantum theory and various related aspects of quantum cosmology. Some of these results are directly associated with two seminal papers of Heinz-Dieter Zeh. Other results are devoted basically to the problem of the definition of the quantum state of the Universe in quantum cosmology—the issue which also belongs to Zeh's scientific legacy. Let us reiterate: one of his works [71] has laid the foundation of such an approach to quantum theory as decoherence, while the other [72] treated at the technical level the problem of the preferred basis in the many-worlds interpretation of quantum mechanics at the time when this interpretation was hardly known yet (it has been published the same year when the monograph edited by DeWitt and Graham has just appeared). No doubt that these works of Zeh will attract the attention of researchers also in the future and it is hard to predict the diversity of contexts in which they can produce groundbreaking effect on expansion of our knowledge about Nature.

Acknowledgements The work of A.B. was supported by the RFBR grant No 20-02-00297 and by the Foundation for Theoretical Physics Development "Basis". The work of A.K. was partially supported by the RFBR grant No 20-02-00411. We are grateful to C. Deffayet, C. Kiefer, I.V. Mishakov, D.V. Nesterov, V.N. Ponomariov, O.V. Teryaev, A. Tronconi, T. Vardanyan and G. Venturi for useful discussions.

References

1. A. Albrecht, Investigating decoherence in a simple system. Phys. Rev. D **46**, 5504–5520 (1992)
2. J.D. Barrow, F.J. Tipler, *The Anthropic Cosmological Principle* (Oxford University Press, Oxford, 1988)
3. A.O. Barvinsky, A.Y. Kamenshchik, V.N. Ponomariov, *Fundamental Questions of the Interpretation of Quantum Mechanics, A Modern Approach* (Publishing House of the Moscow Pedagogical University, Moscow, 1988)
4. A.O. Barvinsky, A.Y. Kamenshchik, V.N. Ponomariov, Anthropic principle and many-worlds interpretation of quantum mechanics, in *Proceedings of the International Seminar "Anthropic Principle in the Structure of Scientific Picture of the World"*, November 28–30, 1989, Leningrad (1989), pp. 48–50
5. A.O. Barvinsky, A.Y. Kamenshchik, Preferred basis in the many-worlds interpretation of quantum mechanics and quantum cosmology. Class. Quantum Grav. **7**, 2285–2293 (1990)
6. A.O. Barvinsky, A.Y. Kamenshchik, One loop quantum cosmology: the Normalizability of the Hartle-Hawking wave function and the probability of inflation. Class. Quant. Grav. **7**, L181–L186 (1990)
7. A.O. Barvinsky, A.Y. Kamenshchik, Quantum scale of inflation and particle physics of the early universe. Phys. Lett. B **332**, 270–276 (1994)
8. A.O. Barvinsky, A.Y. Kamenshchik, Preferred basis in quantum theory and the problem of classicalization of the quantum Universe. Phys. Rev. D **52**, 743–757 (1995)
9. A.O. Barvinsky, A.Y. Kamenshchik, Preferred basis in the many-worlds interpretation in quantum theory and the symmetries of the system. Grav. Cosmol. **1**, 261–265 (1995)
10. A.O. Barvinsky, A.Y. Kamenshchik, I.V. Mishakov, Quantum origin of the early inflationary universe. Nucl. Phys. B **491**, 387–426 (1997)
11. A.O. Barvinsky, A.Y. Kamenshchik, C. Kiefer, I.V. Mishakov, Decoherence in quantum cosmology at the onset of inflation. Nucl. Phys. B **551**, 374–396 (1999)
12. A.O. Barvinsky, A.Y. Kamenshchik, C. Kiefer, Effective action and decoherence by fermions in quantum cosmology. Nucl. Phys. B **552**, 420–444 (1999)
13. A.O. Barvinsky, A.Y. Kamenshchik, C. Kiefer, Origin of the inflationary universe. Mod. Phys. Lett. A **14**, 1083–1088 (1999)
14. A.O. Barvinsky, A.Y. Kamenshchik, Cosmological landscape from nothing: some like it hot. JCAP **0609**, article number 014 (2006)
15. A.O. Barvinsky, A.Y. Kamenshchik, Thermodynamics via creation from nothing: limiting the cosmological constant landscape. Phys. Rev. D **74**, article number 121502(R) (2006)
16. A.O. Barvinsky, Why there is something rather than nothing (out of everything)? Phys. Rev. Lett. **99**, article number 071301 (2007)
17. A.O. Barvinsky, BRST technique for the cosmological density matrix. JHEP **1310**, article number 051 (2013)
18. A.O. Barvinsky, C. Deffayet, A.Y. Kamenshchik, Anomaly driven cosmology: big boost scenario and AdS/CFT correspondence. JCAP **0805**, article number 020 (2008)
19. A.O. Barvinsky, C. Deffayet, A.Y. Kamenshchik, CFT driven cosmology and the DGP/CFT correspondence. JCAP **1005**, article number 034 (2010)
20. A.O. Barvinsky, Thermal power spectrum in the CFT driven cosmology. JCAP **1310**, article number 059 (2013)

21. A.O. Barvinsky, A.Y. Kamenshchik, D.V. Nesterov, Origin of inflation in CFT driven cosmology: R^2-gravity and non-minimally coupled inflaton models. Eur. Phys. J. C **75**(12), article number 584 (2015)
22. A.O. Barvinsky, A.Y. Kamenshchik, D.V. Nesterov, New type of hill-top inflation. JCAP **1601**, article number 036 (2016)
23. A.O. Barvinsky, CFT driven cosmology and conformal higher spin fields. Phys. Rev. D **93**, 103530 (2016)
24. Y. Ben Dov, An observer decomposition for Everett's theory. Found. Phys. Lett. **3**, 383–387 (1990)
25. D. Bohm, B.J. Hiley, *The Undivided Universe?: An Ontological Interpretation of Quantum Theory* (Routledgge, London, 1993)
26. N. Bohr, *Atomic Physics and Human Knowledge* (Wiley, New York, 1958)
27. M. Born, Zur Quantenmechanik der Stoßvorgänge. Zeit. Phys. **37**(12), 863–867 (1926)
28. P. Byrne, *The Many Worlds of Hugh Everett III: Multiple Universes, Mutual Assured Destruction, and the Meltdown of a Nuclear Family* (Oxford University Press, Oxford, 2010)
29. P.D. D'Eath, J.J. Halliwell, Fermions in quantum cosmology. Phys. Rev. D **35**, 1100–1123 (1987)
30. D. Deutsch, Quantum theory as a universal physical theory. Int. J. Theor. Phys. **24**(24), 1–41 (1985)
31. B.S. DeWitt, Quantum mechanics and reality. Phys Today **23**(9), 30–35 (1970)
32. B.S. DeWitt, N. Graham (eds.), *The Many-Worlds Interpretation of Quantum Mechanics* (Princeton University Press, Princeton, 1973)
33. D. Dieks, Resolution of the measurement problem through decoherence of the quantum state. Phys. Lett. A **142**, 439–446 (1989)
34. G. Dvali, Black holes and large N species solution to the hierarchy problem. Fortsch. Phys. **58**, 528–536 (2010)
35. A. Ekert, P.L. Knight, Entangled quantum systems and the Schmidt decomposition. Amer. J. Phys. **63**(5), 415 (1995)
36. H. Everett, Relative-state formulation of quantum mechanics. Rev. Mod. Phys. **29**, 454–462 (1957)
37. H. Everett, The theory of the universal wave function, in DeWitt and Graham, ed. (1973), pp. 3–140
38. D. Finkelstein, The logic of quantum physics. Trans. N.Y. Acad. Sci. **25**, 621–637 (1963)
39. R.N. Graham, The measurement of relative frequency, in DeWitt and Graham eds. (1973), pp. 229–252
40. J.J. Halliwell, S.W. Hawking, The origin of structure in the universe. Phys. Rev. D **31**, 1777–1791 (1985)
41. J. Hartle, Quantum mechanics of individual systems. Amer. J. Phys. **36**(8), 704–712 (1968)
42. J.B. Hartle, S.W. Hawking, Wave function of the universe. Phys. Rev. D **28**, 2960–2975 (1983)
43. G. 't Hooft, M.J.G. Veltman, Regularization and renormalization of gauge fields. Nucl. Phys. B **44**, 189–213
44. M. Jammer, *The Philosophy of Quantum Mechanics?: The Interpretations of Quantum Mechanics in Historical Perspective* (Wiley, New York, 1975)
45. E. Joos, H.D. Zeh, C. Kiefer, D. Giulini, J. Kupsch, I.-O. Stamatescu, *Decoherence and the Appearance of a Classical World in Quantum Theory*, 2nd edn. (Springer, Berlin, 2003)
46. A.Y. Kamenshchik, O.V. Teryaev, Many-worlds interpretation of quantum theory, mesoscopic anthropic principle and biological evolution. NeuroQuantology **11**, 129–148 (2013)
47. A.Y. Kamenshchik, A. Tronconi, T. Vardanyan, G. Venturi, Time in quantum theory, the Wheeler–DeWitt equation and the Born–Oppenheimer approximation. Int. J. Mod. Phys. D **28**(6), article number 1950073 (2019)
48. C. Kiefer, Continuous measurement of minisuperspace variables by higher multipoles. Class. Quant. Grav. **4**, 1369–1382 (1987)
49. C. Kiefer, Wave packets in minisuperspace. Phys. Rev. D **38**, 1761–1772 (1988)

50. C. Kiefer, Continuous measurement of intrinsic time by fermions. Class. Quant. Grav. **6**, 561–568 (1989)
51. C. Kiefer, Decoherence in quantum electrodynamics and quantum gravity. Phys. Rev. D **46**, 1658–1670 (1992)
52. C. Kiefer, *Quantum Gravity*, 3rd edn. (Oxford University Press, Oxford, 2012)
53. O. Kübler, H.D. Zeh, Dynamics of quantum correlations. Ann. Phys. **76**, 405–418 (1973)
54. R. Laflamme, The Euclidean vacuum: justification from quantum cosmology. Phys. Lett. B **198**, 156–160 (1987)
55. E.M. Lifshitz, I.M. Khalatnikov, Investigations in relativistic cosmology. Adv. Phys. **12**, 185–249 (1963)
56. M.A. Markov, V.F. Mukhanov, Classical preferable basis in quantum mechanics. Phys. Lett. A **127**, 251–254 (1988)
57. M.A. Nielsen, I.L. Huang, *Quantum Computation and Quantum Information* (Cambridge University Press, Cambridge, 2000)
58. T. Okamura, On divergence of decoherence factor in quantum cosmology. Prog. Theor. Phys. **95**, 565–576 (1996)
59. D.N. Page, Density matrix of the universe. Phys. Rev. D **34**, 2267–2271 (1986)
60. L. Pauling, E.B. Wilson, *Introduction to Quantum Mechanics* (Addison-Wesley, Reading, MA, 1935)
61. J.P. Paz, S. Sinha, Decoherence and back reaction in quantum cosmology: multidimensional minisuperspace examples. Phys. Rev. D **45**, 2823–2842 (1992)
62. E.G.P. Rowe, The classical limit of quantum mechanical hydrogen radial distributions. Eur. J. Phys. **8**, 81–87 (1987)
63. E. Schmidt, Zur Theorie del linearen and nichtlinearen Integralgleichungen. Mathematische Annalen **63**, 433–476 (1907)
64. E. Schrödinger, Discussion of probability relations between separated systems. Proc. Camb. Philos. Soc. **31**, 555–563 (1935)
65. E. Schrödinger, Probability relations between separated systems. Proc. Camb. Philos. Soc. **32**, 446–452 (1936)
66. A. Sommerfeld, *Wave Mechanics* (Methuen, London, 1930)
67. A.A. Tseytlin, On partition function and Weyl anomaly of conformal higher spin fields. Nucl. Phys. B **877**, 598–631 (2013)
68. G. Veneziano, Large-N bounds on, and compositeness limit of, gauge and gravitational interactions. JHEP **0206**, article number 051 (2002)
69. A. Vilenkin, The birth of inflationary universes. Phys. Rev. D **27**, 2848–2855 (1983)
70. J. von Neumann, *Mathematical Foundations of Quantum Mechanics* (Princeton University Press, Princeton, 1955)
71. H.D. Zeh, On the interpretation of measurement in quantum theory. Found. Phys. **1**, 69–76 (1970)
72. H.D. Zeh, Toward a quantum theory of observation. Found. Phys. **3**, 109–116 (1973)
73. H.D. Zeh, Emergence of classical time from a universal wavefunction. Phys. Lett. A **116**, 9–12 (1986)
74. W.H. Zurek, Pointer basis of quantum apparatus: Into what mixture does the wave packet collapse? Phys. Rev. D **24**, 1516–1525 (1981)
75. W.H. Zurek, Environment-induced superselection rules. Phys. Rev. D **26**, 1862–1880 (1982)

Wave Function First—H. D. Zeh and the Foundations of Physics

Heinrich Päs

1 H. D. Zeh and the Question About "What Is Real?"

On September 21, 2016, together with Claus Kiefer, Erich Joos, Joachim Kupsch, Domenico Giulini and Ion-Olimpio Stamatescu, I received an e-mail from Dieter Zeh, carrying the laconic subject "Altersbeschäftigung", i.e. "retirement activity". The body of the mail contained a brief greeting and an arXiv link to Zeh's essay "The strange (hi)story of particles and waves". In this paper Zeh reviewed how the conception about the nature of matter and light has changed over time. As Zeh described, despite of the usefulness of the concept of atoms in 19th century chemistry and thermodynamics, influential scientists insisted on the view that matter was continuous until around 1900. Light, on the other hand, was understood as composed of particles by Newton and later shown to behave like waves by Thomas Young. In the early 20th century Planck and Einstein then reluctantly concluded the existence of quanta or "particles of light", while De Broglie speculated that the electron, usually understood by now as a particle, could also behave as a wave, a concept that would prove to be seminal soon in the development of atomic models. These arguments continued during the development of quantum mechanics. Heisenberg understood his "matrix mechanics" as a method to obtain probability predictions for classical particles, whereby Schrödinger's competing approach was based on his famous wave equation. At the end of his paper, Zeh took up his stance and concluded that "the particle concept was recognized as a delusion" and that "the observed wave aspects of microscopic objects can be understood only as part of a universal wave function" [47].

For almost five years between 2013 and 2018 Zeh had kept constantly updating the paper, in total 23 versions exist, without much doubt driving the arXiv administrators crazy. But Zeh never hesitated to drive others crazy when it served his cause to reveal the foundations of reality.

Heinz-Dieter Zeh was born in Braunschweig, aka Brunswick, Germany, in 1932. People from Brunswick are described as reserved, humble and outspoken, a charac-

H. Päs (✉)
Fakultät Für Physik, Technische Universität Dortmund, Dortmund, Germany
e-mail: heinrich.paes@tu-dortmund.de

C. Kiefer (ed.), *From Quantum to Classical*, Fundamental Theories of Physics 204,
https://doi.org/10.1007/978-3-030-88781-0_7

terization that seems to fit Zeh himself pretty well. Zeh grew up under the sanguinary Nazi dictatorship that soon instigated the Second World War until, on his 13th birthday, the regime collapsed and the war ended in Europe with Germany's unconditional surrender. In the months before, about ninety percent of Brunswick's inner city, dating back to the medieval ruler Henry The Lion, had been destroyed by Allied air raids, while the Nazi party sent children only three years senior to Zeh himself to the front lines of a lost war. If the Germans couldn't win the war, Hitler reportedly argued, then they deserved to perish. As a consequence of the defeat, Germany was split into a Western and an Eastern state under Soviet influence, and Brunswick, for almost half a century, would suffer economically from its geographical vicinity to the Iron Curtain separating the Western and Eastern zones of influence in Europe.

Did Zeh's background and his childhood coin him in a way that made him particularly unwilling to accept preconceived notions and follow party lines? This is hard to say. As his widow Sigrid Zeh emphasized, her husband "was very reluctant to tell about his life" [50]. Fact is, when Zeh started his career in nuclear physics, the party line for quantum mechanics was the "Copenhagen interpretation", and Zeh was everything but willing to comply. A centerpiece of the Copenhagen interpretation is Max Born's interpretation in which Schrödinger's wave function, in analogy to Heisenberg's matrix mechanics, is understood as a mere tool to obtain probability rules for observations rather than an element of reality. Niels Bohr reportedly went so far as to affirm that "There is no quantum world" beyond what could be measured with classical experiments [21].

"Profoundness or insanity?" [45] Zeh asked rhetorically, only to assert bluntly that such a philosophy constitutes "perhaps... the greatest sophism in the history of science" [47], and a "surrender of consistency" [45].

Until today the discussion about how quantum mechanics has to be interpreted is dominated by the argument about "what is real" [2]. Broadly, a camp of "realists" adopting the wave function as reality struggles with a camp of "epistemologists" or "anti-realists" who in the Copenhagen tradition are understanding the wave function as a tool to describe knowledge about classical objects. Zeh was drawn into this conflict when he discovered the phenomenon of decoherence that allowed to understand how a classical world emerges from an underlying yet unobservable quantum reality [40]. In contrast to other ideas that supplement the formalism of quantum mechanics to achieve this task (such as, for example, a physical collapse of the wave function or Bohm's pilot wave interpretation), decoherence requires no further assumptions beyond the unitary time evolution of the quantum-mechanical wave function. Decoherence arises as a consequence of entanglement between the observed system and its environment combined with the local perspective of an observer lacking information about the exact state of this environment [40], and it is of crucial importance for a wide class of phenomena in physics.

"Decoherence is—in my humble opinion—one of the most important discoveries of the past century" [31], Max Tegmark, MIT cosmologist and—together with John Wheeler—author of the Scientific American feature "100 years of the Quantum" [29] wrote to me in an e-mail in 2017. Tegmark's judgment surely acknowledges the crucial role decoherence plays in the quantum-to-classical transition, including recent developments in quantum computing. But these consequences, that are essentially unrelated to the interpretation of quantum mechanics, don't get even close to exhaust the relevance of decoherence for the foundations of physics.

In fact, decoherence has many facets: First, it is an experimentally proven and well-understood phenomenon without any direct relation to the interpretation of quantum mechanics that describes the loss of coherence when quantum systems interact with their environment. Next, once the wave function is understood as a description of nature, decoherence can be interpreted as the missing link in Hugh Everett's "Many-Worlds-Interpretation" [8]: Decoherence describes how alternative realities such as "particle here" versus "particle there" described by the wave function decouple and give rise to parallel realities, "worlds" or "Everett branches". Finally, decoherence may provide a new paradigm for the foundations of physics. To reach this conclusion, it is necessary to shift our focus away from the question of "what is real" and to concentrate on how this fundamental reality looks like, and how it is related to our everyday, classical experience. In order to do so, we will develop the concepts of "quantum monism" and "ontological priority". Once the view that a quantum description is more fundamental or ontologically prior to any description in terms of classical physics is adopted, a radical consequence entailed by this view is that matter, space and time are derivative concepts emerging via decoherence from a fundamental quantum description.

2 Quantum Monism and the Foundations of Physics

It should be stressed that Zeh wasn't a revolutionary at heart. If Zeh embraced the counterintuitive Many-Worlds-Interpretation of quantum mechanics he did "since he felt forced by Nature to do so", my Dortmund colleague, the philosopher Brigitte Falkenburg once explained to me. "It appears impossible to avoid this consequence as long as one accepts the universal validity of quantum mechanics", Zeh wrote himself [45].

As hard as it is to imagine Zeh to publish in obscure journals such as the Epistemological Letters or to attend the 1983 workshop at the Californian Hippie resort in Esalen where the participants would sit naked in hot pools while discussing parapsychology and the physics of superluminal travel [12], he felt compelled to do so to advocate a rational foundation of reality. According to Zeh, the various essays he collected in his book "Physik ohne Realität?" (i.e. "Physics without reality?") focus exclusively on "the search for a unified and conceptually consistent description of Nature in modern physics" [45]. For Zeh, such a physical reality implied the development of an "observer-independent" description.

Such a notion of "observer-independence" is evocative of the concept of spacetime in Einstein's theory of Special Relativity. In this context, observers moving with different speeds will measure different time and space intervals dt and dx, respectively, between two events, while agreeing on the spacetime distance $ds = \sqrt{dt^2 - dx^2}$. In contrast to dt and dx the spacetime distance ds is, however, not directly observable. Similarly, in quantum mechanics, only a single outcome out of the possibilities described by the wave function is observed in a measurement. While Einstein originally pursued an operationalist approach to derive the laws of Special Relativity, soon after Hermann Minkowski established the notion of an objective spacetime that later would be found to warp and oscillate in the context of General Relativity. Accordingly, Zeh credits Hugh Everett who adopted the unitary evolution of the quantum-mechanical wave function as real to be the "Minkowski of quantum theory" [45]. Such a notion of reality is observer-independent but not directly observable, a "heuristic fiction", as Zeh would characterize it, following philosopher Hans Vaihinger.

But how does this fundamental reality look like? An important aspect that is often overlooked when discussing the Everett interpretation is that Everett's many worlds or branches are not fundamental, but rather constitute an "emergent multiverse", as the American philosopher David Wallace describes it [36]. These worlds or branches describe possible realities experienced by a local observer. But a local observer by definition hasn't access to the complete information about the universe. Max Tegmark has illustrated this local perspective with the terms "frog perspective", in order to contrast it with the hypothetical outside "bird perspective" onto the entire universe [27]. It is the specific perspective of the "frog observer" that defines the quantum factorization into observer, observed and unknown environment as well as the interactions between these partitions and that gives rise to decoherence, emerging Everett branches and a classical experience. Obviously though, the frog perspective provides only an imperfect and incomplete projection of the fundamental bird reality.

Is there a multiverse of many worlds predicted by quantum mechanics? The answer to this question depends on the perspective chosen. As far as our experience goes there is but a single classical universe (although it has been speculated that altered states of consciousness might allow to experience the quantum world [16]). According to how this classical universe arises from a quantum-to-classical transition triggered by decoherence, there exists no reason to consider the classical universe experienced to be more real than alternative Everett branches arising in the same way. Yet, on the most fundamental level of description all these Everett branches are part of a superposition, a single entangled quantum universe.

Zeh approached this idea by entertaining the "wild speculation of a gigantic atomic nucleus, big enough to contain complex subsystems such as measurement devices and even conscious observers" [45] when he first started to think about the foundations of quantum mechanics and the measurement problem—his first foray into quantum cosmology.

This single quantum universe not only alleviates the purported problem of Everett's interpretation featuring a "heavy load of metaphysical baggage" [37], it totally invalidates this criticism as such a fundamental reality isn't only a single universe but a single unique entity comprising matter, space and time as well as all potentially possible events and situations. "In general the state of the individual particle doesn't exist, even in the case where the particles are spatially separated, but only the total state" of the entire system [45], describes Zeh.

So how does the universe look like if experienced from the bird perspective? From this perspective everything that ever has interacted is integrated into a single entangled quantum state, described by the wave function of the universe. This foundational reality entailed by decoherence theory is thus a pure quantum state for the entire universe, an all-encompassing unity: "It seems necessary, therefore, to give up the idea that the world can correctly be analyzed into distinct parts, and to replace it with the assumption that the entire universe is basically a single, indivisible unit", as David Bohm wrote in his 1951 textbook "Quantum Theory" [4]. Approvingly, Zeh writes in his "strange (hi)story": "If this wave function describes reality completely, no local properties can generically 'exist' [47]".

Such a notion strongly reminisces philosophical concepts such as the Tao in Taoism, the Brahma in Hinduism, or Plato's "One": It is a conjunction of opposites accommodating everything that is physically possible. The experienced world, then, emerges from this foundational "One" by fiat of decoherence. According to Zeh, although the frog is not able to fly like a bird and experience this fundamental reality, with "imagination guided by reason" [45] the frog is capable to develop a description of the bird perspective of quantum reality (by solving the Schrödinger equation). Such a notion of reality has been recently described with the term "quantum monism" by University of Geneva philosopher Claudio Calosi [6]., and similar approaches have been advocated by Lev Vaidman as "All is Psi" [33], by Max Tegmark as "physics from scratch" [30], by Sean Carroll as "Mad-Dog Everettianism" [7]. The basic idea is that a state vector or a density matrix and a Hamiltonian are adopted and that everything else, including spacetime and fields propagating in it are derived. An open problem in these approaches is typically to justify the quantum factorization, i.e. the factorization of the initial quantum state into degrees of freedom constituting quasi-classical objects and an environment as a sink for information to trigger decoherence.

Another problem in this argument is that traditionally, reality has two facets: It is supposedly both observer-independent *and* observable. Depending on which aspect is valued as more important, scientists and philosophers tend to subscribe to a realist or anti-realist world view. Yet, on a closer look, this discussion seems oversimplified. A more suitable approach to the foundations of physics should acknowledge that the sciences offer a great variety of contexts to address the various possible questions that can be posed. Depending on the context, for example, a fellow human may be characterized a bunch of atoms or particles, a biological organism, or an integral part of a social construct such as a state, a company or a soccer team. The more complex the chosen agent (in this kind the fellow human) is, the more idealized and substrate-independent it is. A living organism for example is characterized by its function rather than by its constituent particles, see e.g. [25]. In fact, most atoms in a human's body are replaced over the course of a lifetime, and a soccer team remains the same team even if the players change. Each of these contexts can be righteously considered as "real", though not all of these contexts are equally fundamental. Living organisms and soccer teams are still restricted by the laws of physics, while biology or sociology impose no restrictions in the field of particle physics. Such different yet proper ways to describe certain relevant aspects of nature are thus a matter of perspective rather than ontology. Thus a more relevant problem is not so much whether a concept or description is "real" or not but rather which concept is more fundamental or "ontologically prior".

3 What Is Prior?

From Brunswick to New Brunswick: In 2010, Rutgers University philosopher Jonathan Schaffer wrote an essay that begins by describing him musing over the relation of a circle to the pair of semicircles comprised in it: "Are the semicircles dependent abstractions from their whole, or is the circle a derivative construction from its parts?" [24]. Schaffer goes on and explains that "the monist holds that the whole is prior to its parts, and thus views the cosmos as fundamental, with metaphysical explanation dangling downward from the One. The pluralist holds that the parts are prior to their whole, and thus tends to consider particles fundamental, with metaphysical explanation snaking upward from the many."

Schaffer decided to defend the monist view:

"I will claim that monism was never refuted but only misinterpreted. Monism is now usually interpreted as the view that exactly one thing exists. On such a view there are no particles, pebbles, planets, or any other parts to the world. There is only the One. Perhaps monism would deserve to be dismissed as obviously false, given this interpretation. But how uncharitable! The core tenet of historical monism is not that the whole has no parts, but rather that the whole is prior to its parts" [24]. Accordingly, Schaffer argues that the debate should not be so much about what is real but rather about what is fundamentally prior.

This notion of a fundamentally prior whole referring to the bird perspective's fundamental, entangled quantum universe has to be contrasted with the "privileged-perspective realism" of the frog perspective that has been described as "non-naturalistic" and merely psychologically motivated by philosopher Nora Berenstain [3]. As I have described in some detail in [17, 18], it is even possible to quantify how fundamental a chosen context of description is by specifying how much information about a more fundamental layer of description had been discarded, corresponding to the entropy of the macrostate employed in the chosen context of description. The bird perspective's quantum universe then provides the fundamental, entropy-zero description of nature.

It is of utmost importance that such a shift of perspective can have crucial consequences for our approach to identify the foundations of physics. If a description in terms of a universal wave function for the entire universe is more fundamental or "ontologically prior" to any description in terms of isolated, localized constituents such as particles or strings, it will turn our quest for a theory of everything upside down—to build up on quantum cosmology rather than on particle physics or string theory.

In the remainder of this article I will roughly sketch how decoherence in this context may provide the agent that triggers the emergence of everything we experience, including matter, space, time and mind.

4 It from One

"It from Bit" was a slogan John Wheeler came up with to describe his view that "every particle, every field or force, even the spacetime continuum itself—derives its function, its meaning, its very existence entirely—even if in some contexts indirectly—from the apparatus-elicited answers to yes or no questions, binary choices, bits" [38]. "It from Q-Bit" is a pun on this slogan characterizing a recent research project of scientists working on fundamental physics and quantum information theory to discuss relations between the physics of spacetime and quantum phenomena such as entanglement. If, as indicated above, the fundamental description of nature is "The One", an entangled universal quantum state of the universe, a more proper slogan for the endeavor to derive spacetime (as well as matter and mind) from quantum physics would be "It from One".

From the local perspective of a conscious observer, the "One" gets split into quantum system, observer or apparatus and environment [29], and since some information about the complete system is stored in the unknown environment, the timeless "One" appears as many things in space and time. It may well turn out that the most significant aspect of Zeh's decoherence theory is that it at least in principle provides a new paradigm for how to develop a theory of everything, i.e. a theory of space, time and matter in terms of a fully-fledged quantum cosmology, and to derive space, time and matter from this fundamental description via decoherence, i.e. by determining a perspective and thus a quantum factorization (system, observer and environment) and to trace out over the degrees of freedom storing unobservable information in the environment.

In the following we will provide a very brief and vague sketch about how matter, space and time may at least in principle be derived from quantum cosmology, and how this may help to resolve the fine-tuning problems presently distressing particle physicists and cosmologists.

4.1 Matter from One

"There are no quantum jumps, nor are there particles!" [42], Zeh has entitled a paper published in 1993, and in his "Strange (hi)story", he affirmed: "the particle concept was recognized as a delusion" [47]. The basic idea behind this notion is, that matter, or more specifically particles, are not fundamental.

"The spontaneous occurrence of photons as apparent particles (in the form of clicking counters, for example) is then merely a consequence of the fast decoherence caused by the macroscopic detector", Zeh explains [47]: Matter emerges via decoherence from the quantum-mechanical wave function.

Thus, in principle, it should be possible to derive the Standard Model of particle physics from quantum cosmology. This is of course impracticable, since neither the exact wave function, nor the Hamiltonian of quantum cosmology is known. What is known, however, is the Wheeler-DeWitt equation, the Schrödinger-type equation for the wave function of the universe obtained in canonical quantum gravity,

$$H\psi = 0. \tag{1}$$

The Wheeler-DeWitt equation describes a stationary or timeless state fulfilling the Hamiltonian constraint (1) that, as Zeh points out is a "a genuine quantum property, as it reflects the absence of trajectories" [47]. A possible strategy to derive particle physics from quantum cosmology could then at least in principle be to study solutions to the Wheeler-DeWitt equation (presumably constrained by appropriate symmetry constraints) and expose these solutions to decoherence by defining a specific perspective onto this genuine, cosmological quantum state, i.e. by realizing that:

Reality = Universe + Perspective

and that

Fundamental Reality = Universe without Perspective.

The result could then be compared with what we know about the fundamental constituents of matter from particle physics (quarks, leptons, gauge bosons and the Higgs boson including their interactions). If for example the obtained decohered wave function of the universe would respect a certain symmetry that the Standard Model doesn't, this could motivate new particle degrees of freedom that would make that symmetry possible (such as superpartners in supersymmetric field theories). If, on the other hand, the Standard Model would exhibit a symmetry the decohered wave function of the universe wouldn't have, one would search for new degrees of freedom that could yield this symmetry as an emergent feature. The entire approach would be top-down rather than bottom-up guessing of Grand Unified gauge groups comprising the Standard Model. An interesting aspect in this context follows from the "holistic features" of gauge symmetries [10, 23]: The fact that local gauge symmetries allow for distinct symmetry transformations at different space-time locations may betray their origin from a monistic quantum cosmology. Conversely, global symmetries do not exhibit such a property and are presumably broken by quantum gravity effects, suggesting that these symmetries are less fundamental or emergent [39]. These findings may provide an important clue to whether a law of nature is a feature of an emergent layer of description or whether it derives directly from the non-local, fundamental "One".

4.2 Time from One

A curious property of the Wheeler-DeWitt equation is that it describes a quantum state without time evolution. Dieter Zeh encountered this possible feature of entangled quantum systems already when he investigated atomic nuclei in the 1960ies and felt vexed by the success of approximation methods that adopted time-dependent wave functions to approximate stationary states of heavy nuclei. "According to which logic can the solution of a time-independent equation approximately depend on time", Zeh asked himself [45]. The notion to depict the entire universe as a "gigantic atomic nucleus" in combination with the Wheeler-DeWitt equation enabled Zeh to understand time as an emergent, classical concept [41]. "Among the consequences are the fundamental timelessness of quantum gravity, the approximate nature of a semiclassical time, and the correlation of entropy with the size of the Universe," explains Claus Kiefer who has pointed out in detail how time emerges in theories of quantum gravity that respect the universality of quantum mechanics and the superposition principle [13]. The emergence of time so far constitutes the only derivation of a fundamental property of the universe by means of decoherence actually realized. It may thus serve as a proper example inspiring similar approaches to the emergence of space and matter.

4.3 *Space from One*

Already in his 1929 Chicago lectures on "The Physical Principles of the Quantum Theory", Werner Heisenberg explained that quantum mechanics could be understood *either* as an a-causal theory defined in space and time or as a "causal relationship expressed by mathematical laws" that cannot be described in spacetime. From today's perspective, it doesn't seem too far-fetched to adopt an abstract fundamental reality that isn't defined in space and time but rather gives rise to an emergent spacetime. As Zeh emphasized in his "Strange (hi)story", "only if one insisted that reality must be defined in space and time would this very concept of reality (which is essentially equivalent to the existence of a consistent description of Nature) have to be abandoned" [47]. In fact, Zeh's notion of an emergent spacetime may serve as a consistent foundation for recent approaches to derive spacetime from entanglement.

These approaches include Erik Verlinde's "entropic gravity" where "gravity is explained as an entropic force caused by changes in the information associated with the positions of material bodies" [34]. In more recent works, spacetime geometry is derived from the entanglement entropy of a quantum field theory defined on the boundary of the spacetime considered (see for example [22]). Here, gravity emerges within a quantum information theoretic viewpoint as describing the change in entanglement caused by matter [35].

A related idea has been recently explored by Anthony Aguirre, Max Tegmark, Yasunori Nomura, Raphael Bousso, Leonard Susskind, Juan Maldacena and others, sometimes under buzzword "ER=EPR" [1, 5, 14, 15, 26]. The basic idea here is that different Everett branches in the Many-Worlds-Interpretation and different patches of space in the universe should describe the same set of possible configurations of matter. If the many worlds of quantum mechanics and other bubble universes created in cosmological inflation and residing beyond our cosmological horizon are one-and-the-same thing, then entanglement between quantum systems might be similar to wormholes connecting spacelike separated regions in the universe. In this case, argues for example Susskind, the Everett interpretation could be mapped to a Copenhagen interpretation where entangled particles are connected by wormholes. Zeh was very skeptical of such speculations, commenting a Scientific American article by Maldacena discussing the idea with "Its only nontrivial sentence that I could agree with is that 'this picture is still wild speculation'—except that I could not see any justification for the optimism to insert the word 'still' [48]. Yet, Zeh himself had entertained the basic idea that parallel universes are other Everett worlds instead of separated regions in space himself as early as 2002, when he wrote: "A typical example for the appearance of probabilistic quantum events is the decay of an unstable state by means of tunneling through a potential barrier. The decay products (which in quantum cosmology may even be whole universes) are here assumed to enter existence" and goes on criticizing that "even hypothetical 'baby' or 'bubble universes' are defined to be somewhere else in space—in contrast to the 'many worlds' of quantum theory" [43]. While such approaches are still speculative and sketchy, the basic idea that space in its entirety is understood as correlations being traced back to entanglement appears promising and seems to fit very well into a "Wave Function First" approach to quantum cosmology.

4.4 Mind from One

Perhaps the least understood phenomenon in the universe is the nature of consciousness. Among the few facts known about conscious experience is that it generally appears to be accompanied by activity in the cerebral cortex. An interesting approach known as "Integrated Information Theory" [32] pursues a mathematical framework for evaluating the quality and quantity of consciousness based on properties of the corresponding information processing such as the irreducibility into subsets or "integration of information". Max Tegmark has generalized the integrated information framework from neural-network-based consciousness to arbitrary quantum systems and has raised many interesting questions [30].

In fact, it is an unresolved yet highly interesting question whether—and if so how—consciousness might be related to the quantum-to-classical transition described by decoherence. In an early manuscript written in 1966 [45], Zeh describes how localization is a macroscopic feature of the inherently non-local quantum theory. Accordingly, Zeh argues that one has to accept that also "the content of consciousness can only be realized macroscopically as long as it is localized inside a living organism". In another place, Zeh goes on to explain that "therefore, every conceivable subjective observer who is part of the universe requires an extreme "individualization" (multifold localization) in the complete quantum world, and hence of his "Everett world" (his frog's perspective)" [47]. A related aspect is the fact that the spatial localization of an observer's consciousness may be instrumental in defining her specific frog perspective onto the world including for example the preferred basis the fundamental quantum reality decoheres into. As Zeh writes, "the question why there appears to be just one definite measurement outcome would therefore be similar to the question... why you, as an observer, are somebody rather than the whole world!... the relative state with respect to the observer's mind describes the specific observer's "frog's perspective" of the quantum world" [46]. How, for example, would the Universe look like for example for an alien intelligence whose consciousness may be a product of an algorithm running on momentum eigenstates such as plane waves? Quite obviously, in this case decoherence may converge towards a different 'preferred basis' [17]. Although highly speculative, it might be not even considered as too far-fetched to speculate that the change in information processing inside the brain experienced in altered states of consciousness may have consequences for the subject's frog perspective and thus her experience of the quantum-to-classical transition [16, 20].

5 Conclusion

While monist concepts are usually associated with Asian religions these days, it was the monism in Platonism that became most influential in the development of science. In late antiquity Platonism and Christianity competed to become the dominating philosophy of the Roman empire. When Christianity finally prevailed, it adopted core concepts from contemporary Platonism where God and One were identified. But at the same time, early Christians such as St. Augustine also adopted dualist concepts, such as devil, sin and hellfire, for example from the ancient Persian religion of Manichaeism. As a consequence, monism became a battleground in medieval Christianity. Thinkers such as (Pseudo-)Dionysius The Areopagite, Meister Eckhart, Nicholas of Cusa and Giordano Bruno likewise embraced a monist One and very similar philosophies, though they lived quite different lives: While Dionysius was valued as one of the church fathers and Nicholaus became a cardinal and deputy of the pope, Eckhart was convicted to be possessed by the devil, and Bruno was burned at the stake. With the rebirth of ancient philosophy and sciences during the Renaissance, also monism started to flourish again. Ancient and temporary philosophers in the Platonist tradition such as Marsilio Ficino, Baruch Spinoza, Ralph Cudworth and Friedrich Schelling inspired the thinkers that fueled scientific revolutions such as Galileo Galilei, Johannes Kepler, Isaac Newton, Hans Christian Oersted, Michael Faraday and Albert Einstein. Many of these revolutions started with apparently unlikely coincidences: Why do all bodies (in vacuo) fall with the same speed? Why do the same laws govern physics on Earth and in the heavens? Why and how do electricity and magnetism seem to be related? A mindset that adopts nature as a unity is particularly prone to find and exploit such coincidences. It may prove to be seminal also for the present challenges in fundamental physics.

In fact, the fundamental, non-local "It from One" or "Wave Function First" approach as introduced above is particularly timely, as researchers working on the foundations of physics "are now witnessing the beginning of the phase of crisis", "the Dawn of the Post-Naturalness Era" in particle physics [9]. According to Gian Giudice, the head of the CERN Theory Department, naturalness, i.e. the requirement that the laws of physics aren't fine-tuned, "drove the efforts in physics beyond the Standard Model of the last several decades". Nevertheless, persistent fine-tuning problems appear in loop contributions to the Higgs mass, and in the amount of dark energy driving the accelerated expansion of the universe. Ideas that have been pursued to resolve these problems include supersymmetry and large extra space dimensions, both motivated by string theory, an approach to develop a theory of everything in terms of one dimensional local, fundamental degrees of freedom known as strings that provide a natural cutoff in the procedure of renormalization and a candidate for the graviton, the hypothetical force-carrier of quantum gravity.

In his "Strange (hi)story", Zeh has raised the objection, that "because of the unavoidable entanglement of all variables, one cannot expect the effective quantum fields, which are said to describe "elementary particles", to be related to these elusive fundamental variables in a simple way. This conclusion puts in doubt much of the traditional approach to QFT, which is based on concepts of renormalization and 'dressing' [47]". In turn, this implies that a renewed understanding of reductionism based on quantum cosmology rather than on local degrees of freedom such as particles or strings may provide the necessary paradigm shift desperately sought after.

To illustrate this with a trivial example, assume for example the solution fulfilling the Wheeler-DeWitt equation would be the Bell singlet state,

$$|\psi_s\rangle = |\uparrow\downarrow\rangle - |\downarrow\uparrow\rangle, \tag{2}$$

and the known particles of the Standard Model would be described by the spinor $|\uparrow\rangle$. In this case one could conclude that the missing piece for the wave function of the universe would be matter described by the spinor $|\downarrow\rangle$ and by fiat of the global entangled state $|\psi_s\rangle$ the "unlikely coincidence" of correlations such as a cancellation of the total spin of Standard Model and beyond the Standard Model particles would be implied.

On April 13, 2018, only two days before he died, I received a flattering e-mail in which Zeh provided me with encouraging feedback about my essay "A Fundamental Universe" that discussed the ideas reviewed here in an entry for the Foundational Questions Institute's Essay Contest: "Your article has my fullest sympathy. It surely contains some totally new philosophical aspects also concerning my own interpretation..." [50]. It is sad that Zeh won't live longer to witness which important role his discovery of decoherence and decoherence theory may play in the future.

"With every man an entire world is dying", Zeh's widow entitled the notice of death with a quote by the German novelist Gerhart Hauptmann. For Zeh this is true in at least two aspects, as it describes both his individual perspective in the Everettian "many-minds-multiverse" and his world view, his impressive, physical cosmology. We can only hope that at least the latter will live on in the works and views of his students, collaborators and the researchers who had the privilege to discuss with him.

It is consolatory not only for those who miss Zeh, that time in the quantum universe doesn't exist on the most fundamental level and that death and transience are mere illusions of our coarse-grained existence.

References

1. A. Aguirre, M. Tegmark, Born in an infinite universe: a cosmological interpretation of quantum mechanics. Phys. Rev. D **84**, 105002 (2011), arXiv:1008.1066 [quant-ph]
2. A. Becker, *What is Real?* (Basic Books, New York, 2018)
3. [Berenstain 2020] N. Berenstain, Privileged-perspective realism in the quantum multiverse, in *The Foundation of Reality: Fundamentality, Space, and Time*, ed. by D. Glick, G. Darby, A. Marmodoro (Oxford University Press, Oxford, 2020)
4. D. Bohm, *Quantum Theory* (Prentice-Hall, New York, 1951)
5. R. Bousso, L. Susskind, The multiverse interpretation of quantum mechanics. Phys. Rev. D **85**, 045007 (2012), arXiv:1105.3796 [hep-th]
6. C. Calosi, Quantum monism: an assessment. Philos. Stud. **175**(12), 3217–3236 (2018)
7. S.M. Carroll, A. Singh, *Mad-Dog Everettianism: Quantum Mechanics at Its Most Minimal*, arXiv:1801.08132 [quant-ph]
8. H. Everett, Relative state formulation of quantum mechanics. Rev. Mod. Phys. **29**, 454 (1957)
9. G. Giudice, *The Dawn of the Post-Naturalness Era*, arXiv:1710.07663
10. H. Gomes, *Holism as the Empirical Significance of Symmetries*, arXiv:1910.05330 [physics.hist-ph]
11. W. Heisenberg, *The Physical Principles of the Quantum Theory* (University of Chicago Press, Chicago, 1930)
12. D. Kaiser, *How the Hippies Saved Physics* (Norton Paperback, New York, 2012)
13. C. Kiefer, *Does Time Exist in Quantum Gravity?* arXiv:0909.3767 [gr-qc]
14. Y. Nomura, Physical theories. Eternal inflation, and quantum universe. JHEP **1111**, 063 (2011), arXiv:1104.2324 [hep-th]
15. Y. Nomura, Quantum mechanics, spacetime locality, and gravity. Found. Phys. **43**, 978 (2013). ([arXiv:1110.4630 [hep-th]])
16. H. Päs, Can the Many-Worlds-Interpretation be probed in Psychology? Int. J. Quantum Found. **3**(1) (2017), arXiv:1609.04878
17. H. Päs, A fundamental universe, in *Contribution to the FQXi Essay Contest 2018*, https://fqxi.org/community/forum/topic/3096
18. H. Päs, Physics beyond the multiverse - naturalness and the quest for a fundamental theory. Found. Phys. **49**(9), 1051–1065 (2019), arXiv:1809.06326 [physics.hist-ph]
19. H. Päs, *Quantum Monism Could Save the Soul of Physics* (Scientific American Observations, 2019)
20. H. Päs, M. Wittmann, How to set goals in a timeless quantum Universe, in *Contribution to the FQXi Essay Contest 2019*, https://fqxi.org/community/forum/topic/2882
21. A. Petersen, The philosophy of Niels Bohr. Bull. Atom. Sci. **19**, 12 (1963)
22. M. Van Raamsdonk, Building up spacetime with quantum entanglement. Gen. Rel. Grav. **42**, 2323 (2010). [Int. J. Mod. Phys. D **19**, 2429 (2010)], arXiv:1005.3035 [hep-th]
23. C. Rovelli, *Why Gauge?* arXiv:1308.5599 [hep-th]
24. J. Schaffer, Monism: the priority of the whole. Philos. Rev. **119**(1), 31–76 (2010)
25. E. Schrödinger, *What is Life?* (Cambridge University Press, Cambridge, 1944/1967)
26. L. Susskind, Copenhagen versus Everett, teleportation, and ER=EPR. Fortsch. Phys. **64**(6-7), 551 (2016), arXiv:1604.02589 [hep-th]
27. M. Tegmark, Does the universe in fact contain almost no information? Found. Phys. Lett. **9**, 25–42 (1996). (arXiv:quant-ph/9603008)

28. M. Tegmark, The multiverse hierarchy, in *Universe or Multiverse?*, ed. by B. Carr (Cambridge University Press, Cambridge, 2007)
29. M. Tegmark, J.A. Wheeler, 100 years of the quantum. Sci. Am. **284**, 68–75 (2001)
30. M. Tegmark, Consciousness as a state of matter. Chaos, Solitons Fractals **76**, 238–270 (2015). (arXiv:1401.1219)
31. M. Tegmark, Priv. Comm., e-mail to the author (2017)
32. G. Tononi, An information integration theory of consciousness. BMC Neurosci. **45**, 42 (2004)
33. L. Vaidman, *All is Psi*, arXiv:1602.05025 [quant-ph]
34. E. Verlinde, On the origin of gravity and the laws of Newton. JHEP **1104**, 029 (2011). (arXiv:1001.0785)
35. E. Verlinde, Emergent gravity and the dark universe. SciPost Phys. **2**, 016 (2017), arXiv:1611.02269
36. D. Wallace, *The Emergent Multiverse: Quantum Theory According to the Everett Interpretation* (Oxford University Press, Oxford, 2012)
37. J.A. Wheeler, Law without law, in *Problems in the Foundation of Physics, Proceedings of the International School of Physics Enrico Fermi,' Varenna on Lake Como 25th July-6th August 1977* (North-Holland Publishing Company, 1979), p. 396
38. J.A. Wheeler, Information, physics, quantum: the search for links, in *Complexity, Entropy and the Physics of Information*, ed. by W. Zurek (Westview, Boulder, 1990)
39. E. Witten, *Symmetry and Emergence*, arXiv:1710.01791 [hep-th]
40. H.D. Zeh, On the interpretation of measurement in quantum theory. Found. Phys. **1**, 69 (1970)
41. H.D. Zeh, Emergence of classical time from a universal wave function. Phys. Lett. A **116**, 9 (1986)
42. H.D. Zeh, There are no quantum jumps, nor are there particles! Phys. Lett. A **172**, 189–195 (1993)
43. H.D. Zeh, The Wave Function: It or Bit? arXiv:quant-ph/0204088, also, in *Science and Ultimate Reality*. ed. by J.D. Barrow, P.C.W. Davies, C.L. Harper Jr. (Cambridge University Press, Cambridge, 2004), pp. 103–120
44. H.D. Zeh, *The Physical Basis of the Direction of Time* (Springer, Heidelberg, 2007)
45. H.D. Zeh, *Physik ohne Realität: Tiefsinn oder Wahnsinn* (Springer, Heidelberg, 2012)
46. H.D. Zeh, The role of the observer in the Everett interpretation. NeuroQuantology **11**, 97 (2013), arxiv:1211.0196v6
47. H.D. Zeh, *The Strange (Hi)story of Particles and Waves*, arXiv:1304.1003
48. H.D. Zeh, Letter written in response to an article by Juan Maldacena in the November 2016 issue of the Scientific American, http://www.rzuser.uni-heidelberg.de/~as3/ER=EPR.pdf. Accessed Sep. 30 2020
49. H.D. Zeh, Priv. Comm., e-mail on April 13 (2018)
50. S. Zeh, Priv. Comm, e-mail on February 2 (2020)

Geometrodynamics as Functionalism About Time

Henrique Gomes and Jeremy Butterfield

1 Introduction

It is an honour to commemorate Heinz-Dieter Zeh, whose contributions to foundational physics, especially in the areas of quantum decoherence and the nature of time, were so significant. We would like to do so by reviewing three broadly geometrodynamical—and in part Machian or 'relational'—projects. The review is prompted by a recent literature in the philosophy of physics, about a doctrine called 'spacetime functionalism': which, as we will explain, each of the three projects vividly illustrates. Although these projects, and so this paper, are confined to classical physics, the final Section will briefly discuss quantum aspects. Thus we will briefly connect with Zeh's own ideas about how his two main areas were linked: for example, that the matter degrees of freedom could decohere the metric-gravitational degrees of freedom so as to enable, in an Everett-style interpretation, a classical time to emerge.

We first introduce the projects Sect. 1.1. Then we give some details about the physics Sect. 1.2, and the philosophy Sect. 1.3. These details come together in Sect. 1.4.

1.1 Three Projects

Recently, a literature has sprung up about 'spacetime functionalism'. Like all 'isms', it comes in various versions. But the broad idea is that the concept of spacetime,

H. Gomes (✉) · J. Butterfield
Trinity College, University of Cambridge, CB2 1TQ Cambridge, UK
e-mail: jb56@hermes.cam.ac.uk

C. Kiefer (ed.), *From Quantum to Classical*, Fundamental Theories of Physics 204,
https://doi.org/10.1007/978-3-030-88781-0_8

and other more specific chrono-geometric concepts such as 'distance' 'simultaneity' or 'inertial frame', are functional concepts. This means, roughly speaking, that the concept is defined by its pattern of relations to other concepts. The pattern is called the concept's *functional role*; so the idea is that you best understand the concept by looking at 'the role it plays'. Hence the slogan: 'spacetime is as spacetime does'. Here the 'does'-what makes up the functional role—is realized i.e. instantiated by the physics of matter and radiation. So spacetime functionalism is closely related to relational, and specifically Machian, approaches to chrono-geometry and dynamics; and to what has recently been called the 'dynamical approach' to chrono-geometry.[1]

We are sympathetic to spacetime functionalism. But in two previous papers, we argued that the recent literature missed a trick—though there was also good news.

The missed trick was that this literature did not notice that in its original and, we think, best formulation (by Lewis), functionalism is a variety of reduction. In *reduction*, a problematic discourse or theory is vindicated or legitimized by being reduced to (i.e. shown to be a part of) an unproblematic discourse or theory. In *functionalism*, this reduction is achieved by spelling out the functional roles of the problematic discourse's concepts or properties, and arguing that each of them is uniquely realized i.e. instantiated by certain (usually complicated) concepts or properties of the unproblematic discourse.

Thus we urged that spacetime functionalism should say that a chrono-geometric concept is, or several such concepts are, uniquely specifiable in terms of their functional roles (patterns of relations). Then the thrust of the reduction—and the echo of Machian and relationist ideas—is that these functional roles are realized by (usually complicated) features of the physics of matter and radiation, without adverting to other concepts of chrono-geometry.

The good news was that the older literature in foundations of chrono-geometry (before the recent 'functionalism') contained successful examples of this reductive endeavour. We reported four such examples: cases where the unique specifiability of a chrono-geometric concept in terms of matter and radiation, and the corresponding reduction, was secured by a precise theorem.[2]

We can now state this paper's aim: to show that three projects in the physics literature give vivid and impressive illustrations of spacetime functionalism. As it happens, they have hitherto been almost wholly ignored by philosophers of physics; so we submit that they deserve philosophers' attention. The first project is forty years

---

[1] Spacetime functionalism has also recently been a theme in philosophical discussions of the emergence of spacetime from quantum gravity. We will here set this aside, apart from our brief return to Zeh in our final Section. But we thank Nick Huggett for stressing to us the question whether the concepts in a putative quantum gravity theory—the "bottom" theory in Sect. 1.3's jargon—can be grasped without invoking spatiotemporal notions. For more discussion, cf. e.g. [5, Introduction, Sect. 6] and [6].

[2] The 'missed trick' accusation, and our overall account of functionalism and reduction, is in our [7]. The 'good news' examples are in our [8]. Of these examples, the fourth is Barbour and Bertotti's Machian dynamics [9], in which a temporal metric is defined by the dynamics of the point-particles. So this example also illustrates this paper's specific topic, of functionalism about the temporal metric.

old and is well-known in mathematical relativity. But the other two are much more recent and less well-known—so, again: worth advertising.

More specifically, each of these projects illustrates functionalism *about time*. So in broad philosophical terms, they each assume that:

(a) some discourse or theory about matter and radiation, and *space*, is unproblematic; while
(b) a discourse or theory about time is problematic, and is to be vindicated by being reduced to (shown to follow from) the unproblematic. In slightly more technical terms: each provides, within a theory about spatial geometry, a functionalist reduction of the temporal metric and time-evolution: and the reduction is summed up in a theorem that the temporal metric and-or the Hamiltonian governing time-evolution is, in an appropriate sense, unique.[3]

These three lines of work are all 'general-relativistic', in a sense we will make precise. But they differ substantially in exactly what they assume, and in what they deduce. They are, in short:

(1) The recovery of geometrodynamics, i.e. general relativity's usual Hamiltonian, from requirements on deformations of hypersurfaces in a Lorentzian spacetime. This is due to [1].
(2) The programme of Schuller et al.: [2, 3]. They deduce from assumptions about matter and radiation in a 4-dimensional manifold that is *not* initially assumed to have a Lorentzian metric, the existence of a 'generalized metric', with e.g. causal future and past cones and mass hyperbolas; and they deduce a lot about how their generalized metric relates to matter and radiation.
(3) The deduction of general relativity's usual Hamiltonian in a framework without even a spacetime: that is, without initially assuming a 4-dimensional manifold, let alone one with a Lorentzian metric. This is due to [4] (with precursors in [10, 11]).

We will discuss these in this order: in Sects. 2, 3 and 4 respectively. Then in Sect. 5, we give a positive corollary of (3). In short: (3) answers a misgiving you might have about a recent programme in the foundations of classical gravity, viz. shape dynamics. Section 6 concludes with a table summarising the results surveyed, and briefly draws some connections with the work of H-D. Zeh.

[3] The words 'problematic' and 'unproblematic' are our labels: these projects' authors do not use them. We admit they are imperfect: one might prefer less judgmental adjectives such as 'troublesome' and 'okay' (a suggestion we discuss in Sect. 1.3). And whatever the adjectives, one could of course resist the urge to philosophize! One could abjure all ideas about reducing chrono-geometry, and read the uniqueness results as "just" mathematical theorems. But as we read these projects' authors, they *do* sympathize with the endeavour of reduction.

1.2 Appetizer: The Projects Introduced

All three projects are relativistic, indeed general-relativistic. In calling them 'general-relativistic', we mean, more precisely: (a) they either assume or deduce a Lorentzian metric, and a dynamical geometry, i.e. a spatial geometry that varies in time according to an equation of motion; and (b) they focus on this equation of motion being the orthodox general-relativistic one. Thus in this paper, the label 'general relativistic' connotes a notion of evolution, and, indirectly, of global hyperbolicity, that is not present in the Einstein field equations *simpliciter*.

But note our phrases: 'either assume or deduce'; and 'focus on', rather than 'prove'. That is, as these phrases suggest: these projects' details differ considerably. There are contrasts both about what is assumed, and about what is proved. For example, each of the first and third projects deduces the orthodox general-relativistic Hamiltonian in the presence of any of a wide class of matter fields; but the second project does not. And while the first project assumes a 4-dimensional spacetime manifold and Lorentzian metric, and the second project assumes a spacetime manifold: the third manages to avoid assuming either of these concepts. It assumes only a spatial 3-manifold, from which it in effects *builds* a spacetime manifold.

We introduce the three projects, in more detail as follows.

1. The first project [1] assumes a 4-dimensional spacetime manifold with a Lorentzian metric. But Hojman et al. adopt a '3+1' or 'canonical' perspective: so they consider 3-dimensional spacelike hypersurfaces each equipped with its Riemannian geometry, and discuss how such a space, Σ say, can be embedded in spacetime. They show that if we transcribe some natural requirements about how these embeddings mesh—the requirements are encoded in a *deformation algebra*—into a Hamiltonian framework for describing how the spatial geometry on Σ changes, so as to give a constraint algebra, then the Hamiltonian *must be* the usual Hamiltonian in general relativity in its ADM form. This is a remarkable, and rightly lauded, achievement.
In short, and in philosophers' jargon: Hojman et al. show that the ADM Hamiltonian is the unique realizer of the functional role '... is the generator of how spatial geometry changes over time, that reproduces the assumed deformation algebra of hypersurfaces'.
The technical apparatus used by this first project is also important for us, since parts of it will appear again in the second and third projects.
2. On the other hand, the second project [2, 3] assumes a spacetime manifold that is *not* considered as foliated, nor is even required to have a Lorentzian metric. It then proceeds in two stages. The first stage assumes that the dynamics of any matter-or-radiation fields in the spacetime obey some conceptual requirements, e.g. about being 'well-behaved' and 'predictive' in certain senses, and about being deduced from a diffeomorphism-invariant action functional. These assumptions imply a description of a causal structure for the spacetime, and so of spatial hypersurfaces, and of how a generalized, tensorial spatial geometry changes over time, satisfying certain properties. Then the second stage adopts a 3+1

decomposition, and applies the apparatus of Hojman et al., i.e. the first project, to show that this description is consistent, i.e. can be satisfied.

So far, the only specific matter-or-radiation dynamics for which the resulting consistency equations have been solved is classical electromagnetism: for which the resulting evolution of spatial geometry is (like in the first project) exactly as described by general relativity, i.e. by the ADM Hamiltonian. (In saying 'the only known solution', we do not a criticism of the programme: it is a familiar point that a reduction can be of restricted scope, a 'local reduction', yet of great scientific interest.)

This is, we submit, a remarkable achievement. It deserves to be better known by philosophers: not least because of its evident affinities with the 'dynamical relativity' viewpoint of [12], and the spacetime functionalism literature (which was prompted in part by that viewpoint). Thus in this project, the functional role of the temporal metric requires that the equations of motion of the matter-or-radiation fields should be well-behaved and predictive in senses that turn out to define causal cones and make time "the dimension along which we make predictions."[4] In Sect. 3, we will discuss these affinities in more detail.

For the moment, we just note the obvious moral: that here is another uniqueness result *à la* functionalism. In short, it is: given these authors' general assumptions, and classical electromagnetism, general relativity i.e. the ADM Hamiltonian uniquely satisfies the functional role '... describes how spatial geometry changes over time' (where this last phrase gets spelt out in terms of predictivity etc.).

But there is a *caveat*: these authors' deduction of general relativity depends sensitively on their assumption of spacetime diffeomorphism invariance. For we will see (in Sect. 3.4) that without it, there are models of their other assumptions that have the Maxwell field coupled to the Lorentzian metric in the usual way, and obeying Maxwell's equations (in temporal gauge), while the evolution of geometry is completely different from that of general relativity. More generally: it is not clear how broad a class of assumptions within these authors' framework, would yield orthodox couplings between matter or radiation, and a Lorentzian metric with orthodox i.e. general-relativistic evolution of geometry.

3. Finally, the third project [4] is more austere, i.e. self-denying, in its initial assumptions about the 'unproblematic' base, from which to build the reduction of the 'problematic' temporal metric. It does *not* assume, ab initio, a spacetime manifold. It assumes a 3-manifold whose spatial geometry is to vary over time, in a manner that is locally definable in a certain sense: roughly, that local changes of a quantity in distinct spatial regions are required to commute in an appropriate way. This assumption is then set in a Hamiltonian framework, with some

[4] This project's requirement that the dynamics of matter be predictive yields another affinity. Namely, with the proposal [13, Chap. 7] that time is characterised in contrast to space by the fact that data can be specified on spacelike surfaces, but not timelike and null surfaces, in an unconstrained way. He argues that this makes partial differential equations that evolve data from spacelike surfaces 'more informative', in a sense that he derives from a broadly Humean account of laws of nature as summaries of informative patterns in phenomena. We will not pursue details of this imaginative proposal: [14] is a detailed assessment.

technical conditions, such as the Hamiltonian being at most second-order in momenta.

Remarkably, it turns out that these assumptions are enough to *construct* a spacetime, with a Lorentzian metric; *and* to show that the geometry evolves over time according to the general-relativistic Hamiltonian, i.e. the ADM Hamiltonian. Besides, this construction holds good in the presence of any of a wide class of matter and radiation fields, not just e.g. electromagnetism.[5]

We will see in Sect. 4 that this result is a descendant, for relativistic field theories with a dynamical spatial geometry, of Barbour and Bertotti's definition of the temporal metric in their non-relativistic Machian particle dynamics ([9], reviewed in our [8]; cf. footnote 2).

So here again is a uniqueness result, consonant with functionalism. The ADM Hamiltonian is the unique realizer of the functional role '... is the generator of how spatial geometry changes over time, in a locally definable way, without assuming spacetime'. Broadly speaking, this is a stronger result than predecessors since there is no initial assumption of a spacetime.

Besides, we will see in Sect. 5 that this result has a positive corollary for a recent programme in classical gravity, viz. shape dynamics. Namely: this result exonerates shape dynamics from the accusation that, because in a certain regime its Hamiltonian matches that of general relativity, shape dynamics amounts to 'theft' rather than 'honest toil', as Bertrand Russell's much-cited quip puts it. In a bit more detail, the point will be: agreed, shape dynamics can admit to having, as a matter of its own history, 'piggy-backed' on general relativity in the formulation of its Hamiltonian. But this result shows that this 'piggy-backing' need not be 'theft'. For shape dynamics can 'take over'—i.e. an advocate of shape dynamics can transcribe into their formalism—the derivation of Sect. 4's result, and so obtain the desired Hamiltonian—even without initially assuming a spacetime manifold. In short, the result has the merits of 'honest toil', and shape dynamics can invoke that toil, to answer the accusation of 'theft'.

To sum up this appetizer: all three projects have as their punchline, an impressive uniqueness result, that the orthodox Hamiltonian of 3+1 general relativity, the ADM Hamiltonian, is the unique satisfier of a condition. In philosophers' jargon: it is the unique realizer of a functional role. The results vary in their assumptions. In particular: as one passes from (1) to (3), less is assumed about the manifold. One ends with (3) only assuming a 3-manifold with a spatial metric; and while (2) assumes a 4-manifold, it does not postulate ab initio any metric, even a spatial one. Also, the detail of what is deduced varies. For example, both (1) and (3) apply to a wide class of matter fields; while for (2) the uniqueness result is, so far, 'local', i.e. proven only for electromagnetism.

[5] This project, as well as project (1) of Hojman et al., naturally allow for an Euclidean version of geometrodynamics. In both cases, we can impose further conditions that would specialize to the Lorentzian signature.

1.3 Functionalist Reduction: A Review

In Sect. 1.1 we said about reduction and functionalism, only that:

(i) in a functionalist reduction, the functional roles of the concepts or properties of the problematic discourse or theory are argued to be uniquely realized by concepts and properties of the unproblematic discourse or theory; and
(ii) for this paper's three projects, it is time, especially the temporal metric, that is problematic, while spatial geometry, and the physics of matter and radiation, are unproblematic.

In this Section, we will fill out (i). We will begin with how philosophers of science (and indeed, philosophically inclined physicists) usually discuss reduction of one theory to another. This will be a matter of recalling the basic idea of Nagel's account of reduction. Then we will expound functionalism, in the formulation of Lewis. We will introduce functional definition, and then stress how functionalist reduction differs from—we say: improves on—Nagelian reduction. (All this will summarise Sects. 2–5 of our [7]: they contain various developments, defences and references which, for reasons of space, we here forego.) With this review in hand, Sect. 1.4 will then spell out how this paper's three projects illustrate functionalist reduction, thus filling out (ii) above.

Nagelian and functionalist reduction contrasted. In *Nagelian reduction*, one envisages reducing one theory, say T_t ('t' for 'top'), to another, say T_b ('b' for 'bottom'), by adding to T_b a set B of so-called *bridge-laws*. The core idea is that these are sentences that use the vocabularies of both T_b and T_t in such a way that from the conjunction of T_b and B, one can deduce all of T_t ([15, pp. 354–358]; [16, pp. 361–373]). So, assuming that theories are sets of sentences closed under deduction, T_t is shown to be part of, i.e. already contained in, T_b. The traditional idea is that T_b is an improvement on T_t, since it recovers T_t's successes but also says more. So 'b' is also a mnemonic for 'better' and 't' is a mnemonic for 'tainted'.[6]

Several adjustments or variations of this core idea are widely agreed (including by Nagel himself: [17–19]). Most philosophers agree that to fit real-life scientific cases, one must allow that, not all of T_t, but only most of it, or some approximation or analogue, of it, is deduced from the conjunction of T_b and B. And most philosophers agree that bridge-laws do not always (even collectively) 'define' each T_t term, even in the usual minimal sense of logic books: viz. specifying the term's (contingent) extension—nevermind its intension or "meaning". (For a predicate F, the extension is the set of actual instances of F; for a name or other expression referring to an object e.g. a definite description 'the unique F', it is that object.) After all: not determining such extensions is perfectly compatible with our nevertheless being able to deduce all of T_t—as Nagelian reduction demands.

---

[6] Beware: philosophers and physicists use 'reduce' in converse senses. We adopt the philosophical jargon, in which the 'tainted' or 'worse' theory is reduced to the better one; but physicists say that, for example, special relativistic kinematics reduces to Newtonian kinematics for small velocities.

This "looseness" means that there can be a choice of which bridge-laws to postulate: a choice of which set B to add to T_b, in order to deduce T_t, or an approximation or analogue of it. For there might be more than one way to relate all of T_t's terms to those of T_b, while nevertheless securing the deduction. Of course, this is not to say that all such choices are equally good; and one would expect it to be a matter of scientific judgment, relative to perhaps various criteria, which bridge-laws to postulate. Indeed, most philosophers in the Nagelian tradition regard it as a matter of scientific judgment, whether to postulate bridge-laws at all and thereby perform a reduction. Thus their rationale for a reduction is often that it makes for a more unified, or more parsimonious or simpler, picture of the world.

We can now state the contrast with *functionalist reduction*, as formulated by Lewis. It says instead:

(i) Each bridge-law connects just one T_t term with the vocabulary of T_b.

(ii) Each bridge-law 'defines' its T_t term, in logicians' weak sense of specifying its extension. But beware: (a) the extension specified by the bridge-law is contingent, and need not reflect pre-given meanings of the terms involved; and (b) the bridge-law 'definition' is *not* one of the functional definitions emphasised by functionalism—for as we will see in a moment, the functional definition of a T_t term is given *wholly within* T_t.

(iii) The bridge-laws are contingent statements: they are statements of contingent co-extension. (That is: two predicates have the same set of actual instances; two referring expressions actually refer to the same object.) But they should *not* be called 'hypotheses'. For they are *mandatory*, not optional. There is no variety or choice or optionality, of the sort Nagelians envisaged.

Lewis defends these proposals about reduction and bridge-laws as part of his overall view about *functionalism*. But in his classic expositions of his overall view [20, 21], these proposals about bridge-laws come after his main statements about functionalism, i.e. towards the ends of the papers (viz. at [20, pp. 441–445]; [21, 255f]). As a result, in the subsequent literature they have been much less emphasised than the main statements. This has been unfortunate since it has fostered the widespread view that functionalism is opposed to reduction—when in fact Lewis has shown how they can be persuasively combined.[7]

[7] The worthy exception to this unfortunate neglect is 'the Canberra Plan'. This is the nickname for advocacy of Lewisian functionalist reduction, which is familiar in metaphysics, philosophy of mind and ethics (to which Lewis also applied the doctrine): but which is unfortunately almost unknown to philosophers of science, and wholly unknown to recent 'spacetime functionalists'. So our position in this and our other papers is that philosophy of chrono-geometry gives good illustrations of the Canberra Plan. As will be clear, our main difference from the usual Plan will be that we do not require the functional tole of a concept, extracted from the top theory T_t, to be strictly faithful to some pre-given meaning, or to give what philosophers call a 'conceptual analysis'. For details of the Canberra Plan, cf. [22].

For physicist readers, a further word of introduction about Lewis might be helpful. He was a giant of twentieth-century philosophy; and although he never worked in philosophy of physics, his views about topics other than reduction (especially about modality and causation), have close

To explain Lewis' proposals (i) to (iii), the key idea is that while Nagelian reduction contrasts two vocabularies, viz. those of the two theories T_t and T_b: for Lewisian reduction, there are *three* vocabularies at issue—not two. The reason is that *within* T_t, there is a division of vocabulary, into two sorts. Each term in one sort will get *functionally defined* in terms of items in the other sort. So for the next few paragraphs, we set aside reduction, and focus only on T_t.

Functional definition. First, there are the terms[8] that have an agreed interpretation in a certain domain. Lewis calls them O-terms, where O stands for 'old'. Another good mnemonic is 'okay' (suggested by [26, p. 55]). For the idea is that O-terms are fully understood: there is nothing problematic about them. So as Lewis stresses: O does not stand for 'observational', as against 'theoretical' in the sense of 'non-observational'.[9] So there is an agreed, unproblematic or 'okay', notion of truth in that domain for sentences of T_t that contain only O-terms.

Then there are the other terms of T_t: terms that, Lewis proposes, T_t introduces to us for the first time, so that their interpretation is not settled. Lewis calls them T-terms. But again, as he stresses: T does not stand for 'theoretical', as against 'observational'—any more than his O stands for 'observational'. The T_t-terms are just *new* terms, whose interpretation needs to be settled. Another good mnemonic is 'troublesome' (Button and Walsh ibid.): for the T_t-terms are troublesome, problematic, at least in the minimal sense of our needing to interpret them.

Lewis now proposes that the assertions of T_t, taken together, give sufficiently rich information about how all the T-terms are related, both to the (old, interpreted) O-terms in T_t, and to each other, that the truth of T_t (and so advocacy of T_t) implies that each T-term has a *unique* interpretation. That is: advocacy of T_t involves claiming that each such term has a unique extension, that is implicitly determined by the entirety of T_t. Assuming this uniqueness, Lewis then provides a systematic procedure whereby given T_t, one can write down definitions (in the usual logicians' weak sense of: specifications of extensions) of each of the T-terms. So he provides a procedure for getting *simultaneous unique definitions*.

Here we meet the jargon of *functional role*, announced at the beginning of Sect. 1.1. The general idea is that a property or concept of interest can be uniquely specified by a pattern of relations (typically: causal or nomic, i.e. law-like, relations) that it has to other properties. The pattern is called the property's *functional role*. The property is called the *realizer* or *occupant* of the role.[10]

---

connections with foundational issues in spacetime theories, and in mechanics, both classical and quantum. Cf. for example, [23–25].

[8] We concentrate on predicates. Let us assume that names and other expressions referring to objects, e.g. definite descriptions, are eliminated in favour of predicates in the usual way.

[9] So Lewis does not aim to 'reduce theory to observation' in the way that some logical empiricists did. More generally: his treatment of both functionalism and reduction is "realist", not "instrumentalist" or "eliminativist" about T_t.

[10] (1): Nothing here will turn on any distinction one might make between property and concept. So we will mostly say 'property'. But note that accordingly, 'properties' includes also relations, of two or more places. (2): We will mostly say 'specified', 'specification' etc., though often 'defined'

In this paper, we do not need the details of Lewis' procedure for extracting from T_t specifications of the extensions of each of its T-terms: his procedure for getting *simultaneous unique definitions*. The reason is essentially that within each of our three projects, only one T-term, viz. the temporal metric, gets specified. So for us it suffices to note the following six points.

(1) When we envisage a whole set of properties being functional, i.e. specified by their relations to each other and to yet other properties, one naturally worries that there will be a vicious "logical circle" of specification. For if one property X is specified by its functional role that mentions another property Y, *and* vice versa, i.e. Y's functional role mentions X: then surely there is a circle, and both specifications fail. And similarly for circles with more than two members.

(2) But Lewis showed that there need be no logical circle. To avoid such circles, one need only maintain that the body of doctrine mentioning the properties (in our notation: the theory T_t) is sufficiently informative or rich (logically strong) that each of the functional roles is satisfied by just one property. That is: one maintains that each functional role has a *unique* realizer. Then each such property can be specified by its functional role, without any logical circularity. Each T-term has its extension fixed by its functional role; and although the functional role seems to include other T-terms, they are eliminated, so that *au fond* the functional role uses only O-terms.[11]

(3) The main area in which this insight, of simultaneous unique definability, has been discussed is philosophy of mind. Here, mental properties are considered problematic; while bodily and behavioural properties are unproblematic. So the functionalist idea is that each mental property, like being in pain or believing that it is raining, is specified by its functional role: its characteristic pattern of relations to other mental properties, and to bodily/behavioural properties. For example: being in pain is specified by its being typically caused by tissue damage and its typically causing both distress (another mental property) and aversive behaviour. So here, T_t is the non-technical everyday theory of mental lives and behaviour, often called 'folk psychology'. And in this area, Lewis' insight is that by extracting mental properties' functional roles from folk psychology, these properties can all be simultaneously uniquely specified, without logical circularity, in terms of bodily and behavioural properties.

(4) Lewis does not require unique realization in "all possible worlds", but only in the actual world: 'I am not claiming that scientific theories are formulated in such a way that they could not possibly be multiply realized. I am claiming

and 'definition' is used. Saying 'specified' and 'specification' has the advantage of avoiding the connotations of either free verbal stipulation, or of being faithful to a pre-given meaning. This is an advantage since in some examples these connotations do not hold good.

[11] For details, cf.: [20, pp. 428–438], with ancillary material in 438–441; [21, pp. 253–254]; our [7, Sect. 4]. Lewis also gives (1972, 250–253) as a parable, a detective story in which the detective gathers information about three conspirators to a murder, whom he labels X, Y and Z: sufficiently rich information that for it to be true, X, Y and Z must be three people, whom he and the police know by other means—so that they can be arrested.

only that it is reasonable to hope that a good theory will not in fact be multiply realized' [20, pp. 433].

(5) Lewis of course discusses the cases where there is not a unique realizer, or more than one; and also nuanced cases like (a') there is none, but there are near-realizers, of which perhaps one is nearer than the others; and (b') there are many, but some are (perhaps one is) in some way a better realizer than the others. (Cf. also [7, Sects. 3.3–3.5, 4.2 and 5].) But here we can skip details, since this paper's three projects are 'clean'. In each of them, the functional role of the temporal metric has a unique realizer. (But we will briefly meet a nearest realizer in Sect. 5.3.)

(6) Lewis is not suggesting that T_t says all there is to know about the interpretation of a T-term. On the contrary: we might come to know much else about it, after or independently of our acceptance of T_t. This point is all-important for Lewis' account of reduction and bridge-laws—to which we now return.

Functionalist reduction and mandatory bridge-laws. Suppose that after accepting, or independently of accepting, T_t and its simultaneous functional specifications, we come to accept *another* way of specifying the realizers of the functional roles. Then since the realizer of each functional role is unique, we *must* accept an identity statement—we must identify the original realizer with the new specification. In the jargon of reduction: we must accept that the original property is reduced to the new specification.

In terms of our notation of T_t and T_b, and the example of pain (cf. point (3) above):—Suppose we develop a theory T_b of neurophysiology. It has its own specialist vocabulary: 'neuron', 'synapse' etc. But it may well also use some or all of the unproblematic O-terms (bodily and behavioural terms) of T_t, i.e. of folk psychology. Then for the example of pain: we can imagine discovering that a neurophysiological property, say 'C-fibre firing', realizes the functional role of pain, as it was spelt out in the T_t. Then we *must* accept: pain is C-fibre firing. This is the point that we labeled (iii), when we contrasted Nagelian and functionalist reduction. Namely: in the latter, bridge laws are mandatory.

Now we also see why there are *three* vocabularies at issue in reduction; (not two, as discussions of reduction usually assume—one for T_t, one for T_b). For first: the reduced theory's vocabulary divides into two sorts, a division that sets the stage for functionally defining the first in terms of the second; and then there is the reducing theory's vocabulary—in our example, neurophysiological vocabulary ('neuron', 'C-fibre firing').

These ideas are all illustrated in this inference, from two premises:

(i) Being in pain = the inner state that is typically caused by tissue damage and typically causes both distress and aversive behaviour;

(ii) C-fibre-firing = the inner state that is typically caused by tissue damage and typically causes both distress and aversive behaviour;

So, (iii): Being in pain = C-fibre-firing.

We accept (i) because of T_t. This first premise is the functional definition, with 'distress' being, like 'pain', a mental term; whose occurrence in the *definiens* signals the idea (and legitimacy!) of simultaneous definition. We accept (ii) by accepting the later (or independent) theory that we have labelled T_b: for this example, contingent neurophysiology. And we infer (iii) by the transitivity of identity (i.e. of co-extensiveness of predicates). Thus (iii) is the *derived bridge-law*. It is a contingent statement of co-extension (since premise i) is contingent).

To sum up: this example makes clear how we are, as Lewis puts, 'logically compelled to make theoretical identifications' 1970, p. 441), given premises to which we are already committed. To put the premises in general terms:

(1) according to T_t, our T_t-term τ (e.g. 'being in pain') has a unique realizer (extension, referent), that is specified by its pattern of relations, its functional definition; and

(2) according to our later (or independent) theory T_b, ρ ('ρ' for 'reducing') also refers to that realizer.
So we must infer: $\tau = \rho$.

It seems that many scientific examples of reduction fit this account. An obvious, because famous, example is Maxwell's theoretical identification of visible light with electromagnetic waves. To put this with a cartoon-like simplicity: recall that Maxwell calculated within his electromagnetic theory that some solutions were wave oscillations of the electric and magnetic fields, that travelled at ca. 3×10^5 km s^{-1}. This was very close to the measured speed of visible light; and it strained credulity to imagine that two mutually independent phenomena propagated at the same speed. So he inferred that light consisted of these oscillating solutions to his equations. In terms of the notation above: the (rough!) functional role is 'travels at ca. 3×10^5 km s^{-1}'; the T-term 'visible light' in the contemporary theory of optics (τ in our T_t) is functionally defined by this role; and '(suitable) oscillating solution of the Maxwell equations' is the other term ρ drawn from the reducing theory T_b, i.e. Maxwell's electromagnetic theory.

Agreed, this is a *cartoon* of what Maxwell did. And one naturally fears that many, perhaps most, real-life scientific cases that are called 'reductions' will not conform to this neat account. But there are two points to make in its defence; the second is specific to this paper.

(i) Lewis also gives two examples fitting his account (1970, 443–444), argues that other cases do conform to it (1970, 444–445), and also discusses how the account treats later revisions of the theories (1970, 445–446).
(ii) As we said in (5) above: Lewis also discusses what to say in cases where there is not a unique realization, or more than one. (Indeed, we now know that in the Maxwell example, there is more than one realizer of the functional role, 'travels at ca. 3×10^5 km s^{-1}': gravitational influence, as well as light.) But in this paper, we can skip these complications, since our three projects give 'clean' examples—of time having a functional definition, and a reduction.

1.4 How the Projects Illustrate Functionalist Reduction

To prepare for these projects and to summarise this Section, we end by stating the trio of claims that will be the analogues of premises i) and ii) and the conclusion iii) above. Indeed, we will do this twice: first, in general, indeed vague, terms, so as to fit equally well all three projects; and then, in more specific terms that differ between the projects.

Note that our giving both a general trio, and more specific trios, is not a matter of indecision. For our functionalist reductive perspective is a philosophical template that can be brought into contact with the scientific details in various ways. And we think there need be no—there probably is no—fact of the matter about which among the various precise T_t one could diligently extract from e.g. [1] is the best, or the right, T_t to fix on as providing a functional definition of the time parameter (a definition that is then realized by properties mentioned by a corresponding Hamiltonian 'bottom theory' T_b).

So first, the general version. As we have said, the problematic property (the T-term) that is to be given a functional definition and reduction (the analogue of pain) is time. More precisely: it is duration, or the temporal metric. For in the '3+1' framework of this paper, the unification of the temporal metric with the spatial metric in a Lorentz-signature metric—in physical terms: the relativity of simultaneity—will be 'in the background'. In other words: although Lorentz-invariance will hold good (and be postulated or derived in diverse ways: reviewed in Sect. 1.2), it will not be prominent.

The main idea of the temporal metric's functional role, as it would be spelt out in a T_t, is that time is the measure, the parameterisation, of change. Stated as simply as that, the idea has a very long pedigree: from Aristotle through Descartes to McTaggart's insistence that 'time involves change' [27, p. 459]. But for us, the idea is of course more specific: time is to parameterise changes *of spatial geometry*, and its rate of ticking should in some sense mesh appropriately with the amounts of those changes. Thus the analogue of 'being typically caused by tissue damage and typically causing distress and aversive behaviour' will be along the lines: 'being related in an appropriate way to spatial geometry—including to how it changes over time'. Here, the last phrase, 'changes over time', signals—as does the mental term 'distress', in the pain example—that in general, a T-term's functional role involves other T-terms. But as discussed: using other T-terms does not give any problem of a logical circle.

So much by way of T_t (and thus premise i)). For the general version of T_b, i.e. the analogue of neurophysiology (and so premise ii)'s C-fibre firing), we suggest: the general features of a Hamiltonian theory of geometrodynamics. This will characterize time by: (i) its appearing in the denominator of the right-hand-side of equations $\{H, \varphi\} = \frac{d}{dt}\varphi$, governing the relevant fields φ; and thus implicitly by (ii) the symplectic flow of (or Poisson bracket with) the Hamiltonian in the left-hand-side.

We turn to the specific versions, one for each project. Of course, the details mentioned here (for example, the first project's deformation and constraint algebras) will become clearer in the following Sections.

- (1) Hojman et al. [1]: Here one can take T_t as the theory of (globally hyperbolic) Lorentzian spacetimes, and the functional role of time (or rather: of elapsed time, duration) to be defined by the relative evolution of spacelike hypersurfaces, and its being required to obey the deformation algebra. One can take T_b to be the Hamiltonian theory, whose constraint algebra realizes the deformation algebra. Thus in terms of Lewis' notation, τ and ρ, introduced above: the T_t-term τ (the analogue of 'being in pain') equals: the generator of how spatial geometry changes over time, that reproduces the assumed deformation algebra of hypersurfaces. And the T_b-term ρ (the analogue of 'C-fibre firing') equals: the ADM Hamiltonian. In this way, Hojman et al.'s theorem, their uniqueness result, can be written as the Lewisian bridge-law: $\tau = \rho$.
- (2) The programme of Schuller et al.: Here it is clearest to express reduction in two stages, corresponding to the earlier work, reviewed in [3] and then the later work, revised with more modern techniques in [2]. The first stage is the functionalist reduction of a theory about chrono-geometry to a theory about matter and radiation. The second stage develops the theory of gravity, considered as the evolution of a 3-geometry coupled to the first stage's dynamics of matter and radiation. The bridge between the stages is a set of consistency conditions, called 'gravitational closure equations': which Schuller et al. show can be satisfied.
  In the first stage, i.e. the functionalist reduction, one takes T_t as a theory of 4-dimensional manifolds, *not* equipped with any metric, but equipped with matter-or-radiation fields whose equations of motion obey conditions about predictivity, good behaviour and being deduced from a diffeomorphism-invariant action functional. Then T_b is a framework for generalized Lorentzian geometries: geometries that possess a generalized metric (not necessarily a contravariant tensor of rank two), by which we can identify causal cones, a future and past direction, and spacelike (in effect: Cauchy) hypersurfaces. Section 3 will give some details. The broad result is that again, from T_t one could extract the functional role of the time parameter, along the lines that time is "the dimension along which we make predictions": although, owing to the details, it would certainly be more cumbersome to state exactly than is the role extracted from Hojman et al.'s T_t.
  This leads to the consistency conditions, the gravitational closure equations. Thus in their second stage, Schuller et al. adopt the '3+1' framework; and thereby connect to the technical apparatus of project (1), equating the commutation algebra between the Hamiltonian generators with that of the hypersurface deformation. As we announced in Sect. 1.2: using this apparatus, Schuller et al. show that the consistency conditions are indeed satisfied by classical electromagnetism, with the resulting evolution of spatial geometry being described by the ADM Hamiltonian. Thus in this case at least, we can conclude that, with τ expressing the (admittedly cumbersome) functional role gestured at by the slogan that time is "the dimension along which we make predictions" and with ρ being again the ADM Hamiltonian: the functionalist reduction is again summed up by the Lewisian bridge law, $\tau = \rho$.
- (3) Gomes and Shyam [4]: In this project, one begins with only a 3-dimensional Riemannian manifold, and the general ("philosophical") idea above that its geometry changes. So although one assumes that time is the dimension of change, one

does *not* assume that it forms a dimension of a spacetime manifold. Accordingly, T_t is a "pre-relativistic" theory about the time-evolution of causally unrelated quantities: quantities encoding facts about local geometry. Then time, understood as the measure of the local changes of these quantities (as elapsed time, duration), is functionally defined by the different local measures of change being required to commute in the appropriate way. (Details in Sect. 4.) and T_b is an avowedly Hamiltonian theory (as it was in projects (1) and (2)). It is a theory that implements, or realizes, the ideas of T_t in a phase space formalism, subject to certain assumptions such as the Hamiltonian being at most second-order in momenta.
These conditions are rich enough to imply (with hard work!) that the Hamiltonian is the ADM one. So as for projects (1) and (2): with τ expressing the functional role of time given by the T_t and with ρ once again the ADM Hamiltonian, the functionalist reduction can be summed up by the Lewisian bridge-law: $\tau = \rho$.

2 Realizing the Temporal Metric—From Geometrodynamics

We turn to project (1): 'Geometrodynamics regained' by Hojman, Kuchar and Teitelboim, in *Annals of Physics*, 1976; [1]. For us this work is significant, not only as an important illustration of functionalist reduction in a spacetime theory; but also because (as we mentioned in Sect. 1.2) some of its apparatus is also used in projects (2) and (3). Following the 'T_t and T_b' pattern of functionalist reduction, we present: in Sect. 2.1, this project's T_t, essentially the hypersurface deformation algebra; and in Sect. 2.2, its T_b, essentially the constraint algebra. We state the uniqueness result for (the characterization of) the ADM Hamiltonian [28] as Theorem 1.

Of course, we will not display the entire proof here. That is not to our purpose; and anyway, its original exposition is very clear. But we will see, in the second half of Sect. 2.2, that their result can be parsed as amounting to two main achievements. The first, which we have already articulated, finds phase space functions whose algebra recovers the hypersurface commutation algebra. The second, by demanding that there is an algebra homomorphism between the constraints and the deformation vectors, guarantees that these phase space functions are actually constrained. (Cf. Corollary 1 in Sect. 2.2.) We will be interested in the second of these, since the apparatus involved will be deployed again in Sects. 3 and 4.

2.1 Hypersurface Deformations

Hojman et al. assume a 4-dimensional Lorentzian manifold M, but adopt a '3+1' perspective. So they consider a 3-dimensional spatial manifold, which we will here

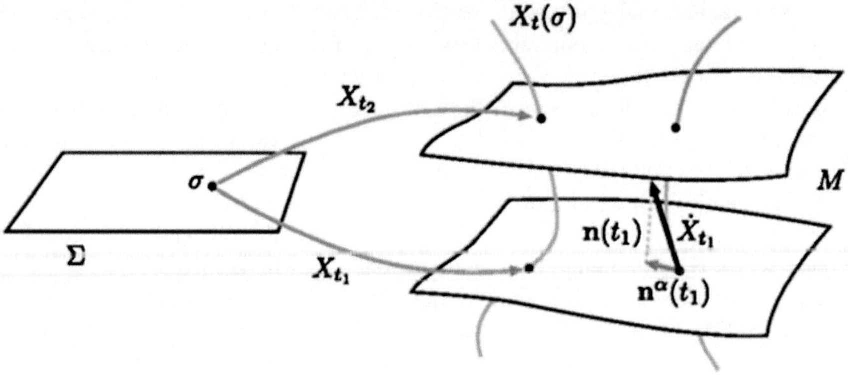

Fig. 1 The foliation of spacetime, induced by a one-parameter family of embeddings of the model spatial manifold, Σ. The coordinates on Σ are denoted by σ^i, and those on M are denoted by x^μ. (Taken from arxiv, with permission; from Düll et al. [2])

call Σ. They assume it is compact and without boundary, and consider its embeddings as spacelike hypersurfaces of M.

Specifically, suppose a foliation of spacetime by spacelike surfaces exists: $X : \mathbb{R} \times \Sigma \to M$. In this notation, X takes a given $\sigma \in \Sigma$ to an entire worldline (these are the points with the same spatial coordinate) and a given t to an entire hypersurface. Given a spacetime metric, $g_{\mu\nu}$, with signature $(-, +, +, +)$, at any point $X(\sigma, t')$, the tangent to this worldline can be decomposed in normal and tangential directions to the hypersurface:

$$\dot{X}_{t'} := \frac{d}{dt}_{|t=t'} X_t(\sigma) = N\mathbf{n} + \mathbf{s} =: (N, \mathbf{s})(x) =: (\dot{X}_{t'\perp}, \dot{\mathbf{X}}_{t'\parallel})(x) \tag{1}$$

where $x = X_{t'}(\sigma)$ denotes the base spacetime point for the vector, N is called *the lapse*, $\mathbf{s}$ is the spatial vector field called *the shift*,[12] and in the last two equalities we represented this spacetime vector in a doublet with its normal and parallel components to the hypersurface (Fig. 1).

There is a representation of Diff(M) onto the embedding maps, Emb(Σ, M) (an infinite-dimensional space of mappings). That is, any spacetime diffeomorphism $d \in \text{Diff}(M), d : p \mapsto d(p)$, defines a map, $\Delta_{d,\Sigma}$, on Emb(Σ, M), as follows. If $p = X(t, \sigma) \in M$, then d defines another embedding (i.e. another member of Emb(Σ, M)), viz. $d \circ X$. Thus:

$$\Delta_{d,\Sigma} : X \mapsto d \circ X$$

[12] We have deviated slightly from the usual notation, in which the shift is also, somewhat confusingly from the 3+1 point of view, denoted by $\mathbf{N}$, or, in components, N^i.

is a diffeomorphism of the infinite dimensional manifold Emb(Σ, M). Correspondingly, a vector field ξ, being an infinitesimal generator of diffeomorphisms on M, defines an infinitesimal generator $D(\xi)$ of diffeomorphisms on Emb(Σ, M); i.e. a vector field on Emb(Σ, M); the map: $\xi \mapsto D(\xi)$ is a Lie algebra homomorphism. We only have good control over the infinitesimal version of the embedding maps (and of the hypersurface deformation algebra below), and we will thus only work at that, infinitesimal, level.[13]

Thus proceeding at the infinitesimal level, on one side of this homomorphism, using the spacetime metric on M, we can decompose a vector field ξ into components normal and parallel to the leaves of the hypersurface, $\xi = (\xi_\perp, \boldsymbol{\xi}_\parallel)$, with $\xi_\perp$ a scalar and $\boldsymbol{\xi}_\parallel$ a spatial vector field.

Then with ξ^1, ξ^2 vector fields, $[D(\xi^1), D(\xi^2)] = D([\xi^1, \xi^2])$ yields [29]:

$$[D(\xi^1_\perp, \boldsymbol{\xi}^1_\parallel), D(\xi^2_\perp, \boldsymbol{\xi}^2_\parallel)] = D(\xi^3_\perp, \boldsymbol{\xi}^3_\parallel) \tag{2}$$

where

$$\xi^3_\perp(\xi^1, \xi^2) = \boldsymbol{\xi}^1_\parallel[\xi^2_\perp] - \boldsymbol{\xi}^2_\parallel[\xi^1_\perp] \tag{3}$$

$$\boldsymbol{\xi}^3_\parallel(\xi^1, \xi^2) = [\boldsymbol{\xi}^1_\parallel, \boldsymbol{\xi}^2_\parallel] - \epsilon\left(\xi^2_\perp h^{-1}(\mathrm{d}\xi^1_\perp) - \xi^1_\perp h^{-1}(\mathrm{d}\xi^2_\perp)\right) \tag{4}$$

where $\epsilon = \pm 1$ is the spacetime signature—it is $-$ for Euclidean and $+$ for Lorentzian—and we have used coordinate-free notation (e.g. $\boldsymbol{\xi}^1_\parallel[\xi^2_\perp] = \xi^{1^i}_{\;\parallel}\partial_i\xi^2_\perp$, and $\xi^2_\perp h^{-1}(\mathrm{d}\xi^1_\perp) = \xi^2_\perp h^{ij}\partial_i\xi^1_\perp$, where h_{ij} is the pull-back of $g_{\mu\nu}$ to the hypersurface).

Essentially, (2) gives the orthogonal decomposition of the commutator of (orthogonally decomposed) vector fields. It has a simple geometrical meaning: one is first deforming the leaves of the foliation one way (ξ^1) and then another (ξ^2), and then comparing that resulting foliation with one in which the opposite ordering of deformations was applied. Suppose that the first vector field deforms the hypersurface Σ into Σ_1, then deforming that surface with ξ^2 we obtain Σ_{12}, i.e. $\Sigma \longrightarrow \Sigma_1 \longrightarrow \Sigma_{12}$. The commutator $[D(\xi^1_\perp, \boldsymbol{\xi}^1{}_\parallel), D(\xi^2_\perp, \boldsymbol{\xi}^2{}_\parallel)]$ is the deformation map between Σ_{12} and Σ_{21}.

The insight of Hojman, Kuchar and Teitelboim was to require that the hypersurface evolution given by D is matched by the geometrodynamical evolution given by the Hamiltonian.

[13] The reason is that it is difficult to control the signature of the embedding after the application of a diffeomorphism. If we apply a diffeomorphism to a given spacelike embedding there is no guarantee that the hypersurface remains spacelike. Nonetheless, under infinitesimal diffeomorphisms (i.e. vector fields) an originally spacelike embedding will remain spacelike, since being spacelike is an open condition: the normal $\mathbf{n}_x$ must be timelike at any point x, i.e. $g_x(\mathbf{n}_x, \mathbf{n}_x) < 0$, which is an open condition. Thanks to Klaas Landsman for this point.

2.2 Commutation Algebras Translated into Constraint Algebras

The standard notation for the phase space variables—the metric on a space-like hypersurface and its conjugate momentum—is $(h_{ab}(\sigma), \pi^{ab}(\sigma))$, (the explicit dependence on space via the label σ reminds us that this is a pair of canonically conjugate variables per space point in an infinite dimensional phase space). So, at each point $\sigma \in \Sigma$ and before the imposition of constraints, we have twelve phase space dimensions as both the metric and its conjugate momentum are symmetric tensors.

We will denote the canonical Poisson brackets as $\{\cdot, \cdot\}$, which, when applied to the fundamental phase space conjugate variables $(h_{ab}(\sigma), \pi^{ab}(\sigma))$ yields

$$\{h_{ab}(\sigma), \pi^{cd}(\sigma')\} = \delta_a^{(c}\delta_b^{d)}\delta(\sigma - \sigma').$$

for $\sigma, \sigma' \in \Sigma$. Note that by $\delta(x - y)$ we really mean $\delta^{(3)}(\sigma - \sigma')$. Thus the action of the Poisson brackets on arbitrary phase space functions (omitting indices) $F(g(\sigma), \pi(\sigma))$ and $P(g(\sigma'), \pi(\sigma'))$ is given by:

$$\begin{aligned}&\{F(h(\sigma), \pi(\sigma)), P(h(\sigma'), \pi(\sigma'))\}\\ &= \int d^3\sigma'' \left(\frac{\delta F(\sigma)}{\delta h_{ab}(\sigma'')}\frac{\delta P(\sigma')}{\delta \pi^{ab}(\sigma'')} - \frac{\delta P(\sigma')}{\delta h_{ab}(\sigma'')}\frac{\delta F(\sigma)}{\delta \pi^{ab}(\sigma')}\right).\end{aligned}$$

To obtain their uniqueness result, Hojman et al. assume the total Hamiltonian for gravity is a sum:

$$\mathscr{H}(\xi_\perp, \boldsymbol{\xi}_\parallel) = \mathscr{H}_\perp(\xi_\perp) + \mathscr{H}_\parallel(\boldsymbol{\xi}_\parallel) \tag{5}$$

with

$$\mathscr{H}_\parallel(\boldsymbol{\xi}_\parallel) := \int d^3\sigma\, \xi^i_\parallel(\sigma)\mathscr{H}_{\parallel i}[h, \pi; \sigma) \quad \text{and} \quad \mathscr{H}_\perp(\xi_\perp) := \int d^3\sigma\, \xi_\perp(\sigma)\mathscr{H}_\perp[h, \pi; \sigma).$$

where we have used DeWitt's mixed index notation: $F[\pi, h; \sigma)$ may depend on the phase space data (and their derivatives) at each spatial point σ. Hojman et al. then require that the constraint algebra reproduces the deformation algebra:

$$\left\{\mathscr{H}(\xi^1_\perp, \boldsymbol{\xi}^1_\parallel), \mathscr{H}(\xi^2_\perp, \boldsymbol{\xi}^2_\parallel)\right\} = \mathscr{H}(\xi^3_\perp, \boldsymbol{\xi}^3_\parallel) \tag{6}$$

with $(\xi^3_\perp, \boldsymbol{\xi}^3_\parallel)$ obeying (3) and (4).

They also assume:

(1) The generator of spatial diffeomorphisms is (with ∇ the Levi-Civita covariant derivative associated to h_{ij})

$$\mathscr{H}_{\parallel i}[h, \pi; \sigma) := \left(h^{kj}\nabla_k\pi_{ij}\right)(\sigma). \tag{7}$$

That is:

$$\left\{h_{ij}(\sigma), \int d\sigma' \xi^\ell \left(\nabla^k \pi_{\ell k}\right)(\sigma')\right\} = \mathcal{L}_\xi h_{ij}(\sigma) ,$$

with the same holding for π_{ij}. They also assume

(2) The scalar generator $\mathscr{H}_\perp[h, \pi; \sigma)$ is such that:

$$\left\{\int d^3\sigma\, \xi_\perp(\sigma)\mathscr{H}_\perp[h, \pi; \sigma) \, , h_{ij}(\sigma)\right\} \equiv \left\{\mathscr{H}_\perp(\xi_\perp), h_{ij}(\sigma)\right\} = 2\epsilon(\xi_\perp K_{ij})(\sigma) \quad (8)$$

where K_{ij} is the extrinsic curvature of the embedded hypersurface in M. This is motivated by requiring $\mathscr{H}_\perp[h, \pi; \sigma)$ to generate diffeomorphisms normal to the leaves of the foliation.

With these assumptions, they then prove:

Theorem 1 (HKT)*Under assumptions* (6)*,* (7)*, and* (8)*, the unique choice of* $\mathscr{H}_\perp[h, \pi; \sigma)$ *is:*

$$\mathscr{H}_\perp[h, \pi; \sigma) = \mathscr{H}_{\text{ADM}\perp}[h_{ab}, \pi^{ab}; \sigma) := \left(a\epsilon\frac{\pi^{ab}\pi_{ab} - \frac{1}{2}\pi^2}{\sqrt{h}} - a^{-1}(R - 2\Lambda)\sqrt{h}\right)(\sigma) \quad (9)$$

where a and Λ *are arbitrary constants.*

This is a remarkable uniqueness result. It entirely fixes the form of the generators of tangential and normal diffeomorphisms.

But condition (8) requires some justification, both technically and conceptually. First, technically: it requires the absence of terms containing derivatives of the momenta in the pure-normal part of the Hamiltonian. In fact, $\mathscr{H}_\perp[h, \pi; \sigma)$ is assumed to be a polynomial, ultralocal functional of π^{ab}. Conceptually, the assumption has a physical meaning inherited from spacetime intuitions. Namely, it encodes the fact that $\mathscr{H}_\perp[h, \pi; \sigma)$, or equivalently, $\mathscr{H}(\xi_\perp, 0)$, should represent an infinitesimal diffeomorphism normal to the leaves, i.e. a normal deformation in the spacetime embedding. Of course, this expectation is alien to the Hamiltonian vocabulary, and thus must be imported from the spacetime domain.

But arriving at (9) does not yet show that these generators are constrained, i.e. that they are generators of symmetries. The total Hamiltonian of General Relativity is a sum of four constraints, the scalar Hamiltonian constraint (one per point) and the vector diffeomorphism constraint (three per point). These constraints, together with their Poisson algebra, encapsulate the initial value problem, the gauge symmetries and the evolution of the theory. But here we have so far only described $\mathscr{H}(\xi_\perp, \boldsymbol{\xi}_\parallel)$ as a symplectic generator of evolution, or of the hypersurface deformations. We have not yet discussed the reason for constraining these generators to be zero on-shell.

This further restriction can still be extracted by imposing not only that the constraint algebra is the same as the deformation algebra, but that the algebra homomorphism works also at the level of generators of symplectic transformations. That is: the geometrodynamical evolution matches the geometrical deformation of surfaces.

For that, we want to demand that:

$$\{\{\bullet, \mathscr{H}(\xi_1)\}, \mathscr{H}(\xi_2)\} - \{\{\bullet, \mathscr{H}(\xi_2)\}, \mathscr{H}(\xi_1)\} \approx \{\bullet, \mathscr{H}\}(\xi_3) \tag{10}$$

where: ξ_3 is the commutator given in (3) and (4), the last equality need hold only on-shell, $\bullet$ stands for any phase space functional, and we abbreviate, e.g.

$$\begin{aligned} \{\bullet, \mathscr{H}_\perp\}(\xi_\perp) &= \int d^3\sigma\, \xi_\perp(\sigma)\, \{\bullet, \mathscr{H}_\perp[h, \pi; \sigma)\} \\ \left\{\bullet, \mathscr{H}_\parallel\right\}(\boldsymbol{\xi}_\parallel) &= \int d^3\sigma\, {\xi_\parallel}^i(\sigma)\left\{\bullet, \mathscr{H}_{\parallel i}[h, \pi; \sigma)\right\}. \end{aligned} \tag{11}$$

For phase-space-independent smearings, there is no difference according to where we place the Poisson brackets. But since the commutation algebra produces field-dependence, one must distinguish the two. What we mean by an evolution by a parameter ξ is the latter, i.e. (11), where the parameter is placed outside the Poisson bracket.

Using the Jacobi identity and (6) we have:

$$\left\{\left\{\bullet, \mathscr{H}(\xi^1_\perp, \xi^1_\parallel)\right\}, \mathscr{H}(\xi^2_\perp, \xi^2_\parallel)\right\} - \left\{\left\{\bullet, \mathscr{H}(\xi^2_\perp, \xi^2_\parallel)\right\}, \mathscr{H}(\xi^1_\perp, \xi^1_\parallel)\right\} = \left\{\bullet, \mathscr{H}(\xi^3_\perp, \xi^3_\parallel)\right\}. \tag{12}$$

But as can be seen from (4),

$$\boldsymbol{\xi}^3_\parallel = [\boldsymbol{\xi}^1_\parallel, \boldsymbol{\xi}^2_\parallel] - \epsilon\left(\xi^2_\perp h^{-1}(\mathrm{d}\xi^1_\perp) - \xi^1_\perp h^{-1}(\mathrm{d}\xi^2_\perp)\right)$$

clearly contains the metric h_{ij}; (Eq. (3) does not).

Therefore, from (12), we would obtain:

$$\left\{\bullet, \mathscr{H}(\xi^3_\perp, \boldsymbol{\xi}^3_\parallel)\right\} = \{\bullet, \mathscr{H}\}(\xi^3_\perp, \boldsymbol{\xi}^3_\parallel) + \mathscr{H}_\parallel(\{\bullet, \boldsymbol{\xi}^3_\parallel\}). \tag{13}$$

From this, we can clearly see that the last term in (13) is the obstruction to obtaining (10). Note also that the last term is *not* an infinitesimal diffeomorphism of $\bullet$. In fact, it will only have an effect on the momenta of $\bullet$, since $\boldsymbol{\xi}^3_\parallel$ only contains the metric. Therefore, for the Hamiltonian evolution to *mesh* in the right way, we must demand that, on-shell (in practice, 'on-shell' means, implementing equality after taking all functional derivatives, i.e. calculating all the Poisson brackets):

$$\mathscr{H}_\parallel(\boldsymbol{\xi}_\parallel) \approx 0,\ \forall \boldsymbol{\xi}_\parallel \in C^\infty(T\Sigma) \qquad \text{or equivalently} \qquad \mathscr{H}_\parallel^i[h, \pi; \sigma) \approx 0\ ; \tag{14}$$

where on-shell equality is denoted by a $\approx$. But to be conserved, this condition will also imply the vanishing of $\mathscr{H}_\perp$:

$$\left\{\mathscr{H}_\perp(\xi^1_\perp), \mathscr{H}_\parallel(\boldsymbol{\xi}^2_\parallel)\right\} = \mathscr{H}_\perp(\boldsymbol{\xi}^2_\parallel[\xi^1_\perp]) \approx 0,\ \forall \boldsymbol{\xi}^2_\parallel \in C^\infty(T\Sigma), \forall \xi^1_\perp \in C^\infty(\Sigma).$$

Therefore, the final condition required to obtain general relativity in its ADM form from Theorem 9 is (10); or, assuming (6) (and using the Jacobi identity), the required condition is:

$$\{\bullet, \mathscr{H}(\xi^3_\perp, \xi^3_\parallel)\} \approx \{\bullet, \mathscr{H}\}(\xi^3_\perp, \xi^3_\parallel) \tag{15}$$

where ξ may have phase-space-dependence (in the example above, $\xi = \xi^3$ had a specific dependence on the metric, as a result of the commutation algebra).

Mathematically, this condition is necessary to ensure that we can interpret the mapping of the deformation vectors into a lapse and shift as a homomorphism from the algebraic structure of hypersurface deformations into the derivations of phase-space functions, i.e. as Hamiltonian evolution. That is:, we want to guarantee that the successive evolution by a pair of parameters stays within the same phase-space solution curve, irrespective of their ordering. Hojman et al. interpret this demand as encoding the "path-independence" of the evolution of geometrical quantities—the irrelevance of the intermediate foliations [1, Sect. 5].

Thus we can now state:

Corollary 1 *Under assumptions* (10), (7), *and* (8) *(that is, replacing* (6) *by the stronger* (10)*) we obtain ADM Hamiltonian geometrodynamics.*

Once one has recovered foliation invariance, and all that comes with it, the closure of the brackets translates back in covariant language to the Bianchi identities. The geometrodynamical coupling of matter then follows suit: it corresponds to the Legendre transform of a generally covariant matter theory coupled to a general relativistic spacetime.[14]

Note that, since it is only the normal components of the two diffeomorphisms that form the obstruction in (13), we can state path-independence employing only equivalence through spatial diffeomorphisms. That is, for 'pure' time evolution, setting the parallel components of ξ to zero, one must demand only that *the geometry* of Σ_{12} is the same as that of Σ_{21} (i.e. that they differ by the action of a spatial diffeomorphism). Of all of the principles of Hojman et al.'s reconstruction, only this weaker 'path-independence', or meshing condition, will be imported into the non-spacetime-based theorems of Sect. 4.

3 Realizing the Temporal Metric—From Matter Without Geometry

We turn to project (2): the programme of Schuller and his collaborators [2, 3] to recover chrono-geometry, specifically, geometrodynamics, from postulates about the physics of (massive or massless) matter fields defined on a 4-dimensional manifold,

[14] Nonetheless, the consistent propagation of matter can also be obtained in the canonically 3+1 language, as above.

but without assuming *ab initio* a Lorentzian metric. This is clearly a programme that is both ambitious and avowedly 'relationist'. So it deserves the attention of philosophers of physics, especially those sympathetic to relationism (e.g. the dynamical approach of [12]).[15] All the more so, since the programme has been developed in detail and has achieved a great deal. But our discussion of it must be limited: we will confine ourselves to showing how the programme illustrates, not just reduction, but our preferred functionalist reduction.

As we said at the end of Sect. 1.4, it is best to think of the reduction in two stages. We give an overview of both stages in Sect. 3.1, then give their details (Sects. 3.2 and 3.3 respectively). Finally, in Sect. 3.4, we will sketch an example showing the programme's dependence on assuming diffeomorphism invariance (as we announced in Sect. 1.2).

3.1 Overview of the Two Stages of Reduction

The broad idea is that the various familiar roles that the Lorentzian metric plays in relativistic physics are to be replaced by, indeed reduced to, conditions on matter and radiation: conditions that are less problematic from a relationist viewpoint. However, Schuller et al. do assume from the start: a 4-dimensional manifold, and that the equations of motion for matter and radiation are deduced from a fully diffeomorphism-invariant action functional. (So this precludes certain theories that take a fundamentally 3+1 perspective—theories such as shape dynamics [31, 32] and Horava-Lifschitz gravity [33].)

Their first stage then consists of adding some reasonable assumptions about these equations (details below), and deducing that spacetime has a causal structure and, indeed, a generalized geometry. More precisely: they deduce that there are well-defined future and past causal cones (limiting the speed of causal propagation) at each point, and this yields definitions of timelike trajectories and spatial—or rather, Cauchy—hypersurfaces, as well as mass hyperbolas. We will call this a *kinematical geometry*.[16] Then the second stage requires that both the matter and the geometry of a spacetime canonically evolve together, starting and ending on shared Cauchy surfaces, and independently of the intermediate foliation. This requirement means a set of consistency conditions, called 'gravitational closure equations', must be satisfied. Schuller et al. then deploy a version of the apparatus of Hojman et al., i.e. project (1) of Sect. 2—suitably generalized to their kinematical geometries—to show that classical electromagnetism satisfies these consistency conditions. Thus they recover orthodox geometrodynamics, i.e. the orthodox general relativistic canonical co-evolution of the Maxwell field and spatial geometry—without having assumed *ab initio* a Lorentzian

[15] But so far as we know, the only philosophical discussion of it is [30], which we commend.

[16] Although this is our label, not these authors', it is prompted by some of their phrases like 'kinematical meaning' in the next quotation.

metric! Clearly, this is a major achievement, even though it is restricted to classical electromagnetism; (so that in philosophers' jargon, it is a local reduction).

These two stages are summarised successively in the introduction to [2]; and this is followed by a remark which reminds us of the dynamical approach of [12]:

> [We first prescribe] the matter dynamics on a spacetime. The dynamics of the underpinning spacetime geometry are then shown to follow from the matter dynamics, essentially by a sufficiently precise requirement of common canonical evolution ... the matter dynamics crucially determines the kinematical meaning of their geometric background [i.e. first stage], and it is precisely this information that directly funnels into the structure of the gravitational dynamics [i.e. second stage]. ... And [this is] not a new perspective either, considering that it was the dynamics of matter, namely the classical electromagnetic field, that led Einstein to the identification and kinematical interpretation of Lorentzian geometries and finally the field equations for their dynamics [cf. the dynamical approach]. (2018: Introduction, p. 084036–1)

We end this Section with a couple of details and quotations, about the first stage. At the start, one assumes three conditions about the matter: (technically, conditions on the principal polynomial of the corresponding field equations). In the authors' words [2]:

> Classically [these three conditions] correspond, in turn, to: the existence of an initial value formulation for the matter field equations; a one-to-one relation between momenta and velocities of massless particles; the requirement of an observer-independent definition of positive particle energy. It is interesting to note that if one insists on the matter field equations being canonically quantizable, these three properties are directly implied. (2018: Introduction, p. 084036–2)

Moreover, one assumes that the equations of motion are obtained from extremising a diffeomorphism-invariant action functional S_{matter}. As we will see, this second assumption is central. For one thing, it means assuming that there is some spacetime density or volume-form, i.e. some way to count infinitesimal volume—which will be a "seed" for metrical relations. And one assumes that additional to the matter degrees of freedom, there is a tensor G on the spacetime, that is constrained only by how it enters the action—through the volume-form for example, and perhaps by how it contracts indices of derivatives or tensor indices of the matter fields. (So nothing about the signature, or tensor valence, of G is assumed: it could be given by a symplectic-form, or by a four-valence tensor contracting area-elements, etc.)

G is called a *geometric tensor*; though in the absence of the other assumptions, it will in general not define a causal structure or other geometrical notions. (So while 'G' naturally stands for 'geometry', one might also think of it as standing for 'gadget'!) But the three conditions on matter fields are then translated into conditions on dispersion relations which, in their turn, are encoded in a totally symmetric contravariant even-rank tensor field, satisfying three simple algebraic conditions. One can then identify much of the structure, such as a causal (cone) structure, usually associated with there being a Lorentzian metric—even though one has not assumed such a metric. In Schuller's words:

> These physically inevitable properties single-handedly ensure that the entire kinematical apparatus familiar from physics on a Lorentzian manifold is defined in precisely the same

way for any spacetime; causality, in particular, is perfectly compatible with superluminal propagation in spacetimes, but one only learns this from a subtle interplay of convex analysis, real algebraic geometry and the modern theory of partial differential equations. [3, p. 2]

Then it turns out that whatever the geometric tensor G, and whatever the matter fields (as long as they satisfy the assumed conditions), a set of consistency conditions on the deduced spacetime structure must be satisfied—leading into the second stage.[17]

3.2 *Kinematical Geometry from Matter Dynamics*

As we have admitted, a detailed account of this programme would go far beyond this paper: a summary of the general ideas must suffice. The starting assumption is the action for matter or radiation, described by a tensorial field B on the spacetime manifold M, which is ultralocally coupled to a "geometric" tensor field, G. G is of arbitrary tensor valence. It is needed at least to form scalar densities of weight one (i.e. ensuring diffeomorphism invariance for S_{matter}), but it can also be used to contract indices of the same valence[18]:

$$S_{\text{matter}}[B; G) = \int d^4x \mathcal{L}_{\text{matter}}(B, \partial^{(K)} B; G(x)) \tag{16}$$

where $\partial^K B$ denotes dependence on an arbitrary but finite number of derivatives of B, i.e. from $\partial_i B$ to $\partial_{i_1} \cdots \partial_{i_F} B$. From this action, it is assumed that the equations of motion for B are of the form:

$$Q^{i_1 \cdots i_n}_{\mathcal{AB}}(G(x))(\partial_{i_1} \cdots \partial_{i_F} B^{\mathcal{A}})(x) + \mathcal{O}(\partial^{F-1} B) = 0 \,. \tag{17}$$

In this equation, $\mathcal{A}$, $\mathcal{B}$ denote the tensor indices of the matter field. $Q^{i_1 \cdots i_n}_{\mathcal{AB}}(G(x))$ is a tensor density of weight one by itself, functionally dependent on the geometric tensor. Despite the appearance of partial—not covariant—derivatives, by ignoring the lower order terms (where the non-covariant terms would appear), (17) is made appropriately covariant. It is important that the equations of motion are linear in the

[17] In general terms, it is unsurprising that there should be such consistency conditions. Recall that in the covariant framework, Lovelock's theorem [34] severely restricts the possible Lagrangian densities for the gravitational part of the action. Agreed: that theorem assumes that the "felt" spacetime metric is the standard Lorentzian metric $g_{\mu\nu}$: without that assumption, the theorem does not apply, and one must seek more general theorems constraining the gravitational part of the action. But instead of seeking a generalized version of Lovelock's theorem, Schüller et al. adopt the alternative, canonical framework, and so can invoke known consistency requirements, drawing on Sect. 2.

[18] Here we are supposing a single geometric tensor. In [2], it is shown how to incorporate multiple G's.

field B, for otherwise one cannot isolate the implications of the dynamics for the causal structure.[19]

While the deduction of the "felt" spacetime properties from Q is not at all straightforward, we can illustrate the gist of it with a simple example: consider the Wentzel-Kramers-Brillouin expansion of the matter field:

$$B^{\mathcal{A}}(x) = \exp(iS(x)/\lambda)(b^{\mathcal{A}} + \mathcal{O}(\lambda)) .$$

Then, from applying the equations of motion, to lowest order we obtain:

$$Q^{i_1 \cdots i_n}_{\mathcal{AB}}(G(x))k_{i_1} \cdots k_{i_F} b^{\mathcal{A}} = 0$$

which precisely captures the infinite frequency limit of the fields. Of course, these equations need not define anything like the causal cones of standard Lorentzian metrics: for starters, here the k_i's are momenta, and must be (Legendre) transformed into vectors to constitute elements of spacetime. It is part of the merit of [3] to show that in fact they do define causal cones—when (17) satisfy the necessary conditions.

Generalizing the manipulation of Q above, one defines the *principal polynomial*, which carries most of the dynamical significance of the equations of motion[20]:

$$P(x,k) = \pm\rho \det_{A,B}\left(Q^{i_1 \cdots i_n}_{\mathcal{AB}}(G(x))k_{i_1} \cdots k_{i_F}\right)$$

If the (linear, partial differential) equations of motion have a well-posed initial value-problem, then $P(x,k)$ defines a hyperbolic homogeneous polynomial in each cotangent space, $T^*_x M$, where ρ is a scalar density obtained from G. Here, the polynomial is called 'hyperbolic' if there is a covector $k \in T^*_x M$ such that $P(x,k) \neq 0$ and the equation $P(x, p + \lambda k) = 0$ has only real solutions for λ, for any further covector $p \in T^*_x M$.

This way of describing P leads directly to the definition of hyperbolicity cones, which are the seeds for defining the causal structure of the theory directly from the equations of motion, and which in fact generalize the familiar causal cones in Lorentzian geometry.[21] Namely, for each such k, there is always an open and convex cone $C_x(P;k)$ that contains all hyperbolic covectors that lie together with k in one connected set. It is a cone not in the geometric sense *per se*, but algebraically: for every $k' \in C_x(P;k)$, we have $C_x(P;k') = C_x(P;k)$, and for $k', k'' \in C(P,k)$,

[19] The terminology 'probing matter', as introduced in [3, pp. 12–13], signifies matter with linear equations of motion. As he describes it, with non-linear equations "it would often be impossible to disentangle the properties of the matter field from properties of the underlying geometry".

[20] More care is needed in the presence of gauge symmetries, but that case too is encompassed by the framework of [2].

[21] For higher order polynomials, one gets an even number of distinct hyperbolicity cones, but not necessarily two, as in Lorentzian geometry. Importantly, this discrepancy does not block general dynamical considerations; and in many cases, there are work-arounds. For instance, for reducible polynomials (i.e. forming many null directions, since $P(x) = P_1(x) \cdots P_n(x)$), one obtains the overall hyperbolicity cone by intersection.

$\lambda k' \in C(P,k)$ and $k' + k'' \in C(P,k)$. To render it geometrical, one must convert cotangent vectors to vectors.

Another distinguishing feature (which we will not delve into), is that in the generalized case, the dual polynomial (e.g. defining the causal structure through vectors instead of co-vectors) may not even be completely well-defined.[22] In fact, finding the dual polynomial is instrumental for a proper Legendre transformation of the matter-induced dynamical structure (such as 'observer worldlines'). A Legendre map L is essentially the spacetime gradient (at the base point) of $\ln P(x,k)$, that is: $L_x : C_x \to T_x M : k \mapsto -\mathrm{grad}_x(\ln P)(x,k)$. It is well-defined and exists when P is bi-hyperbolic and energy-distinguishing. Such a Legendre map will take elements of the hyperbolicity cone C_x to elements of $T_x M$. In this case, one can move from dynamical structures in the cotangent spaces T^*M to structures in TM and, eventually, to structures in spacetime M. For instance, for 'an observer' curve, whose tangent $\dot{\gamma}$ is always within the observer cone $C_x^\#$, the vector space $V_x(\dot{\gamma}) := \{X \in T_x M : L^{-1}(\dot{\gamma})(X) = 0\}$ defines the purely spatial directions as seen by such an observer.[23]

Proceeding in this manner—though we should note that this construction does not directly provide a Lorentzian spacetime metric—one is ultimately able with hard labour to use the dynamics to identify spatial hypersurfaces, temporal directions, momenta, observer trajectories, observer frames, particle energies, etc.

In sum, the great merit of this programme [2, 3, 35] is that it builds a causal spacetime notion directly from the dynamics of matter. But it is not true that 'anything goes'. For even after satisfying the conditions on the matter equations, consistency conditions emerge once one couples the dynamics of matter to the auxiliary geometrical structures, G—leading us to the second stage.

3.3 Geometrodynamics from Kinematical Geometry

The equations of motion also bequeath spatial geometric structure to the spacelike surfaces and temporal geometric structure to the remaining, causal directions. More precisely: the existence and uniqueness of the way to associate normal directions along a hypersurface with its canonical normal co-directions by means of a Legendre map requires all three algebraic properties: the hyperbolicity, time-orientability and energy-distinguishing properties.

---

[22] A dual principal polynomial to P at x is any principal polynomial $P_x^\#$ that acts on the tangent space and annihilates tangents to the bases of null cones of P. That is, let the causal cones (in cotangent space) be defined as $N_x := \{k \in T_x^*M : P(x,k) = 0\}$, with smooth subcones defined by $N_x^{\mathrm{smooth}} := \{k \in N_x : \mathrm{grad}_x P(x,k) \neq 0\}$. Then $P_x^\# : T_x M \to \mathbb{R}$ is defined by $P_x^\#(\mathrm{grad}_x(N_x^{\mathrm{smooth}})) = 0$. Then P is bi-hyperbolic if both P and $P^\#$ are hyperbolic; and similarly: for C_x, one can define $C_x^\#$.

[23] Although Schuller in [3] uses different maps for the null cone and the time-like trajectories—the Gauss map and the Legendre map, respectively—this is not strictly necessary to build a geometry in the sense above, and the distinction is indeed dropped from the later papers in the series.

But once we have such surfaces and their orthogonal directions, one can deform the hypersurfaces, either orthogonally to the surfaces, or parallel to them. Applying such deformations in succession, one can construct a purely geometric *hypersurface deformation algebra*, as seen in Sect. 2, and Eq. (2) (see [29]).

With these algebras in hand, we can apply a generalized form of the Hojman-Kuchar-Teitelboim (HKT) procedure [1], also described in Sect. 2, by which one defines Hamiltonian generators in the gravitational phase space. These generators are designed to recover the action of the hypersurface deformations, and are thus thought, in some sense, to generate spacetime diffeomorphisms.

In sum, the procedure consists in the following steps. First, we use the dynamically-induced causal structure to define not only the spatial hypersurfaces, but also its associated normal vectors in the time direction. Then, one obtains a hypersurface deformation algebra by decomposing general spacetime vector fields in terms of this foliation. This need not match the standard metric spacetime hypersurface commutation algebra; the difference is due to the kinematical information in the construction of the hypersurfaces. Thus, upon computing the hypersurface deformation algebra, only the pure time-time component can differ from (3) and (4):

$$[D_\perp(\xi^1_\perp), D_\perp(\xi^2_\perp)] = \mathscr{H}_\parallel \left(K[Q]^{ij}(\xi^1_\perp \partial_i \xi^2_\perp - \xi^2_\perp \partial_i \xi^1_\perp)\right) \tag{18}$$

The tensor $K[Q]^{ij}$ is defined implicitly by Q (see Eq. 45 and 48 in [2]). It is only in this dynamical relation that the equations of motion make an appearance.

In parallel, using the 'gadget' geometrical tensor G, one can find the induced geometrical structure on these spatial hypersurfaces—call these $\phi(\sigma)$ for $\sigma \in \Sigma$, the spatial hypersurface—and a corresponding phase space for these quantities—labelling as $\pi(\sigma)$ the associated conjugate momenta to $\phi(\sigma)$ (the actual definition is also laborious, and involves again the principal polynomial). Finally, one requires that the hypersurface deformation generators be emulated in the phase space, by respective Hamiltonian generators. That is, equating the commutation algebra between the Hamiltonian generators with that of the hypersurface deformation, one obtains further consistency conditions—this time on the geometrodynamical part of the action. In a nut-shell, the matter dynamics in this way restricts the geometrodynamics of any model embedded in a spacetime manifold.

Finally, two last conditions, in the spirit of Hojman et al., are then assumed in order to obtain the compatible gravitational dynamics. They are called in [2]: (i) "local phase space avatars of deformation operators"—corresponding to a matching of the Hamiltonian constraint algebra and the "felt geometry" deformation algebra, and (ii) "spacetime diffeomorphism invariance"—corresponding to (10).

The two conditions together require the existence of constraint functions on phase space:

$$\mathscr{H}_\perp(\xi_\perp) = \int d^3\sigma\, \xi_\perp(\sigma)\mathcal{H}_\perp(\phi,\pi)(\sigma) \approx 0 \quad \text{and} \tag{19}$$

$$\mathscr{H}_\parallel(\boldsymbol{\xi}_\parallel) := \int_\Sigma d^3\sigma\, \xi^i_\parallel(\sigma)\mathcal{H}^\parallel_i(\phi,\pi)(\sigma) \approx 0$$

satisfying the same algebra as in (18) and (3). As before, the demands of (i) spacetime embeddability and (ii) homomorphism of deformations to symplectic generators imply that the total Hamiltonian for gravity be a fully constrained sum:

$$\mathscr{H}_Q(\xi_\perp, \boldsymbol{\xi}_\parallel) = \mathscr{H}_{Q\perp}(\xi_\perp) + \mathscr{H}_\parallel(\boldsymbol{\xi}_\parallel) \tag{20}$$

where we have inserted a subscript Q to remind ourselves that the Hamiltonian could in principle be different from the standard ADM Hamiltonian, depending on the matter dynamics.

Thus we can finally state:

Theorem 2 (Schuller et al.) *For any diffeomorphism-invariant matter action whose integrand depends locally on some tensorial matter field B and ultralocally on a geometric background described by some tensor field G of arbitrary valence,*

$$S_{matter}[B; G) = \int \mathrm{d}^4x\, L_{matter}(B(x), \partial B(x), \ldots, \partial_{finite} B(x); G(x)) \tag{21}$$

and that satisfies the three matter conditions (the existence of an initial value formulation for the matter field equations; a one-to-one relation between momenta and velocities of massless particles; the existence of an observer-independent definition of positive particle energy):—

one can calculate four geometry-dependent coefficients that enter the equations that guarantee that the geometrodynamical commutation algebra emulates the hypersurface deformation algebra (i.e. the gravitational closure equations), which can be posed as a countable set of linear homogeneous partial differential equations. Then a solution of these equations, in turn, provides a consistent action coupling the matter to the geometry:

$$S_{closed}[B; G) = S_{matter}[B; G) + \int \mathrm{d}^4x\, L_{geometry}(G(x), \partial G(x), \ldots, \partial_{finite} G(x)) \tag{22}$$

Note that the theorem does not state the existence of such a solution for any given B and G and equations of motion. Apart from known examples (such as classical electromagnetism) which recover GR: so far as we know, no other geometrodynamical theory has been found by explicitly solving the gravitational closure equations.

Ideally, the uniqueness theorem we seek would answer the following question: allowing for different matter dynamics in the space of possible derived geometries, is the dynamics of the standard model unique in singling out a meaningful dynamical spacetime? Or are there other families of dynamical fields which would imply different types of (generalized) geometries? Alas, no such theorem is known. In this sense, the programme does not as yet inform us whether chrono-geometry as embodied by a Lorentzian spacetime can or cannot be separated from the dynamics of matter. All we know so far is that for the dynamics of matter that we have, e.g. classical electromagnetism, Lorentzian geometry is what we get. But so to speak: we already knew that. In that sense, unfortunately, the programme does not yet help settle whether matter dynamics or chronogeometry should have explanatory priority over the other. But setting aside this limitation, the programme has the great merit of providing a framework in which the matter dynamics poses strict consistency conditions on any gravitational theory which is to accompany it; and in the future, more powerful statements might well be extracted from this framework.

3.4 A Counter-Example Without Spacetime Diffeomorphism Invariance

We have already seen that this programme's assumptions include the action functional being diffeomorphism invariant. In this final Subsection, we bring out the programme's need for this assumption, by pointing to an example that in its absence, exhibits electromagnetism with the standard coupling to a Lorentzian metric, but also a *non*-orthodox geometrodynamics, i.e. without an Einstein spacetime, or the strong equivalence principle.

This counter-example [36, p. 317] is predicated on a symmetry principle distinct from spacetime diffeomorphism invariance: namely, spatial conformal and diffeomorphism invariance. With this principle, one can stipulate an action that is radically different than the orthodox Einstein-Hilbert one,[24] and one can still couple spin-1 fields and gravity within a consistent dynamics. And when the conformal curvature is small, the equations of motion obtained by the action up to second order in derivatives recover precisely the Maxwell dynamical equations in temporal gauge. Nonetheless, the spacetime obtained from the dynamics need not satisfy the Einstein equations.[25]

Thus, without assuming spacetime diffeomorphism invariance from the outset, the dynamical equations—even Maxwell's—seem insufficient to determine the orthodox spacetime geometry, i.e. orthodox geometrodynamics.

[24] Without a reference density, one can build a conformally invariant 3+1 action by using the square root of the Cotton tensor contracted with itself, cf. [36, p. 314]. Such an action has a completely different structure than the Einstein-Hilbert one, even restricting to the simplest second-order in time-derivatives.

[25] The consistency conditions between the geometric tensor G and geometrodynamics found in [3], ensue only if one further assumes the matter dynamics generates *local* hypersurface deformations. This is not at all necessary: the matter dynamics may define the translation of one hypersurface to the next in a global manner.

4 Realizing the Temporal Metric—From only Spatial Geometry

Let us introduce project (3) by sketching how it relates to projects (1) and (2). More precisely, we will sketch how those projects prompt an expectation about the role of spacetime—which project (3) defeats.

In Sect. 2, we saw that [1] found a direct path from spacetime properties to a Hamiltonian formalism: namely by emulating the algebra of embedded surfaces as a constraint algebra in the Hamiltonian formalism. Then in Sect. 3, we reported on work [3] tying (certain types of) matter dynamics to the dynamics of spacetime. We also pointed to obstacles to a unique specification of standard gravitational dynamics, once one no longer assumes spacetime and its associated requirement of diffeomorphism invariance of the action from which the dynamics are to be derived.

So in the light of these projects, one might expect that a 4-dimensional spacetime needs to be assumed in order to arrive at a consistent dynamics of gravity and matter. That would be a blow to relational approaches such as shape dynamics: which, as we will see in Sect. 5, aims to base its ontology not on spacetime, but on conformal superspace. More specifically, it would suggest that shape dynamics must somehow "piggy-back" on a spacetime framework.

But this is not so. In this Section, we will show that weaker requirements—which do not assume spacetime or spacetime diffeomorphism invariance, at least in its usual form—suffice to deduce general relativistic dynamics. That is: without presupposing spacetime, we will deduce precisely the Hamiltonian of general relativity, i.e. of the form (9), for the appropriate scalar and vector constraints of the ADM formalism. In short: while Hojman et al. found a path from spacetime to a Hamiltonian formalism, here we will describe a path in the opposite direction.

4.1 Constraints on Duration

We begin with the 'top theory' T_t, a theory of Riemannian geometry on a 3-manifold. More precisely, we envisage a configuration space, Φ, whose elements φ are representations of the instantaneous state of the Universe. By 'instantaneous' here, we do not mean any particular spacelike embedding of a hypersurface into a spacetime—there is no spacetime yet! We mean only that the fields at different points are dynamically unrelated. Thus we also assume that the state of the Universe can depend on location. That is, we assume that the theory is not altogether topological. Since we are dealing with field theories, this means that the theories in question will have an infinite number of degrees of freedom (at least one per point of space).

In this picture, since change can be measured locally, it seems natural to also associate duration with change *locally*. But duration is required to mesh in the right way; we will require that the end result after evolution is independent of our choice of intermediate states. That is, a final state is independent of the order in which two different changes are successively applied to an initial state. At this point, we follow a slightly altered version of Hojman et al.'s demand of "path-independence". Their demand was that one falls on the same integral curve of the Hamiltonian, irrespectively of the order of evolution. Ours will be a version specialized to pure time evolution, and applying an equivalence relation based solely on the 'instantaneous' field content of the theory.

More precisely, denoting two evolutions for φ, that may be distinct at distinct locations, by $\Delta^1_\perp\varphi$ and $\Delta^2_\perp\varphi$, the consecutive evolution of the two should be independent of the intermediate state. Thus we demand: $\Delta^{12}_\perp\varphi = \Delta^{21}_\perp\varphi$. (We could omit the $\perp$, since here we do not have embedded surfaces; but it is an intuitive label.)

4.2 Constraints on Evolution

Now we move to sketching the formulation of a Hamiltonian geometrodynamics, as our T_b. As announced, the main result (Theorem 3 below) will be that in this theory, the ADM Hamiltonian (9) is the unique realizer of the functional role of $\Delta_\perp$. Since we seek a Hamiltonian theory, the 'instantaneous state' of the Universe will be a phase space doublet, (h_{ab}, π^{ab}). And invoking the idea of spatial geometry, we require that the theory should depend on the doublet (h_{ab}, π^{ab}) only up to diffeomorphisms. That is, the theory should be invariant under spatial diffeomorphisms.

Our definition of the Hamiltonian generators of change will, therefore, take the form of constraints that the momenta and the metric have to satisfy. But here, we will not demand that the symplectic generators of the diffeomorphisms reproduce the algebra of hypersurface deformations—for in the present picture, we have neither spacetime diffeomorphisms nor hypersurfaces. We will not even demand Sect. 2's full meshing relation, (10): we will only need a similar relation for the pure time evolution, that is, the equivalent of $\Delta^{12}_\perp\varphi = \Delta^{21}_\perp\varphi$, up to the equivalence relation of the theory, namely, spatial diffeomorphisms. This equivalent is (23) below.

We will also assume that, whatever constraint we find for time evolution, it must be a scalar. This is based on the demand on T_t that the theory contain local degrees of freedom. The requirement is not mentioned by Hojman et al. because the evolution normal to a hypersurface has only one direction (per point), and is thus already scalar. Here, without the aid of the spacetime picture, we cannot appeal to that argument.

Here, the argument for a scalar Hamiltonian (i.e. that the Hamiltonian can depend at most on a scalar Lagrange multiplier) just counts degrees of freedom. Let us rehearse how this goes. In three dimensions, h_{ij} has six degrees of freedom; diffeomorphisms—being generated by vector fields (Lagrange multipliers with three indices per space point)—take away three of these. Since the Hamiltonian must preserve spatial diffeomorphism symmetry, as a constraint it should be first class with respect to the generator of spatial diffeomorphisms. Therefore, if it was also generated by a vectorial Lagrange multiplier, we would have zero degrees of freedom left![26] Thus, one further assumption that goes into our general derivation of geometrodynamical evolution is that the theory is not topological.

Lastly, we require the Hamiltonian to depend at most quadratically on the momenta. This can be justified by the need to avoid Ostrogradski instabilities (and thereby, an unphysical theory) that higher-than-quadratic momenta would engender in the corresponding Lagrangian theory (by inducing higher time derivatives). But for reasons of space, we will not give details; for a review, cf. [37].

These assumptions suffice to completely specify the Hamiltonian generating evolution. We now sketch the proof presented in [4].

Statement of the results. First, we stipulate the same generator of spatial diffeomorphisms as in project (1), i.e. Hojman et al.'s (7): given the geometrodynamical phase space, this generator, $\mathscr{H}_\parallel$, is unique.

The limited meshing relation we will require, namely, that the pure change part of the evolutions mesh in the right way, can be mathematically represented as follows:

$$\Delta_\perp^{12}\varphi = \Delta_\perp^{21}\varphi + \{\varphi, \mathscr{H}_\parallel\}(F(\xi_1, \xi_2)), \tag{23}$$

where F is some fixed phase space function—about which we presuppose nothing—with the ξ as arguments, and where we have identified:

$$\Delta_\perp^{12}\varphi := \{\{\varphi, \mathscr{H}_\perp(\xi_\perp^1)\}, \mathscr{H}_\perp(\xi_\perp^2)\}, \quad \text{and} \quad \Delta_\perp^{21}\varphi := \{\{\varphi, \mathscr{H}_\perp(\xi_\perp^2)\}, \mathscr{H}_\perp(\xi_\perp^1)\}\,. \tag{24}$$

Equation (23) guarantees that $\Delta_\perp^{12}\varphi \approx \Delta_\perp^{21}\varphi$ because we have taken $\mathscr{H}_\parallel$ to be the generator of spatial diffeomorphisms, and thus its action defines equivalence classes in terms of spatial geometry.

Using the Jacobi identity, we find Eq. (23) holds if and only if

$$\{\mathscr{H}_\perp(\xi_\perp^1), \mathscr{H}_\perp(\xi_\perp^2)\} = \mathscr{H}_\parallel(F(\xi_1, \xi_2)) \tag{25}$$

for some phase space function F; (about which we again presuppose nothing, unlike Hojman et al., who demand that $\mathscr{H}_\parallel(F(\xi_1, \xi_2))$ reproduce the hypersurface deformation algebra). And moreover, if F turns out to be phase space dependent, we will require a second condition, namely that $\mathscr{H}_\parallel \approx 0$.

We first write a general scalar Hamiltonian generator as a sum of a kinetic and a potential term (both assumed to be non-identically zero)[27]:

[26] Agreed: this argument rejects the possibility that the Lagrange multiplier itself be constrained; e.g. a divergence-free vector field. The main reason is that this would introduce a Lagrange multiplier for a Lagrange multiplier. Such symmetries are "reducible" and lie outside the scope of this investigation. If one wishes to completely obviate the need for this argument, one could simply require that the gravitational field have two propagating physical degrees of freedom; this already requires the remaining part of the Hamiltonian to be scalar.

[27] In fact a bit more is assumed. If $V \equiv 0$, a slight generalization of T ensues in Theorem 3 (still no derivatives of the momenta, but more general contractions, i.e. an arbitrary DeWitt parameter);

$$\mathscr{H}_\perp[\pi^{ab}, h_{ab}; \sigma) = T[\pi^{ab}, h_{ab}; \sigma) + V[h_{ab}; \sigma) \, . \tag{26}$$

Since the possible terms linear in the momentum can be generated by a canonical transformation of the variables, T depends quadratically on the momenta.

The actual calculations are painstaking and cannot be meaningfully summarized here. For the purpose of illustration, we present the general form of the kinetic and potential terms, considered in [4, p. 112503-3]. The general kinetic term is of the form:

$$T[h, \pi; \sigma) = \sum_{r,m,n} \mathcal{B}_{abcd}^{i_1 \cdots i_n \, j_1 \cdots j_m} \left(\nabla_{i_1} \cdots \nabla_{i_n} \pi^{cd}\right) \left(\nabla_{j_1} \cdots \nabla_{j_m} \pi^{ab}\right) (\sigma) \, . \tag{27}$$

The scalar function $V[h, x)$ and the tensor $\mathcal{B}_{abcd}^{i_1 \cdots i_n \, j_1 \cdots j_m}$ depend on the metric h_{ab} and its derivatives through $R_{kl}, \nabla_{a_1} R_{kl}, \cdots, \nabla_{a_1} \cdots \nabla_{a_r} R_{kl}$ where the number of explicit covariant derivatives, r, is arbitrary. Note that in all generality, the spatial derivative order of $\mathcal{B}_{abcd}^{i_1 \cdots i_n \, j_1 \cdots j_m}$ and $V[h, \sigma)$ differ.

We can now state the theorem, proven in [4, p. 112503-23]:

Theorem 3 (Rigidity) *Equation* (25) *is satisfied by a unique generator in the family* (26) *(with T as defined in* (27)*). Namely:*

$$\mathcal{H}_{ADM\perp}[h, \pi; \sigma) := \left(a\epsilon \frac{\pi^{ab}\pi_{ab} - \frac{1}{2}\pi^2}{\sqrt{h}} - a^{-1}(R - 2\Lambda)\sqrt{h} \right) (\sigma) \tag{28}$$

where a and Λ are arbitrary constants.

Thus, since (25) is obtained with an F that is the same as given by the hypersurface deformation algebra of (4) (and therefore F is phase-space-dependent). Thus (23) requires $\mathscr{H}_\parallel \approx 0$. And since $\{\mathscr{H}_\perp, \mathscr{H}_\parallel\} \propto \mathscr{H}_\perp$, dynamical preservation of $\mathscr{H}_\parallel \approx 0$ implies also that $\mathscr{H}_\perp \approx 0$.

Thus we can now state:

Corollary 2 *Given an equivalence relation supplied by spatial diffeomorphisms,* (7)*, and a generator of time evolution of the form* (26)*, the only theory that enforces the meshing,*[28] *or path-independence condition, $\Delta_\perp^{12}\varphi \approx \Delta_\perp^{21}\varphi$, given in equation* (23) *is the ADM form of general relativity.*

---

if $T \equiv 0$, any V is trivially propagated. If $V \not\equiv 0$ but contains no derivatives of the metric (i.e. it is proportional to the scalar density $\sqrt{g}$), a linear term in the momenta in T arises, but this change can be accommodated by the purely quadratic term through a canonical transformation which eliminates the potential term.

[28] Of course, as in ADM, these definitions are unique only up to the relevant class of isomorphism of the theory in question: here, symplectomorphisms (or canonical transformations).

5 A Corollary: Shape Dynamics Exonerated

We turn to our positive corollary of Sect. 4. It is that another programme in foundations of classical gravity, shape dynamics, is innocent of an accusation that one might be tempted to make against it. The accusation arises from the fact that within a certain regime of shape dynamics, its Hamiltonian matches that of general relativity i.e. is the ADM Hamiltonian. This prompts the question, like I. Rabi's quip when in 1936 the muon was discovered with its surprising properties: 'who ordered that?' That is: what is the motivation, from within the perspective of shape dynamics, for this form of Hamiltonian? Or more accusingly: is there none? Has shape dynamics 'simply grabbed' the ADM Hamiltonian from its illustrious rival, general relativity?

We will begin by recalling how the dialectic of such an accusation is often treated in philosophy: namely, in terms of a much-cited quip by Bertrand Russell condemning 'theft' and praising 'honest toil' (Sect. 5.1). This review will not be just a philosophical reminiscence. Russell's quip is unfair in a way that the 'top theories' T_t in this paper's three projects illustrate. Then we will turn to shape dynamics. We first introduce it (Sect. 5.2); and then in Sect. 5.3 we explain how, thanks to Sect. 4's theorem, it is innocent of the accusation. It has indeed undertaken, not theft, but honest toil.

5.1 *Russell's Quip*

Much of Russell's writings in the period 1910–1930 proposed reductions in what Sect. 1.3 called the Nagelian sense: deduction of a 'top theory' T_t from a 'bottom theory' T_b, augmented with extra premises we there labelled B (for 'bridge-laws'). Russell's motivation, indeed inspiration, was of course the logicist programme of Frege, himself and Whitehead to try to reduce all of pure mathematics to logic: that is, to deduce pure mathematics from logic, augmented with appropriate definitions of mathematical words (e.g. 'real number') in terms of logic notions.

In the light of that endeavour, especially his and Whitehead's *magnum opus, Principia Mathematica*, Russell aimed to similarly reduce whole swathes of empirical knowledge, both everyday and technical, to theories about sense-experience. And in the logical empiricist school of the 1920s onwards, which was much influenced by logicism, other philosophers, such as Carnap, articulated similarly ambitious philosophical programmes to reduce the 'problematic' realm of 'the theoretical' to 'observation': programmes that led directly to fellow-empiricist Nagel's formulation of reduction, as reviewed at the beginning of Sect. 1.3.

We need not linger on these philosophers' aim of finding a firm foundation of empirical knowledge in sense-experience, or 'pure observation': an aim that was much criticized from about 1950 onwards.[29] What matters for us is these philoso-

[29] Let alone endorse it: cf. footnote 9. Nor need we linger on the shortcomings of logicism: in short, that the most it could claim to achieve is the deduction of pure mathematics from a single part of it, viz. set-theory, not from *logic*.

phers' admitting the need to formulate judicious definitions that enable them to *deduce* the 'problematic' doctrine at issue. They are thus opposed to so-called *implicit definition*, i.e. to thinking it is enough, for justifying a concept or discourse, to give a set of postulates ('axioms') containing the concept and from which one's claims about it can be deduced. How do you know—they might say—that your deduction sets out from safe ground?

Hence Russell's famous quip about 'theft over honest toil'. It is in his *Introduction to Mathematical Philosophy* (1919), in his discussion of deducing the truths of arithmetic from logic, i.e. from set-theory (cf. footnote 29):

> The method of 'postulating' what we want has many advantages; they are the same as the advantages of theft over honest toil. Let us leave them to others and proceed with our honest toil. (1919, p. 71)

Thus 'theft' is here the dogmatic postulation of entities, as being those things that obey certain axioms or postulates; and 'toil' is here the work of finding judicious definitions so that the *definienda* can be shown using logical inference alone to satisfy the claims made about them.

Indeed, a *bon mot*; and much quoted. But it is unfair to dismiss making postulates at the level of the 'top theory' as 'theft'. In many cases, it is obviously hard work, even when one is given an agreed stock of words or concepts, to formulate axioms whose consequences express all that we wish to assert using those words or concepts. Think of the work involved in axiomatising Euclidean geometry: whether by Euclid and his contemporaries, or in our own era by Hilbert (in his *Grundlagen der Geometrie* of 1899) and his successors. Formulating an axiom system that implies all the desired theorems is an achievement; (especially if one requires, as Hilbert did, that the axioms be mutually independent). This is so, regardless of whether one has the philosopher's or logician's interest in giving underlying 'bottom theory' definitions of the objects in geometry's subject-matter; e.g. of what a point in the plane *is* in terms of set-theory.

We see this point also in the top theories T_t of this paper's three projects (and in our other spacetime examples in [8]). For example, the formulation by Hojman et al. of their deformation algebra is an achievement, regardless of whether one goes on—as indeed they did—to give an underlying 'bottom theory' realization of it as a Hamiltonian constraint algebra. Similarly, the formulation by Schuller et al. of generalized geometries is an achievement, regardless of whether one goes on—as indeed they did—to give an underlying 'bottom theory' invoking predictivity and good behaviour conditions for matter and radiation, and showing a solution of the resulting consistency conditions. Lastly, the formulation by Gomes and Shyam of path independence of evolution of spatial quantities is an achievement, regardless of whether one goes on to specify a Hamiltonian that has these properties.[30]

---

[30] Fortunately, the point is also noticed by philosophers of logic. [38, p. 272] call Russell's *bon mot* 'one of the shoddiest slogans in philosophy'. And their reason is exactly this point: that writing down the right analysis of a concept, *before* any 'reduction' begins, often requires honest toil. Thus they point out that (i) Russell originally aimed it, unfairly, at Dedekind's treatment of continuity, and (ii) 'it assumes [wrongly] that we already know what we want . . . the examples from Dedekind show just how much honest toil it takes to discover-to formulate precisely—just what it is that we want'.

So much by way of general philosophy. We turn to shape dynamics: to urge that thanks to Sect. 4's theorem, its use of the ADM Hamiltonian is *not* a case of theft.

5.2 A Brief Introduction to Shape Dynamics

This will be a very brief, and so incomplete, introduction to shape dynamics. For more technical details, see [39, Chap. 12] and [31, 32].

The relational motivation for shape dynamics. Conceptually, shape dynamics is an offspring of relationism. Relationism starts from the idea that all measurements are comparisons: that doubling the size of all the rods and of the universe alongside them, will change nothing measurable; and that speeding up all clocks and physical phenomena alongside them, will change nothing measurable. No one has ever concretely measured (nor ever will measure) a fundamentally dimensionful quantity.[31]

But even if one accepts that fundamentally only relations are physically measured, general relativity associates a dimensionful quantity to trajectories (curves) in space-time, irrespectively of relations to any other object: namely, proper time (or distance). The "experienced time" along a trajectory in space-time depends only on the trajectory itself, without reference to anything else. (This time is of course determined only up to an affine transformation, reflecting a choice of unit and zero.) So *prima facie*, general relativity fails to be relational, because of its physically significant but conformally *non*-invariant "experienced time" (i.e. proper time and distance). Although we believe that proper time should be a physical correlate of something relational, such as a ratio of numbers of oscillations of two atoms, the theory does not *explicitly* require such an understanding.

So it is tempting to impose scale-invariance at the fundamental, theory-building level, as a criterion: to require theories to build all physical quantities from relations. Over the years, prompted by this sort of relational rationale, many consistent theories of gravity with 4–dimensional space-time scale-invariance have been proposed.[32] But, so far as we know, none has overcome this shortcoming of general relativity, while keeping its explanatory power.

But if we accept the ontological '3+1' split of spacetime into space and time, we can perhaps break this *impasse*. For we now gain access to a different kind of symmetry: those that act on each spatial field configuration. Spatial diffeomorphisms enforce the notion that location is relational; i.e. positions only make sense relative to other positions. Analogously, spatial conformal invariance is a relational symmetry: it means that local scale can only be made sense of relationally, i.e. by comparing one size to another.

The technical argument for shape dynamics. In brief, the technical idea behind shape dynamics, is that the ADM formulation of general relativity, with its refoliation

---

[31] See [40] for an eloquent defence of these ideas.

[32] Such models come under different names (e.g. conformal gravity and Weyl gravity). Their appeal lies in their renormalization properties, see e.g. [41, 42].

redundancy (represented by $\mathscr{H}_{\text{ADM}\perp} \approx 0$), is dual to a theory that possesses spatial conformal invariance. The reason such a dual formulation is possible, is that the constraint that is the symplectic generator of conformal transformations, is first class with respect to diffeomorphisms, and can be seen as a gauge-fixing of the symmetries generated by $\mathscr{H}_{\text{ADM}\perp}$. Thus we can represent the ADM Hamiltonian in an alternative (non-local) form in the space of conformal geometries. The details of this duality are spelled out in [32]. Here, for brevity, we will only describe how the theory is closely related to what is known as the 'York method' for solving general relativity's initial value constraints.

We start from the fact that in the ADM formulation, not all initial data (h_{ab}, π_{ab}) are acceptable. To find initial data, we must first solve the two ADM constraints, $\mathscr{H}_\perp = \mathscr{H}_\parallel = 0$. The York method [43, 44] is the only known mathematical tool for generically attacking the initial value problem of general relativity [45]. Indeed, most of the formal proofs of existence and uniqueness of solutions of general relativistic dynamics require the use of a particular choice of foliation, namely that the mean (of the trace of the) extrinsic curvature (CMC) be constant throughout the given leaf. And as we shall see in a moment, this foliation is the one that is compatible with the existence of a shape dynamics solution. As a matter of fact, every test that has ever been passed by GR is known to be consistent with a CMC foliation.[33]

More importantly for us, it turns out that this choice has special properties: it allows a description of the evolution of spatial geometry without reference to local spatial scale. That is, it allows one to describe gravitational evolution in the auxiliary time τ,[34] in terms of spatially 'conformally invariant' geometries. This is because the constraints that these foliations are obliged to satisfy in order to be CMC, namely,

$$\mathcal{C} = \operatorname{tr}\pi - \tau\sqrt{h} = 0, \tag{29}$$

where $\operatorname{tr}\pi = h_{ab}\pi^{ab}$ and τ is a constant, generate, via Noether's theorem, canonical transformations. Such transformations are exactly changes of spatial scale, i.e. the local conformal transformations.[35] These results show that surprisingly, although general relativity is not fundamentally concerned with spatial conformal geometries, it is deeply related to them.

Schematically, the trade can be represented, in the jargon of Dirac constraint analysis, as two first class constraint surfaces being fully second-class with respect to each other. Being 'first class' means that constraints weakly commute, that is, that

---

[33] The foliation does break down inside known solutions with event horizons. But, in the generic case where event horizons form, Marsden and Tipler conjecture that "the [CMC] foliation does fill out at least the exterior of the event horizon of the singularity, and that the foliation fills out the entire spacetime if it is geodesically complete" [46, p. 126].

[34] Which, from (29), is just the average of the trace of the momenta, $\tau := \frac{\int \operatorname{tr}\pi}{V(h)}$, where $V(h) := \int \sqrt{h}$.

[35] In fact, the momenta are set to be constant trace, not trace-free; which would be the true generator of conformal transformations. However, one can interpret the constant trace part as defining an auxiliary quantity, the York time. The transverse traceless choice sets the gravitational momenta to be purely tensorial (spin-2), with no scalar (spin-0) or vector (spin-1) components.

they close on-shell and thus can be taken as symmetry generating; 'second class' means that they do not weakly commute, indeed, their Poisson bracket forms an invertible matrix, and thus one phase space function serves as a gauge-fixing for the other. See Fig. 2.

These properties mean that we can split $\mathscr{H}_\perp$ into a local and a global part. The local constraint generates a symmetry that can be traded with local conformal invariance, whereas the global part generates a non-local Hamiltonian in conformal superspace, that we can write, in the unconstrained phase space as:

$$\mathscr{H}_{\text{global}} = \int (\Omega^6(h, \pi) - 1)\sqrt{h} \tag{30}$$

where Ω is the York conformal factor (described in Appendix 7).

Thus $\mathscr{H}_\perp - \mathscr{H}_{\text{global}}$ is entirely second class with respect to $\mathcal{C}$, and we have two separate first class systems of local constraints: the standard ADM one, given by $\mathscr{H}_\perp \approx 0$ and $\mathscr{H}_\| \approx 0$, and the shape dynamics one,

$$\mathcal{C} \sim 0, \qquad \mathscr{H}_\| \sim 0 \ . \tag{31}$$

which are both conserved by the Hamiltonian $\mathscr{H}_{\text{global}}$. Moreover, the dynamics of this system matches that of ADM general relativity (whenever a CMC foliation is available, as discussed above).

In sum, shape dynamics amounts to taking seriously these hints: hints from the initial value formulation of general relativity and from considering symmetries compatible with instantaneous physical configuration states. The result is a theory about the dynamics of space that is based on different symmetry principles than the standard diffeomorphism or Lorentz invariance of spacetime. Namely, the symmetries of the theory are taken to be conformal diffeomorphisms. Indeed, starting from the broad idea that shape dynamics is to be a theory about space, it is natural to focus on the space Riem, of unconstrained spatial 3-metrics on a 3-manifold, Σ. Then one can justify the symmetry group of spatial conformal diffeomorphisms as being the maximal local group of symmetries acting on Riem [36]. Thus we get the idea of conformal superspace, which has been shown to have a well-defined symplectic reduction [47] (Fig. 2).

But as we have also seen, it has one peculiar feature: a non-local generator of evolution (Hamiltonian). And if we focus on conformal superspace, there are admittedly many Hamiltonians on it that look more natural than the non-local one that shape dynamics officially adopts: viz. the Hamiltonian obtained by extending general relativity's ADM Hamiltonian to this space. Agreed: this official choice enables shape dynamics to match (in many circumstances) the successes of general relativity.[36] But the choice looking unnatural triggers the accusation of 'theft' ...

[36] Fortunately, we can in many circumstances find explicit solutions of shape dynamics, albeit at the cost of simplifying the problem by assuming it is highly symmetric: see [39, Chap. 13], and references therein.

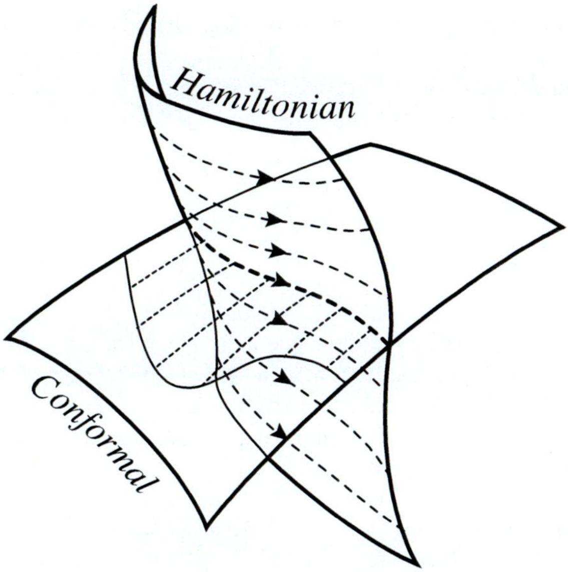

Fig. 2 A schematic representation of the phase space of general relativity. In it, two constraints coexist, which are good gauge-fixings for each other and are both first-class with respect to the diffeomorphism constraint. One is the Hamiltonian constraint and the other is the conformal (Weyl) constraint. The Hamiltonian constraint is completely gauge-fixed by the conformal constraint except for a single residual global constraint. It Poisson-commutes with the conformal constraint and generates a vector flow on the Hamiltonian constraint surface (represented in the figure), which is parallel to the conformal constraint surface. This vector flow generates the time evolution of the system in the intersection between the two surfaces. Any solution can then be represented in an arbitrary conformal gauge by lifting it from the intersection to an arbitrary curve on the conformal constraint surface. All such lifted curves are gauge-equivalent solutions of a conformal gauge theory with conformally-invariant Hamiltonian

5.3 Shape Dynamics Exonerated

But Sect. 4's Theorem 3 exonerates shape dynamics. For from most such natural Hamiltonians on conformal superspace, we would *not* be able to deduce appropriate relations to 'local change', or duration in the sense of Sect. 4.1.

To put the point more specifically and more positively, in terms of functionalist reduction: the shape dynamics Hamiltonian is the nearest realizer on (the cotangent bundle of) conformal superspace of a Hamiltonian possessing a representative that—that is: that in some gauge—yields the notion of duration which we formulated in Sect. 4.1's T_t. That is: this Hamiltonian gives, in some nearest realization, a reduction, in the vocabulary of conformal superspace (T_b), of the T_t notion of time.

That, through Theorem 3, this can be done without invoking spacetime ideas—in particular, without invoking the relativity of simultaneity, or its more sophisticated cousin, refoliation invariance—is, we submit, remarkable. And it is essential for the basic outlook of shape dynamics, i.e. an ontology based on conformal superspace, to be tenable.

In sum: shape dynamics is exonerated as the unique theory in conformal superspace that can be lifted to match the dynamics of a theory of spatial metrics with a local notion of duration.

6 Conclusions

We shall not here summarise the three geometrodynamical projects, (1) to (3), that we have reviewed and connected to functionalist reduction. For Sect. 1.2 summarised them in themselves, and Sect. 1.3 summarised them as illustrations of functionalist reduction. Let us instead give a table, allowing one to compare at a glance these three projects, and also shape dynamics.

The projects: their assumptions, results and uniqueness				
	Spacetime manifold?	Spacetime metric?	Recover Einsteinian dyn.?	Uniqueness theorem?
Hojman et al.	✓	✓	✓	✓
Schuller et al.	✓	×	?	×
Gomes-Shyam	×	×	✓	✓
Shape dynam.	×	×	✓	✓*

This table, with its brief labels, summarizes in the first two columns the *assumptions*, and in the last two columns the *results*, of functional reduction. It is of course a simplification of the situation. For example: although project (2) (by Schuller et al.) does not assume a Lorentzian metric, it does use some (dynamical) notion of a coupling metric, or geometry (the tensor G). On the other hand, the purely spatial ontology of project (3) [4] has a spatial Riemannian metric in its fundamental variables. We have also put an asterisk in the last item (uniqueness theorem for shape dynamics), because (as discussed in Sect. 5.3) it is the unique theory in conformal superspace that can be lifted to match the dynamics of a theory of spatial metrics with a local notion of duration. That is: the asterisk signals the fact that it does not have a uniqueness theorem solely in terms of conformal superspace. The question mark in the third column, for project (2), is correlated to a lack of uniqueness: for we do not know if there are matter fields which would allow a different sort of spacetime (and spacetime dynamics) through the constructions of that project.

In this paper, we have used the philosophical idea of functionalist reduction to characterize time in three broadly geometrodynamical—but purely classical—theories. We thus arrive, in effect, at the threshold of one of H-D. Zeh's main endeavours: to understand time in quantum geometrodynamics. That is of course a vast endeavour, in which Zeh, his many collaborators such as Kiefer [48], and of course many others, have done a great deal of work. We obviously cannot go into this here: (for masterly introductions to the programme, cf. [49, Chap. 6] and [48, Chap. 5]). In closing, let us just make a hopeful remark on the possible use of our philosophical notions of functional role, and functionalist reduction.

We recall that in this endeavour, one main aim is to extract a time variable, with respect to which the matter degrees of freedom would obey a time-dependent Schrödinger equation, from a WKB-like approximation to the (fundamentally timeless) Wheeler-DeWitt equation. We can think of this as an internal time being identified by its functional role: the matter degrees of freedom are functionally differentiated, and identifying the functional coefficient with time yields a Schrodinger-like equation for these degrees of freedom. Accordingly, our discussion suggests that one might explore defining time by other possible functional roles. But of course, whether any such definition will prove useful in the context of quantum geometrodynamics remains to be seen. We can only hope that the functionalist reductive perspective we have offered here for the classical case may prove useful in people's effort to better understand the quantum case ... long may Heinz-Dieter Zeh's scientific creativity and craftsmanship inspire that effort!

Acknowledgements We thank: (i) audiences in seminars at the Black Hole Initiative, Harvard, LSE, MIT and Muenchen, for comments; (ii) Laurenz Hudetz, Nick Huggett, Tushar Menon, Brian Pitts, and Mike Schneider for correspondence; and (iii) Claus Kiefer, for his generosity and comments.

7 A The York Method

Let us briefly review the York method, by which general relativity's constraints decouple and turn into elliptic equations. On a closed spatial hypersurface, we start with the (by-now standard!) metric h_{ab} and tensor density π^{ab}; but also with a real number τ. The York method then consists in the following steps:

1. Given a traceless tensor, one can find its projection to a transverse component. That is, upon the substitution $\pi^{ab} \to \pi^{ab} - \nabla^{(a}\chi^{j)}$ (where round brackets signify index symmetrization, $A_{(ab)} := A_{ab} + A_{ba}$) and ∇ is the Levi-Civita covariant derivative with respect to h_{ab}, one can solve the transversality condition (with respect to the vector field χ^a):

$$\nabla_b(\nabla^{(a}\chi^{b)} - \frac{2}{3}h^{ab}\nabla_c\chi^c) = \nabla_b(\pi^{ab} - \frac{1}{3}\mathrm{tr}\,\pi\; h^{ab})\;.$$

Thereby we obtain the transverse-traceless momentum π^{ab}_{TT}:

$$\pi^{ab}_{TT} = (\pi^{ab} - \frac{1}{3}h^{ab}\mathrm{tr}\,\pi) - (\nabla^a\chi^b + \nabla^b\chi^a - \frac{2}{3}h^{ab}\nabla_c\chi^c)\;\;.$$

This means that for any constant τ, the momentum $p^{ab} := \pi^{ab}_{TT} + \frac{1}{2}\tau\, h^{ab}\sqrt{h}$ satisfies $\mathscr{H}_{\|} = 0$. Importantly, it represents a *CMC slicing* (constant mean curvature), since its trace is necessarily constant:

$$h_{ab}p^{ab} = \frac{3}{2}\tau\sqrt{h} = const.\sqrt{h}\,.$$

2. Upon *conformally transforming* the canonical variables, as $(h_{ab}, \pi^{ab}_{TT}) \to (\Omega^4 h_{ab}, \Omega^{-4}\pi^{ab}_{TT})$, from the scalar constraint $\mathscr{H}_\perp = 0$ one obtains the Lichnerowicz–York equation (LY) with respect to the scalar field Ω:

$$\frac{\Omega^{-6}}{\sqrt{h}}\pi^{ab\,2}_{TT} - \frac{3}{8}\sqrt{h}\,\Omega^6\tau^2 - \sqrt{h}\left(R\,\Omega^2 - 8\,\Omega\Delta\Omega\right) = 0\,.$$

This is an elliptic equation which can, under certain additional assumptions, be solved for Ω: thereby satisfying all of the ADM constraints on a CMC slice $\left(\mathrm{tr}\,p = \frac{3}{2}\,\tau\,\sqrt{h}\right)$. The metric and momenta satisfying the initial value problem are then given by p^{ab} and $\gamma_{ab} = \Omega^4 h_{ab}$.

One important point hidden in the analysis is that the LY equation and the transversality condition are conformally invariant:

$$\begin{cases} h_{ab} \to e^{4\phi}h_{ab} \\ \pi^{ab} \to e^{-4\phi}\pi^{ab} \end{cases} \quad \Rightarrow \quad \begin{cases} \gamma_{ab}[e^{4\phi}h_{ab}, e^{-4\phi}\pi^{ab}] = \gamma_{ab}[h, \pi] \\ p^{ab}[e^{4\phi}h_{ab}, e^{-4\phi}\pi^{ab}] = p^{ab}[h_{ab}, \pi^{ab}]\,. \end{cases}$$

References

1. S.A. Hojman, K. Kuchar, C. Teitelboim, Geometrodynamics regained. Annals Phys. **96**, 88–135 (1976). https://doi.org/10.1016/0003-4916(76)90112-3
2. M. Dull, F.P. Schuller, N. Stritzelberger, F. Wolz, Gravitational closure of matter field equations. Phys. Rev. D **97**(8), 084036
3. F.P. Schuller, All spacetimes beyond Einstein (Obergurgl Lectures). arXiv:1111.4824 (2011)
4. H. Gomes, V. Shyam, Extending the rigidity of general relativity. J. Math. Phys. **57**(11), 112503 (2016). https://doi.org/10.1063/1.4967951
5. N. Huggett, C. Wüthrich, *Out of Nowhere: The Emergence of Spacetime in Quantum Theories of Gravity*, (Oxford University Press, Oxford, 2021)
6. V. Lam, C. Wüthrich, Spacetime functionalism from a realist perspective. Synthese. (Published online 17 April 2020)
7. J. Butterfield, H. Gomes, *Functionalism as a Species of Reduction Current Debates in Philosophy of Science*, (In Honor of Roberto Torretti, Springer, New York, 2020)
8. J. Butterfield, H. Gomes, 2020 Spacetime functionalism *avant la lettre* submitted
9. J. Barbour, B. Bertotti, Mach's principle and the structure of dynamical theories. in *Proceedings of the Royal Society of London A: Mathematical, Physical and Engineering Sciences*, vol. 382, no. 178, pp. 3295–306. http://rspa.royalsocietypublishing.org/content/382/1783/295. https://doi.org/10.1098/rspa.1982.0102 (1982)
10. D. Giulini, What is the geometry of superspace? Phys. Rev. D **51**, 5630–5635. https://doi.org/10.1103/PhysRevD.51.5630
11. J. Barbour, B.Z. Foster, N. O'Murchadha, Relativity without relativity class. Quant. Grav. **19**, 3217–3248 (2002). https://doi.org/10.1088/0264-9381/19/12/308
12. H. Brown, *Physical Relativity* (Oxford University Press, Oxford, 2006)

13. C. Callender, *What Makes Time Special* (Oxford University Press, Oxford, 2017)
14. L. James, A new perspective on time and physical laws. Forthcom. Br. J. Philos. Sci. http://philsci-archive.pitt.edu/17107/ (2020)
15. E. Nagel, *The Structure of Science: Problems in the logic of scientific explanation* (Harcourt, New York, 1961)
16. E. Nagel, Issues in the logic of reductive explanations, emergence: contemporary readings. in *MIT Press*, ed. by M. Bedau, P. Humphreys (Massachusetts, Cambridge, 1979, 2008)
17. J. Butterfield, Emergence, reduction and supervenience: a varied landscape. Found. Phys. **41**, 920–959 (2011)
18. F. Dizadji-Bahmani, R. Frigg, S. Hartmann, Who's afraid of Nagelian reduction? Erkenntnis **73**, 393–412 (2010)
19. K. Schaffner, Ernest Nagel and reduction. J. Philos. **109**, 534–565 (2012)
20. D. Lewis, How to define theoretical terms. J. Philos. **67**, 427–446 (1970)
21. D. Lewis, Psychophysical and theoretical identifications. Australas. J. Philos. **50**, 249–258 (1972)
22. D. Braddon-Mitchell, R. Nola, *Conceptual Analysis and Philosophical Naturalism MIT Press: Bradford Books* (Massachusetts, Cambridge, 2009)
23. J. Butterfield, The hole truth. Br. J. Philos. Sci. **40**(1), 1–28. https://doi.org/10.1093/bjps/40.1.1 (1989)
24. J. Butterfield, David lewis meets john bell. Philos. Sci. **59**, 26–43 (1992). https://doi.org/10.1086/289652
25. J. Butterfield, David lewis meets Hamilton and Jacobi. Philos. Sci. **71**(5), 1095–1106. https://www.jstor.org/stable/10.1086/428013 (2004)
26. T. Button, S. Walsh, *Philosophy and Model Theory* (Oxford University Press, Oxford, 2018)
27. J.E. McTaggart, The unreality of time. Mind **17**, 457–474 (1908)
28. R. Arnowitt, S. Deser, C. Misner, The dynamics of general relativity. in *Gravitation: An Introduction to Current Research*, ed. by L. Witten (Wiley, New York, 1962), pp. 227–264
29. C. Teitelboim, How commutators of constraints reflect the spacetime structure. Ann. Phys. **79**(2), 542–557. http://www.sciencedirect.com/science/article/pii/0003491673900961. https://doi.org/10.1016/0003-4916(73)90096-1 (1973)
30. T. Menon, Why is the metric of general relativity Lorentzian? In preparation (2021)
31. H. Gomes, S. Gryb, T. Koslowski, Einstein gravity as a 3D conformally invariant theory. Class. Quant. Grav. **28**, 045005 (2011). https://doi.org/10.1088/0264-9381/28/4/045005
32. H. Gomes, T. Koslowski, The link between general relativity and shape dynamics. Class. Quant. Grav. **29**, 075009 (2012). https://doi.org/10.1088/0264-9381/29/7/075009
33. P. Horava, Quantum gravity at a Lifshitz point. Phys. Rev. D **79**, 084008 (2009). https://doi.org/10.1103/PhysRevD.79.084008
34. D. Lovelock, The Einstein tensor and its generalizations. J. Math. Phys. **12**(3), 498–501. https://doi.org/10.1063/1.1665613 (1971)
35. K. Giesel, F.P. Schuller, C. Witte, Gravitational dynamics for all tensorial spacetimes carrying predictive, interpretable and quantizable matter. Phys. Rev. D **85**, 104042 (2012). https://doi.org/10.1103/PhysRevD.85.104042
36. H. Gomes, Local gravity theories in conformal superspace. Ann. Phys. http://www.sciencedirect.com/science/article/pii/S0003491618301507. https://doi.org/10.1016/j.aop.2018.05.014 (2018)
37. R.P. Woodard, Ostrogradsky's theorem on Hamiltonian instability. Scholarpedia **10**(8), 32243 (2015)
38. A. Oliver, T. Smiley, *Plural Logic* (Oxford University Press, Oxford, 2016)
39. F. Mercati, *Shape Dynamics: Relativity and Relationalism Oxford University Press* (Oxford, 2017)
40. J. Barbour, *The End of Time: The Next Revolution in Physics Oxford University Press* (Oxford, 1999)
41. E. Fradkin, A. Tseytlin, Conformal supergravity. Physics Reports **119**(4), 233–362. http://www.sciencedirect.com/science/article/pii/0370157385901383. https://doi.org/10.1016/0370-1573(85)90138-3 (1985)

42. K.S. Stelle, Renormalization of higher-derivative quantum gravity. Phys. Rev. D **16**, 953–969. https://link.aps.org/doi/10.1103/PhysRevD.16.953. https://doi.org/10.1103/PhysRevD.16.953 (1977)
43. J.A. Isenberg, N.J.W. Murchadha, J. York, Initial value problem of general relativity. 3. Coupled fields and the scalar-tensor theory. Phys. Rev. D **13**, 1532–1537 (1976)
44. J.W. York, Gravitational degrees of freedom and the initial-value problem. Phys. Rev. Lett. **26**, 1656–1658 (1971). https://doi.org/10.1103/PhysRevLett.26.1656
45. E. Gourgoulhon, 3+1 formalism and bases of numerical relativity. in Lecture notes in Physics, vol. 846 (Springer, Berlin, 2012)
46. J.E. Marsden, F.J. Tipler, Maximal hypersurfaces and foliations of constant mean extrinsic curvature in general relativity. Phys Rep. **66**, 109 (1980)
47. A. Fischer, J. Marsden, The manifold of conformally equivalent metrics. Can. J. Math. **29**, 193–209 (1977)
48. C. Kiefer, *Quantum Gravity* (Oxford University Press, Oxford, 2012)
49. H.D. Zeh, *The Physical Basis of the Direction of Time* (Springer, Berlin, 1992)

Quantum Hierarchical Systems: Fluctuation Force by Coarse-Graining, Decoherence by Correlation Noise

Bei-Lok Hu

–Dedicated to the memory of Professor Heinz Dieter Zeh who made invaluable contributions to the foundational issues of quantum mechanics [1] and pioneered the environment-induced quantum decoherence program [2].
–From Quantum to Classical: Essays in Memory of Dieter Zeh, edited by C. Kiefer, in "Fundamental Theories of Physics" (Springer, Cham, 2021).

1 Democratic, Autocratic and Hierarchical Systems

A more familiar set of nomenclature is, respectively, "Closed, Open and Effectively Open" systems. These two sets differ in their characterization but there are correspondences between them. The title set refers to the system's constituents: democratic means all constituents are equal, autocratic means there is a distinguished party (or a few in the case of oligarchic) and the rest are irrelevant (the 'peasants'). The latter nomenclature refers to the degree the system is influenced by its environment: a closed system has no environment, an open system has [3]. In an autocratic system, it is easy to identify the chosen (few) as the distinguished and the rest acting as its environment, in the sense that the distinguished does not care about the details of the 'irrelevant' individuals, perhaps only their collective opinion in a very coarse way, as noise (maybe not completely white, but somewhat colored). Thus in this sense an autocratic system can easily be dealt with in the manner of an open system and the distinguished party's behavior described by the Langevin dynamics. For a closed democratic system, all constituents are identical. Consider for example a mole of monatomic gas. The question is, how to describe the macroscopic features of this gas, made up of a huge number of identical particles, from the microscopic dynamics of each of these particles, knowing that the micro and macro behaviors are fundamentally different. Even possessing the details about the positions and momenta of

B.-L. Hu (✉)
University of Maryland, College Park, MD, USA
e-mail: blhu@umd.edu

C. Kiefer (ed.), *From Quantum to Classical*, Fundamental Theories of Physics 204,
https://doi.org/10.1007/978-3-030-88781-0_9

each particle one cannot straightforwardly predict the dissipative macro-dynamics and the appearance of an 'arrow of time' from the unitary micro-dynamics of these particles.

This is a daunting challenge Boltzmann took upon himself and resolved with brilliant ideas: First, introduce a hierarchical representation of the nth order correlation functions of the molecules. Then focus on a lower order correlation function—the lowest being the one-particle distribution function. It is driven by the higher order correlation functions—the lowest being the two point function, which gives a reasonable description of a dilute gas. Now introduce the molecular chaos assumption, namely, assume that the colliding partners are uncorrelated initially—we shall refer to this 'causal factorization condition' as 'slaving'. This leads to the Boltzmann equation which depicts the dissipative dynamics of a system of (dilute) molecular gas.

Using a hierarchical representation of this closed system and focusing attention on a certain sector (the lower order correlations) while including the influence of another sector (the higher order correlations) under certain physical conditions, we can treat this otherwise closed system in an open system manner. It is in this sense that we can call it an 'effectively open' system. How to extract useful information about a closed democratic system with no obvious way to distinguish one constituent from another, nor are there conservation laws governing certain kind of special variables (e.g., the hydrodynamic variables)—this is the question we pose in this essay, a study of closed quantum systems by hierarchical representations. With the inspiration of Boltzmann and Langevin, we shall study how a hierarchical order introduced to a closed democratic system—defined either by substance or by representation—can indeed facilitate our understanding of how macro-behaviors arise from micro-dynamics.

The remainder of this section summarizes the two major paradigms of nonequilibrium (NEq) statistical mechanics, a la Langevin and Boltzmann (see Chap. 1 of [4]), and the key ideas behind open quantum systems: how persistent structures emerge from judicious choices of coarse-graining and non-Markovianity from self-consistent backreaction. This will set the stage for the central themes of this paper, a hierarchical representation of a closed system rendering it an effectively open system, and hierarchical coarse-graining in understanding natural phenomena in a systematic way.

In the hierarchical conceptual framework, we shall give two examples showing how ordered coarse-graining in a systematic manner offers ways to (1) calculate the quantum fluctuation force [5], the Casimir-Polder force [6] in our example, between an atom and a dielectric material; (2) derive the effective noise from quantum fluctuations of the higher order correlations in the Schwinger-Dyson hierarchy of equations for an interacting quantum field theory which can act to decohere the lower order correlations. Sections 2 and 3 are about closed quantum systems, and Sect. 5 is devoted to this second task. Beyond the questions of dissipation and time-arrow Boltzmann asked of closed classical systems, an equally fundamental issue is, how does classical behavior appear in a closed quantum system? This issue, often referred to as decoherence [7], has been explored extensively from the 80s and 90s under the consistent/decoherent history conceptual framework and the environment-induced decoherence paradigm. To motivate these issues we shall focus in Sect. 2 on the special roles of the collective and hydrodynamics variables. Section 3 discusses hierarchical systems in both structural and representational senses. We show their particularly adapt characteristics for addressing these issues in closed quantum systems. The derivation of the correlation noise from the nPI (master) effective action is left to Sect. 5.

Section 4 introduces hierarchical coarse-graining (CG) and discusses how, for the same set of variables or components in a closed system, different ordering of CG yields different results. We use the calculation of a fluctuation force, the Casimir-Polder force, as example. In the atom-field-medium system of interest we see five variables associated with the three parties involved: the atom has two and the medium has two. The primary action is between the atom's internal degree of freedom (dof) interacting with a quantum field which is modified by the presence of a dielectric. We show how an orderly coarse-graining over these five variables (see Table 1) by means of the *graded influence action* yields the Casimir-Polder force. We use this exactly controllable example with unambiguous results to illustrate the idea of hierarchical coarse-graining[1] which has deeper implications and broader scope of applications.

1.1 Open Systems: Coarse-Graining and Backreaction

The first order of business in treating an open system is to *identify* which subsystem or which variables (the relevant sector) are of special interest and whether the remainder (the irrelevant sector) constitutes an environment whose details are of lesser importance, and only a coarse-grained description is sufficient to summarize its overall effects on the system of special interest. To make a sensible distinction involves recognizing and devising a set of criteria to *separate* the relevant from the irrelevant variables. This procedure is simplified when the two sets of variables possess very different characteristic time or length or energy scales or interaction strengths. An example is the separation of slow-fast or heavy-light variables as in the Born-Oppenheimer approximation in molecular physics where the nuclear variables are assumed to enter adiabatically as parameters in the electronic wave function. Similar separation is possible in quantum cosmology between the 'heavy' gravitational sector characterized by the Planck mass and the 'light' matter sector [9]. In statistical physics this separation can be made formally with projection operator techniques [10]. This usually results in a nonlinear integro-differential equation for the relevant variables, which contains the causal and correlational information from their interaction with the irrelevant variables.

[1] Looking up the web, I found this terminology appears most often in relation to biological systems. Ideas akin to this, likely broader in scope and in variety, for the description of real and perceived structures in the physical world and adaptive systems can be found in this interesting paper [8].

Apart from finding some way of separating the overall closed system into a 'relevant' part of primary interest (the open system) and an 'irrelevant' part of secondary interest (the environment) in order to render calculations possible, one also needs to devise some *averaging* scheme to reduce or reconstitute the detailed information of the environment such that its effect on the system can be represented by some macroscopic functions, such as the transport or response functions. This involves introducing certain **coarse-graining** measures. It is usually by the imposition of such measures to varying degrees that an environment is turned into a bath, and certain macroscopic characteristics such as temperature and chemical potential can be introduced to simplify its description.

How the environment affects the open system is determined by the **back-reaction** effects. By referring to an effect as backreaction, it is implicitly assumed that a system of interest is preferentially identified, that one cares much less about the details of the other sector (the 'irrelevant' variables in the 'environment'). The backreaction can be significant, but should not be too overpowering, so as to invalidate the separation scheme. To what extent one views the interplay of the two sectors as *interaction* (between two subsystems of approximately equal weight) or as *backreaction* (of a less relevant environment on the more important system) is reflexive of and determined by the degree one decides to keep or discard the information in one subsystem versus the other. It also depends on their interaction strength. Through reaction and backreaction the behavior of each sector is linked to the other in an inseparable way, i.e., by their interplay.

Self-consistency is thus a necessary requirement in backreaction considerations. This condition is imbued in the fluctuation-dissipation theorem (FDT). When the environment is a bath, for systems near equilibrium, their response can be depicted by linear response theory. Even though such relations are usually presented in such a context, its existence in a more general form can be shown to cover nonequilibrium systems. Indeed as long as backreaction is included, such a relation can be understood as a corollary of the self-consistency requirement, which ultimately can be traced to the unitary condition of the original closed system.

A familiar example of self-consistent backreaction process is the time-dependent Hartree-Fock approximation in atomic physics or nuclear physics, where the system could be described by the wave function of the electrons obeying the Schrödinger equations with a potential determined by the charge density of the electrons themselves via the Poisson equation. In a cosmological backreaction problem, one can view the system as a classical spacetime, whose dynamics is determined by Einstein's equations with sources given by particles produced by the vacuum excited by the dynamics of curved spacetime and depicted by the appropriate wave equations in this spacetime.

Much of the physics of open systems is concerned with the appropriateness in the devising and the implementation of these procedures, namely, *separation, coarse-graining* and *backreaction*. (1) the identification and separation of the physically interesting variables which make up the open system, (2) the 'averaging' over of the environment or the projecting out of the irrelevant variables—how different coarse-graining measures affect the final result is important (we give an example in Sect. 4), and (3) the evaluation of the averaged effect—the backreaction—of the environment on the system of interest.

These considerations surrounding an open system are common and essential not only to well-posed and well-studied examples of many-body systems like molecular, nuclear and condensed-matter physics, they also bear on some basic issues at the foundations of quantum mechanics and statistical mechanics, such as decoherence and the emergence of classical behaviors, with profound implications on the emergence of time, spacetime, even quantum mechanics as an emergent theory [11, 12].

1.2 From Closed to Effectively Open Systems

There are many systems in nature which are apparently closed (to the observer), in that there is no obvious way to define a system which is so much different from an environment. These systems do not possess any parameter which can enable the observer to distinguish possible heavy-light sectors, high-low frequency behavior or slow-fast dynamics. Boltzmann's theory of molecular gas is a simple good example: All molecules in the gas are on equal footing, in that no one can claim to be more special than the others. Because of the lack of separation parameters which marks the discrepancy of one component from the other, these systems do not lend themselves to an obvious or explicit system—environment divide like open systems would, and appear like closed. However, usually in their effective description a separation is introduced implicitly or operationally because of their restricted appearance or due to the imprecision in one's measurement. These are called effectively open systems. Boltzmann's treatment of molecular gas, how he "effectively opened up" a closed democratic system, is a sterling example we shall follow to address a number of important issues in closed quantum systems.

1.3 Paradigms of Nonequilibrium Statistical Mechanics

We can highlight the distinction between open and effectively open systems by comparing the two primary models which characterize these two major paradigms of nonequilibrium statistical mechanics: the Boltzmann- Bogoliubov-Born-Green-Kirkwood-Yvon (BBGKY) hierarchy theory [13] of molecular kinetics, and the Langevin (Einstein-Smoluchowski) theory [14] of Brownian motions. The difference between the two are of both formal and conceptual nature.

To begin with, the *setup* of the problem is different: As we remarked above, in kinetic theory one studies the overall dynamics of a system of gas molecules, treating each molecule in the system on the same footing, while in Brownian motion one (Brownian) particle which defines the system is distinct, the rest is relegated to be the environment. The terminology of 'relevant' versus 'irrelevant' variables not so subtly reflects the discrepancy.

The *object* of interest in kinetic theory is the one-particle distribution function (or the nth-order correlation function), while in Brownian motion it is the reduced density matrix. The emphasis in the former is the behavior of the gas as a whole (e.g., dissipative dynamics) taking into account the correlations amongst the particles, while in the latter is the motion of the Brownian particle under the influence of its environment.

The nature of *coarse-graining* is also very different: In Brownian motion it is in the integration over the environmental variables. In kinetic theory coarse-graining is a bit more subtle. It resides in confining ones attention to one particle distribution functions, assuming a factorization condition for the two point functions under the molecular chaos assumption. This corresponds formally to (a) a truncation of the BBGKY hierarchy, and (b) an introduction of a causal factorization condition, the combination of these two steps we refer to as "slaving", the pivotal process which renders a closed system effectively open. Note that truncation would only yield two disjoint partitions, each one being a smaller closed system separated from the complete hierarchy. The part that is largely 'ignored' is where the noise comes from, while its main physical effect on the 'system' is to impart a dissipative component in its dynamics and for a quantum system, acting as the source of decoherence [15].

Finally the *philosophies* behind these two paradigms are quite different: In problems which can be modeled by Brownian motion, the separation of the system from the environment is prescribed: it is usually determined by some clear disparity between the two systems. In closed systems like a molecular gas where there is no clear discrepancy in sight, making a convenient yet physically unjustified separation often leads to wrong predictions. Coarse-graining in Boltzmann's kinetic theory is also very different from that of Brownian motion. The latter is explicit and thus easily identified while the former is more subtle, as the system has its own innate systematics [16]. However, as we shall see in Sect. 5, one effective way of coarse-graining a la Boltzmann lies in the 'slaving' procedures, where information attached to higher correlation orders is not kept in full, but represented by the so-called 'correlation noise', which turns a Boltzmann equation into a stochastic Boltzmann equation [17]. Now, just what correlation order is sufficient for the physics under study is an objectively definable and verifiable act, which ultimately is determined by the degree of precision in a measurement and judged by how good it depicts the relevant physics.

2 Collective and Hydrodynamic Variables

In treating physical systems containing many micro degrees of freedom a meaningful challenge is to design a small set of meso or macro variables to render the problem at the micro level technically tractable while preserving its physical essence. Instead of variables one can also seek to reformulate the theory such that in the domain of interest, e.g., at low energy, an effective theory gives an accurate enough description. In order to construct a simple and accurate enough theory valid at a specified range (of mass, energy, length or time scales) the challenge comes first and foremost in the design of such sets of variables. Familiar examples abound, e.g., thermodynamics from statistical mechanics, hydrodynamic limit of kinetic theory, collective dynamics in condensed matter and nuclear physics. Bear in mind that a coarse-grained description in terms of thermodynamic or hydrodynamic variables of a collective nature and their associated response functions is qualitatively very different from the detailed description in terms of the underlying microscopic variables and dynamics. E.g., the notion of temperature and pressure in thermodynamics makes no sense for each molecule. A theory of collective variables which emerges from the microscopic theories may bear little resemblance to them.

We comment on two such kinds of variables and the effective theories they bring out, namely, the collective and the hydrodynamic variables. We begin with the collective variables which are general and broadly used, we then look at a special class, the hydrodynamic variables, which has been studied in great detail in the 1990s in the context of consistent/decoherent histories. Because of the special properties they possess, namely, obeying conservation laws, they have a clear advantage in becoming 'habitually' decohered in the quasi-classical domains which emerge.

2.1 Collective Variables

Collective variables are derived variables, in the sense that they are not elemental (even quarks and gluons may one day be shown to be composites), and the theories these variables describe are effective theories. When we refer to one branch of physics as particle physics and some other branch of physics as molecular physics, we are already invoking the concepts of collective variables. In studying molecular forces we rarely need to include quark-gluon processes, much less string theory. Molecular theories are effective theories constructed from atomic wavefunctions. Knowing QED and the atomic composition of a molecule helps to construct viable molecular theories for the properties of molecular forces and the dynamics of molecular interactions. Deriving nuclear forces from QCD faces the same kind but more difficult challenge.

In order to construct a certain set of collective variables to serve a specific purpose, one often needs to have some idea in what type of physics are of special interest, and in what parameter ranges. E.g., the many 'ons'—phonon, roton, plasmon etc.—in condensed matter physics are collective excitations, thus it is the quantized normal modes of vibration or rotation of many atoms or the excitation modes of a plasma. Distinguishing these specific excitation modes by different collective variables serves targeted functions. These variables obey equations of motion or possess symmetry properties very different from that of the basic constituents (atoms) they are constructed from. In fact, the collective physics they describe often ceases to make sense when one examines it at the atomic scale. This is true of the hydrodynamic variables we shall discuss in the next subsection.

Consider one very common type of collective variable, the center of mass (CoM) of an assembly of particles. It is defined in such a way that when acted upon by some external force the CoM of a system of particles with total mass M moves like a single particle of mass M. It is a collective variable, not for the description of some type of collective excitation, but as a configuration variable constructed to simplify the description of the dynamics of the collective. Let us examine the role this CoM variable plays and the specific conditions which make it special playing this role in two well-known processes: (1) decoherence of a dust particle in a gas—why does it decohere so fast? and (2) the special role of the CoM in representing the quantum properties of a macroscopic object.

2.1.1 Decoherence in a Gas Environment

A quantum object collides with many particles in its ambient environment. We are interested in how fast this object decoheres and begins to behave classically. After the pioneering work of Joos and Zeh (in [2]), many authors have studied this problem, e.g., [18–22]. Take the calculation of Hornberger as example. In calculating the S-matrix he made a reasonable assumption that, in general, a collision keeps the center-of-mass invariant, and only the relative coordinates are affected. Results become simplified in the limit $m/M \to 0$, namely, when the dust particle's mass M is much bigger than that m of the ambient particles. Putting in the typical mass of a dust particle in the interstellar medium and as background the cosmic microwave radiation, even assuming a very long relaxation time (the age of the universe) the width of a solitonic wave packet describing the center of mass is as small as 2 pm. He uses this example to show "the remarkable efficiency of the decoherence mechanism to induce classical behavior in the quantum state of macroscopic objects".

With the assumption that the mass M of the quantum object is much greater than m, that of the ambient particles, one can describe this object's dynamics as a kind of quantum Brownian motion, and view its decoherence in an open quantum system framework, as environment induced. Emphasizing the complex role an environment plays in decoherence Riedel and Zurek [23] used the same problem to illustrate the so-called quantum Darwinism idea [24]. Note, in contrast, this problem can be dealt with in the closed quantum system framework. No environment need be brought in to see the dust particle's decoherence. E.g., Dodd and Halliwell [21] cast the master equation for the dust particle in a form emphasizing the role of local densities. Being hydrodynamic variables they decohere habitually because of the conservation laws they obey. In contrast, in the decoherence More of this in the next subsection. Let us continue with how the collective variables can serve the function of depicting macroscopic properties from the micro-constituents, even quantum properties.

2.1.2 Macroscopic Quantum Phenomena: Center of Mass Axiom

What are the conditions upon which the quantum mechanical and statistical mechanical properties of a macroscopic object can be described adequately in terms of the kinematics and dynamics of its center-of-mass (CoM) variable? This is an implicit assumption made in many macroscopic quantum phenomena (MQP) investigations, namely, that the quantum mechanical behavior, such as quantum decoherence, fluctuations, dissipation and entanglement, of a macroscopic object, like the nanoelectromechanical oscillator, a mirror [25], or a C^{60} molecule [26], placed in interaction with an environment can be captured adequately by its CoM behavior. For convenience we refer to this as the 'CoM axiom'. This assertion is intuitively reasonable, as one might expect it to be true from normal- mode decompositions familiar in classical mechanics, but when the constituent particles interact with each other in addition to interacting with their common environment, it is not such a clear-cut result. We need to better define the conditions which make this possible and in most cases, not.

With the aim of assessing the validity of the CoM axiom the authors of [27] consider a system modeled by N harmonic oscillators interacting with an environment consisting of n harmonic oscillators and derive an exact non-Markovian master equation for such a system in a environment with arbitrary spectral density and temperature. These authors outlined a procedure to find a canonical transformation from the individual coordinates (x_i, p_i) to the collective coordinates $(\tilde{X}_i, \tilde{P}_i)$, $i = 1, ..., N$ where $\tilde{X}_1$, $\tilde{P}_1$ are the center-of-mass coordinate and momentum respectively. They considered a class of couplings between the system and the environment in the form $f(x_i)q_j$ (instead of the ordinarily assumed $x_i q_j$) and examined if and when the CoM coordinate dynamics separates from the reduced variable dynamics. They note that if the function $f(x)$ has the property $\sum_{i=1}^{N} f(x_i) = \tilde{f}(\tilde{X}_1) + g(\tilde{X}_2, ..., \tilde{X}_N)$, for example $f(x) = x$ or $f(x) = x^2$, one can split the coupling between the system and environment into couplings containing the CoM coordinate and the relative coordinates. They proceed by tracing over the environmental degrees of freedom q_i to obtain the influence action which characterizes the effect of the environment on the system.

However, as the authors of [27] emphasized, the coarse-graining made by tracing over the environmental variables q_i does not necessarily lead to the separation of the CoM and the relative variables in the effective action. When they are mixed up and can no longer be written as the sum of these two contributions, the form of the master equation will be no longer be so clean-cut—the dynamics of both the relative variable and the center-of-mass variable will be mixed-in.

For this CoM axiom to hold what these authors find is, at least for the quantum Brownian motion of a system of N harmonic oscillators in an n-oscillator bath, that the coupling between the system and the environment need be bi-linear, in the form $x_i q_j$, that the potential $V_{ij}(x_i - x_j)$ is independent of the center-of-mass coordinate. In that case, one can say that the quantum evolution of a macroscopic object in a general environment is completely described by the dynamics of the center-of-mass canonical variables $(\tilde{X}_1, \tilde{P}_1)$ obeying a non-Markovian master equation of the Hu-Paz-Zhang [28] type.

What is the role of collective variables in MQP? At least for the N harmonic oscillator model it says: (1) For certain types of coupling the center of mass (CoM) variable of an object composed of a large number of constituents does play a role in capturing the collective behavior of this object (2) For other types of coupling the environment-induced quantum statistical properties of the system such as decoherence and entanglement are generally more complicated.

There is much to be explored in MQP. An interesting related issue is the decoherence by internal degrees of freedom [29–31], especially if the internal dynamics is chaotic [32].

2.2 *Hydrodynamic Variables*

In the chronicle of the development of theories of decoherence, it is well-known that Joos and Zeh [2] and Zurek [33] in the 80s pioneered the environment-induced decoherence program which form the cornerstone for many important theories and experiments in quantum information and foundation developed in the following decade. What was lesser known is the theories of consistent histories of Griffiths [34] and Omnes [35], until Gell-Mann and Hartle [36] in the 90s greatly developed the decoherent histories formalism. The histories formulation of quantum mechanics is a fundamental advance: quantum histories are described by amplitudes, and they interfere. Only when they are decohered sufficiently can probabilities be used to describe a quasi-classical reality. The decoherent histories formulation of quantum mechanics is particularly adapt in treating closed quantum systems [37].

Following this vein and asking what conditions would be conducive to histories getting decohered, comes another important discovery in the 90s: hydrodynamical variables. Reason: they obey conservation laws. We shall give a quick summary of this because it is important for understanding the appearance of quasi-classical domains in quantum closed systems. I'll mention in passing that the same principle, symmetry, underlying the conservation laws, which singles out the hydrodynamic variables, is what allows for the existence of decoherence-free subspaces [38] in quantum information, albeit the latter is in the space of quantum states.

2.2.1 Hydrodynamic Variables and Conservation Laws

A group of papers addressing this issue appeared in the later half of the 90s [39–41]. In the context of the decoherent histories approach to quantum theory, Halliwell showed that a class of macroscopic configurations consisting of histories of local densities (number, momentum, energy) exhibits negligible interference. This follows from the close connection of the local densities with the corresponding exactly conserved (and thus exactly decoherent—exact decoherence implies exact conservation) quantities (which commute with the Hamiltonian), and also from the observation that the eigenstates of local densities (averaged over a sufficiently large volume) remain approximate eigenstates under time evolution. Histories of these variables will be approximately decoherent and that their probabilities will be strongly peaked about hydrodynamic equations.

Decoherence of hydrodynamic modes is considered by Calzetta and Hu [42] using the influence action. They show that long wavelength hydrodynamic modes are more readily decohered, a robust result following from general principles, independent of the details of the microscopic dynamics.

2.2.2 Harmonic Chain

The linear quantum oscillator chain is a solvable example where these principles and criteria can be tested out. Halliwell [43] showed that decoherence arises directly from the proximity of these variables to exactly conserved quantities which are exactly decoherent, and *not* from environmentally-induced decoherence. He further explored the approach to local equilibrium and the subsequent emergence of hydrodynamic equations for the local densities.

Another noteworthy aspect related to our central themes is coarse graining—the difference in outcomes from coarse-graining a local group versus a nonlocal group. For a quantum chain Brun and Hartle [44] found that noise, decoherence, and computational complexity favor locality over nonlocality for deterministic predictability.

3 Hierarchical Systems

Recall the challenge we posed: How to extract interesting physics at the meso or macro scale from a closed quantum system of democratic microscopic constituents. As we mentioned the hydrodynamic variables stand out, aided by the conservation laws. A more familiar approach is mean field theory, which is often used as a first order approximation to the full theory. It assumes that each independent particle interacts with an effective potential depicting the averaged effects of all the other particles. Vlasov equation for a plasma is of this nature. Mean field theory respects democracy in a nice way, all particles are treated equally, each contributes equally to and experiences the same potential. The mean field dynamics remain time-reversal invariant. A mean field description of quantum field theory incorporates the effects of quantum fluctuations through renormalized interaction potential and renormalized coupling constants in the system. However at this level of approximation the dynamics of the system is without dissipation and without noise. Thus it cannot help with the decoherence issue—in fact the background field decomposition commonly used in field theory skips around this issue as it is often assumed to be classical without asking how that comes about (see, e.g., [45]).

Because a closed democratic system has no environment, and no obvious scale differentiation, it is not easy to come up with some parameters to characterize a quantum closed democratic system. But we know of some creative ways to serve specific purposes. The large N expansion is very useful, starting with the Hartree approximation, it produces a pretty good semiclassical theory. The $\epsilon = 4 - D$ (D being the dimension of spacetime) expansion worked well in critical phenomena, and dimensional regularization (identifying ultraviolet divergences by targeting terms proportional to powers of 1/(D-4)) was invented to provide a proof of the renormalizability of non-Abelian gauge theories in a covariant way. Here, we wish to study how a hierarchical organization or representation can provide clues to understanding the behavior of a closed democratic system.

3.1 Structural Versus Representational Hierarchy

Many physical objects have well-defined or easily identifiable levels of structure. Physicists for centuries have been probing these levels of structure, layer by layer, from molecules to atoms to nucleus to nucleons to quarks and gluons, and more. What at one time was considered as elementary turned out to be composites. With this structural hierarchy it is easy to introduce approximations to focus on the activities of those particles of interest at a specific level (from nuclear/particle to atomic/molecular physics) and pretty much ignore the complex underlying structures. Ignore is an exaggeration of course, theoretically speaking, but in essence this is implemented operationally. By invoking the Born-Oppenheimer approximation to separate the nuclear motion from the electronic motion one carries a small electron to nucleus mass ratio. Or, to derive a low energy effective field theory, there are terms proportional to the light to heavy mass sector ratio we need to be mindful of [46]. Similarly for the gauge hierarchy problems or extra dimensional (Kaluza-Klein) unification theories. Physical systems with an innate structural hierarchy are relatively easy to handle.

Another kind of hierarchical ordering is more in the nature of representation than structural. Representation is created to facilitate a better organization of the information about the system which we can extract in an easier way. Consider the molecular gas. It is a closed democratic system—all particles are the same, there is no distinguished party in the system and there is no environment. How do we get a handle on it to perform some analysis? We need to introduce some scale of physical relevance. At this point we may immediately hear objections to the effect that any scale one interjects into the system is subjective, and the description which one gives depends on one's preferred choice in how to represent this system. When the perceptions are different how can everyone agree on a unique objective physical reality?

These are important issues to bear in mind. Indeed this issue is even more acute when we dwell on physical reality from quantum histories which interfere constantly, or, in quantum measurement, where the measuring device not just bear witness to, but also partakes of the 'collapse of the wave function' and shapes the observed reality. That is why we need to appeal to hydrodynamic variables, they give our classical minds some peace. Putting quantum worries aside for now, let us just focus on a classical closed democratic system. It is reassuring that there are types of representation which have realistic physical meanings and capable of producing verifiable or falsifiable consequences. Correlation (BBGKY) hierarchy is an example and a good reason why Boltzmann's arduous journey remains inspirational today.

3.2 Correlation Hierarchy

At the microscopic level (of molecular dynamics) all molecular movements are time-reversal invariant, not so at the macroscopic level—dissipation and violation of time-reversal invariance are observed. To reconcile this difference and understand the origin of dissipation Boltzmann came up with the idea that if only one particle distribution functions observed, and the molecular chaos assumption is imposed for binary collisions, then he can explain the origin of dissipation as a rather common macroscopic phenomena. The BBGKY hierarchy of succesive higher order correlation functions is a suitable way to systemize the information in the gas. It is only upon the truncation of the hierarchy and the imposition of a causal factorization condition—what we called 'slaving' (the re-expression of the higher correlation functions in terms of the lower ones, assuming that colliding partners are uncorrelated initially)—that the closed system represented by the full hierarchy is rendered effectively open, and with it enters dissipation. We shall show that dissipation is accompanied by noise accounting for the contributions of the higher order correlation functions. Thus the name 'correlation noise'.

The reason why we say a representation by the correlation hierarchy produces physically unambiguous results is the following: The 'freedom' in choosing which order of correlation one wishes to measure is entirely up to the experimenter. Oftentimes she is not quite free to choose because her observation is limited by the precision of her measuring devices. If her instrument can only measure the one particle distribution function, well, all information about two particle correlations is lost to her. This is so only for specific processes taking place under specific conditions. What is observed is not subjective, because the same results can be reproduced under the same conditions by anyone. Yes, observers with different degrees of precision will report on different phenomena. If I don't have access to a particle accelerator whether I believe the nucleons are elementary or are composites is totally irrelevant. Subjective views have no place here.

Now we can describe at least conceptually how decoherence in a closed quantum system is implemented in a hierarchical representation. Decoherence of the sector of lower (up to the nth) order correlation functions by the quantum fluctuations associated with the sector of higher (n+1th and above) order correlation functions can be represented by a correlation noise of the nth order. It is in this sense that a closed democratic system (say of N molecules) can be treated as an effectively open system, with the lower sector acting as the system of interest and the higher order sector its environment. The effective environment thus changes with the correlation order stipulated by the experimenter or observers at the upper limit of precision in her measurements. The coarse-graining carried out in a graded environment by correlation order is an example of representational hierarchical coarse-graining.

In the next section we shall consider another kind of hierarchy which has both structural and representational elements. We will see that not only the way we choose to organize the structural components in our system, but also the order we coarse-grain different components, lead to very different results. This is a type of structural hierarchical coarse-graining.

4 Hierarchical Coarse-Graining

We mentioned earlier that almost all macroscopic or mescoscopic objects in nature have hierarchical levels of structure, and as such, theories operating at a particular level are effective theories, results of coarse-graining over the underlying substructures. How good an effective theory is in its depiction of physical phenomena at a particular scale is usually determined by the appropriateness in the choice of the collective variables, the correctness in the choice of coarse-graining for the targeted goal, its extent in relation to the probing scale and the precision of measurement. We talked about the first (collective variables) and the third elements (measurement precision). Here we shall focus on the second element, coarse-graining, which to me is probably the most subtle yet important factor. The issues involved are quite extensive. I listed them in an essay for the 1993 Time-Asymmetry conference [47] (reproduced partly in Chap. 1 of [4]). One key issue concerns *persistent emergent structures* in a closed quantum system: How does the degree of persistence depend on (a) the choice of collective variables, (b) the CG chosen for each layer of substructure, (c) stability with respect to successive CG, and (d) robustness with respect to variations of CG measures? I can only partially address a subset of these issues here. (Similar concerns seem to be shared in [48] from the angle of 'quasi-classical coarse-graining'.)

When the basic constituents or the components of a system are known it is easier to tackle this problem because the substructures are already persistent. For concreteness let us take as example the closed quantum system with three basic components: an atom, a quantum field and a dielectric medium. When we say the correctness in the choice of coarse-graining for the targeted goal, we ask, which components shall be coarse-grained, and in what order. That depends on what we want. In the example below our target is the Casimir-Polder force between the atom and the dielectric. We know the whole atom feels the force but it is only its internal degrees of freedom (the electronic levels) which interact with the field, in fact, the fluctuations in the quantum field. Thus we need two variables for the atom: its external or mechanical dof, usually the center of mass (CoM). For the quantum field, we know its vacuum state is changed when there is a medium present, thus we need to know how the medium modifies the field. As for the medium, the dielectric modeled by oscillators on a lattice also has two sets of variables: the polarization vector and the bath variables which damp the motion of the medium oscillators, preventing run-away behavior. The polarizability is one of the many dynamical susceptibility functions describing the electromagnetic responses of different materials in different capacities. Ordinarily one may choose a shortcut which lumps the first two steps by starting with a phenomenological response function as is done in Lifshitz's macroscopic electrodynamics [49]—there sits an example of CG at the innermost layer.

To illustrate how different targets demand different CG, or the ordering of CG, let's leave the medium aside and just consider an atom in a quantum field. So in this case we have only three variables. Suppose we want to see how the motion of an atom affects its spontaneous emission rate—some may consider this effect as *motional decoherence* [50]. Of these three components, which one should we CG over? The field. The simple reason is because we are asking a question which involves only the two variables of the atom: how does the atom's CoM (edf) motion affect its electronic activities (idf). But more importantly, let's reason out the physics. When an atom moves, it picks up a different field configuration at a different time and space. This changing field configuration is what the atom's internal degrees of freedom respond to, resulting in motional decoherence.

An easy example for this moving atom- quantum field set up: Suppose we wish to calculate the *entanglement* between a moving two-level atom and a quantum field, which variable should be CG over? The atom's edf.

Going in the opposite, somewhat more complex, direction is *quantum friction* [51]. A neutral atom moving very close to the surface of some dielectric material disturbs the field in the medium. Unlike in the Casimir-Polder force where it is sufficient to consider a stationary medium modifying the field, in quantum friction, one can at best assume the whole system is in a *nonequilibrium* steady state. The field in the dielectric back-reacts in a highly non-Markovian manner on the atom and results in slowing the atom down. It is interesting to consider how CG should be carried out in an ordered manner, in stages, capturing the interplay of all the variables of the participating entities.

4.1 An Example: Atom-Field-Medium Interaction

The material in this section is excerpted from the work of Behunin and Hu [52] where the Casimir-Polder force is derived from a microscopic model with three entities: the atom, the quantum field and the medium, represented by five micro/meso-variables: two for the atom, two for the medium. We want to use the procedure laid out there to illustrate how hierarchical coarse-graining plays an essential role in the emergent physics. The ordering of coarse-graining procedures is embodied in the graded influence action. For readers who don't care about the technical details they can skip from the next subsection on, and just have a look at Table 1. For those who want more, you can find a similar derivation of the atomic forces in [53].

The action describing the entire system $S[\vec{z}, \vec{Q}, A^\mu, \vec{P}, \vec{X}_\nu]$ is the sum of eight terms: (We adopt the Einstein summation convention and natural units throughout $\hbar = c = k_B = 1$.)

$$S[\vec{z}, \vec{Q}, A^\mu, \vec{P}, \vec{X}_\nu] \equiv S_Z + S_Q + S_E + S_M + S_X + S_{int}^{AF} + S_{int}^{PF} + S_{int}^{PX}, \quad (1)$$

with five free actions pertaining to the five dynamical variables and three interaction actions. Here (1) S_Z is the action for the motion of the atom's center of mass with coordinate $\vec{z}$ and total mass M under the influence of an external potential $V[\vec{z}(\lambda)]$:

$$S_Z[\vec{z}] = \int_{t_i}^{t_f} dt \left[\frac{M}{2}\dot{\vec{z}}^2(t) - V[\vec{z}(t)]\right] \quad (2)$$

where t is the atom's worldline parameter. (2) The internal degrees of freedom of the atom are modeled by a three-dimensional harmonic oscillator with coordinate $\vec{Q}$ and natural frequency Ω with action

$$S_Q[\vec{Q}] = \frac{\mu}{2}\int_{t_i}^{t_f} dt \left[\dot{\vec{Q}}^2(t) - \Omega^2\vec{Q}^2(t)\right] \quad (3)$$

where μ is the oscillator's reduced mass. (3) The dynamics of the free photon field is described by S_E where 'E' stands for the electromagnetic field,

$$S_E[A^\mu] = \frac{1}{4}\int d^4x\ F^{\mu\nu}F_{\mu\nu}. \quad (4)$$

where $F_{\mu\nu} = \partial_\mu A_\nu - \partial_\nu A_\mu$ is the field strength tensor, with A_μ the photon field's vector potential and $\int d^4x = \int_{t_i}^{t_f} dt \int d^3x$. (4) The matter may be described by a continuous lattice of harmonic oscillators with natural frequency ω which is meant to model the polarization of the medium, the coordinate of each oscillator is described by the vector field, $\vec{P}$, with action

$$S_M[\vec{P}] = \frac{1}{2}\int_V d^4x[\dot{\vec{P}}^2(x) - \omega^2\vec{P}^2(x)]. \quad (5)$$

The subscript V on the integration denotes that the spatial integration is restricted to the volume containing the matter.

(5) Each oscillator comprising the matter field is coupled to a reservoir. The reservoir is composed of a collection of oscillators at each point with frequency dependent mass $I(\nu)$ and coordinates $\vec{X}_\nu$, with natural frequency ν

$$S_X[\vec{X}_\nu] = \frac{1}{2}\int_V d^4x \int d\nu\ I(\nu)[\dot{\vec{X}}_\nu^2(x) - \nu^2\vec{X}^2(x)]. \quad (6)$$

The interaction of these parties is specified by the three interaction actions. (6) The interaction between the internal degree of freedom (dof) of the atom and the field is

$$S_{\text{int}}^{AF}[A^{\mu}, \vec{Q}, \vec{z}] = q \int d\lambda\, Q^{i}(\lambda) E_{i}(\lambda, \vec{z}(\lambda)) \tag{7}$$

where q represents the electronic charge. For the case of the matter the dipole moment of each oscillator is coupled with the local electric field with a coupling Ω_P

$$S_{\text{int}}^{PF}[A^{\mu}, \vec{P}] = \Omega_P \int_V d^4x\, P_{\nu}^{i}(x) E_{i}(x). \tag{8}$$

For our model, which considers only the coupling of the electric field to the local polarization of the matter, there is no magnetic response for the medium and so the permeability can assume its vacuum value, μ_o, throughout. It should be noted that we have *not* included interactions among the elements of the dielectric which will provide a spatially local form for the permittivity in the macroscopic Maxwell's equations.

Finally, each oscillator composing the matter is coupled to a reservoir with frequency dependent 'charge' $g(\nu)$ which will provide dissipation and noise

$$S_{\text{int}}^{PX}[\vec{P}, \vec{X}_{\nu}] = \int_V d^4x \int_0^{\infty} d\nu\, g(\nu) P_i(x) X_{\nu}^{i}(x). \tag{9}$$

We can divide our considerations into two stages: In Stage I we will calculate the effect of the medium on the field. We then make a rest stop, show where the popular Lorentz, Drude and plasma models of the medium arise from, and where the semi-phenomenological 'Macroscopic QED' approach of Lifshitz [49] resides. In Stage II, we first find out how the medium-modified field interacts with the atom's internal degrees of freedom, and then derive the Casimir-Polder force the atom's center of mass (mechanical or external degree of freedom) experiences.

4.2 *The Medium and the Medium-Modified Field*

The 'medium' refers to the combined reservoir + matter system. A reservoir for the network of oscillators modeling the dielectric is needed to damp out their motion and keep them in a stationary state. Thus two steps need be taken in Stage I: (IA) how the reservoir alters the medium, (IB) how the medium alters the quantum field.

After step IA, namely, after coarse-graining over the reservoir the authors of [52] obtain the semiclassical equation of motion for the polarization vector of the matter,

$$\ddot{P}^k + \omega^2 P^k - \int_{t_i}^{t_f} dt' \mathfrak{G}^{ret}(t, t') P^k(t') = \xi_X^k(t). \tag{10}$$

where $\mathfrak{G}^{ret}$ is the retarded Green function of the reservoir field, and $\vec{\xi}_X$ is the classical stochastic force or noise driving the matter with probability distribution described by $\mathcal{P}[\vec{\xi}_X]$. Due to the Gaussianity of $\mathcal{P}[\vec{\xi}_X]$ all of its moments are specified by the mean $\langle \xi_X^j \rangle_{\xi_X} = 0$ and the variance

$$\langle \{\xi_X^j(x), \xi_X^k(x')\} \rangle_{\xi_X} = \delta^{jk} \mathfrak{G}_H(x, x'), \tag{11}$$

where $\mathfrak{G}_H$ is the Hadamard function of the reservoir field, and

$$\langle ... \rangle_{\xi_X} \equiv \int \mathcal{D}\vec{\xi}_X \mathcal{P}[\vec{\xi}_X](...). \tag{12}$$

Thus Behunin and Hu show how coarse-graining over the medium degrees of freedom leads to a permittivity and a classical stochastic source responsible for an additional induced component of field fluctuations. For the particular case of local fluctuations they found that the stochastic semiclassical action for the field takes the same form as the noninteracting field (with frequency dependent velocity) driven by an external current (stochastic force). (A detailed description of the field's microstate is unimportant in the description of the atom-surface force.) From the action we obtain the stochastic semiclassical equations of motion for the medium-modified field

$$\nabla \times \nabla \times \vec{A}(\omega', \vec{x}) - \omega'^2 (1 + \Omega_P^2 \tilde{g}_{ret}(\omega', \vec{x})) \vec{A}(\omega', \vec{x}) = i \Omega_P \omega' \tilde{g}_{ret}(\omega', \vec{x}) \vec{\xi}_X(\omega', \vec{x}), \tag{13}$$

where $\tilde{g}_{ret}$ is the (Fourier-transformed) retarded Green function of the medium and $\vec{\xi}_X(x)$ is the stochastic force from the fluctuations of the medium. By comparing with the macroscopic Maxwell's equations one can readily identify the permittivity as $\varepsilon(\omega', \vec{x}) = 1 + \Omega_P^2 \tilde{g}_{ret}(\omega', \vec{x})$ ($\vec{\xi}_M(x)$ drives the field in addition to the reservoir when the system is not in a steady-state).

We can take a rest stop now at the end of Stage I and take stock of what our micro-physics model delivers: For Ohmic spectral density of the reservoir the frequency-dependent permittivity corresponds with the *Lorentz oscillator model*

$$\varepsilon(\omega', \vec{x}) = 1 - \frac{\Omega_P^2}{\omega'(\omega' + i\gamma) - \tilde{\omega}^2} \qquad \vec{x} \in V. \tag{14}$$

When the restoring force of the matter vanishes we obtain the *Drude model* and the *plasma model* when $\gamma \to 0$ in addition. In this form it's clear that the matter-field coupling Ω_P can be interpreted as the medium's plasma frequency.

One can see from Eq. (13) a striking similarity with *Lifshitz's Macroscopic QED*. Due to the linearity of our theory this is not surprising. Indeed if one were to proceed from this point treating the field semiclassically and choosing all space to be filled with a dielectric material, albeit in vacuum this dielectric is fictitious, one would exactly reproduce the predictions of MQED using (13) and choosing the dielectric to be in a thermal state.[2]

4.3 Medium-Modified Quantum Field's Effect on the Atom

For the second stage, the task is to find out how this medium modified field affects the atom. This also involves two steps: (IIA) how the field interacts with the atom's internal degrees of freedom (here using an oscillator model), (IIB) calculate the force exerted on the atom (its mechanical or external degree of freedom) by coarse-graining over the atoms' internal degrees of freedom.

4.3.1 Medium-Modified Quantum Field

By coarse-graining over the medium-influenced field we can incorporate the averaged effect of the field on the atom's trajectory without a specific knowledge of the final field state, leading to the reduced density matrix for the atom

$$\rho_r(z_f, z'_f, \vec{Q}_f, \vec{Q}'_f; t_f) = \int_{CTP}^{\vec{z}_f, \vec{z}'_f} \mathcal{D}\vec{z} \int_{CTP}^{\vec{Q}_f, \vec{Q}'_f} \mathcal{D}\vec{Q} e^{\{i(S_Z[\vec{z}]-S_Z[\vec{z}']+S_Q[\vec{Q}]-S_Q[\vec{Q}'])\}}$$
$$\times \int \mathcal{D}A^\mu_f \int_{CTP}^{A^\mu_f, A^\mu_f} \mathcal{D}A^\mu \, e^{\{S^{AF}_{int}[\vec{z},\vec{Q},A^\mu]-S^{AF}_{int}[\vec{z}',\vec{Q}',A^{\mu'}]\}} \mathcal{F}_M[A^\mu, A^{\mu'}] \tag{15}$$

(CTP denotes that it is a closed-time-path [54] integral) where the first integral in the second line traces over the final field configurations.

For the assumed initially factorized state the path integrals over the field can be evaluated exactly if the initial state is Gaussian yielding the field-reduced influence functional,

$$\mathcal{F}_\xi[J^\mu, J^{\mu'}, \xi] = \int \mathcal{D}\vec{\xi}_X \mathcal{P}[\vec{\xi}_X] \int \mathcal{D}\vec{\xi}_M \mathcal{P}[\vec{\xi}_M] \exp\{i S^E_{IF}[J^\mu, J^{\mu'}, \xi]\} \tag{16}$$

expressed in terms of the influence action, S^E_{IF}, given by

$$S^E_{IF}[J^\mu, J^{\mu'}, \xi^j] = \int d^4x \int d^4x' J^{\mu-}(x)$$
$$\times \left[\tilde{D}^{ret}_{\mu\nu}(x, x')\big(J^{\nu+}(x') - \kappa^\nu_i \xi^i(x')\big) + \frac{i}{4} \tilde{D}^H_{\mu\nu}(x, x') J^{\nu-}(\vec{x}') \right] \tag{17}$$

[2] However, it is important to note that in this case the stochastic field ξ^j_X represents the fluctuations of the medium only and does not include the intrinsic fluctuations of the field. After the next level of coarse-graining described below the intrinsic quantum fluctuations of the field will enter which is different from the induced fluctuations from interaction with the dielectric medium.

where $\vec{\xi}(x) \equiv \vec{\xi}_M(x) + \int_{t_i}^{t_f} dt' \ \tilde{g}_{ret}(t,t')\vec{\xi}_X(t',\vec{x})$. The current density J^μ in (17) comes from the atom-field interaction and takes the explicit form

$$J^\mu(x) = -q \int d\lambda \ Q^i(\lambda)\kappa_i^\mu \delta^4(x^\alpha - z^\alpha(\lambda)) \tag{18}$$

where the derivative operator $\kappa_i^\mu = -\partial_0\eta_i^\mu + \partial_i\eta_0^\mu$ yields the electric field when contracted with the vector potential $E^i = \kappa_\mu^i A^\mu$ and also enforces current conservation $\partial_\mu\kappa_i^\mu f(x) = (-\partial_0\partial_i + \partial_i\partial_0)f(x) = 0$. The integral kernels $\tilde{D}_{\mu\nu}^{ret}$ and $\tilde{D}_{\mu\nu}^H$ are the retarded Green's and Hadamard function for the medium-altered electromagnetic field which result from solving the semiclassical equations of motion (for the retarded Green's function sourced by a delta function). The retarded Green's function describes the classical electrodynamical propagation of the field in the presence of the dielectric material, and the Hadamard function describes the field's intrinsic quantum fluctuations.

In the temporal gauge the semiclassical equation of motion for the field's retarded Green's function in the presence of a dielectric medium satisfies

$$\begin{aligned} &\epsilon_i^{ab}\epsilon_b^{mn}\partial_a\partial_m \tilde{D}_{nk}^{ret}(x,x') + \frac{\partial^2}{\partial t^2}\tilde{D}_{ik}^{ret}(x,x') \\ &+ \frac{\partial}{\partial t}\int_{t_{in}}^{t_f} dt_2 \ \tilde{g}_{ret}(t,t_2;\vec{x})\frac{\partial}{\partial t_2}\tilde{D}_{ik}^{ret}(t_2,\vec{x},x') = \delta_{ik}\delta^4(x-x') \end{aligned} \tag{19}$$

where ϵ_{abc} is the Levi-Civita symbol (Roman indices refer to spatial components). The solution to (19) gives the particular solution, A_j^P, to the semiclassical equation of motion (13) for the electromagnetic field

$$A_j^P(x) = -\int d^4x' \tilde{D}_{jk}^{ret}(x,x')\frac{\partial}{\partial t'}\xi^k(x'). \tag{20}$$

As was noted previously the Heisenberg equations of motion for the field operators take the same form as the semiclassical equation of motion because of the linear coupling

$$\epsilon_i^{ab}\epsilon_b^{mn}\partial_a\partial_m \hat{A}_n(x) + \frac{\partial^2}{\partial t^2}\hat{A}_i(x) + \frac{\partial}{\partial t}\int_{t_{in}}^{t_f} dt_2 \ \tilde{g}_{ret}(t,t_2;\vec{x})\frac{\partial}{\partial t_2}\hat{A}_i(t_2,\vec{x}) = \frac{\partial}{\partial t}\xi_i(x). \tag{21}$$

Therefore we can use the homogeneous solution to (21) to construct the symmetric two-point function of the field operators to give the Hadamard function

$$\tilde{D}_H^{ij}(x,x') = \langle\{\hat{A}_o^i(x), \hat{A}_o^j(x')\}\rangle. \tag{22}$$

Finally, the field-reduced density matrix takes the form

$$\rho_r(z_f, z'_f, \vec{Q}_f, \vec{Q}'_f; t_f) = \int_{CTP}^{\vec{z}_f, \vec{z}'_f} \mathcal{D}\vec{z} \int_{CTP}^{\vec{Q}_f, \vec{Q}'_f} \mathcal{D}\vec{Q}\; e^{i(S_Z[\vec{z}] + S_Q[\vec{Q}] - S_Z[\vec{z}'] - S_Q[\vec{Q}'])} \times \mathcal{F}_\xi[J^\mu, J^{\mu'}, \xi]. \tag{23}$$

4.3.2 C-P Force by Coarse-Graining the Atom's Internal Dof

At this point we have the density matrix that describes the dynamics of the atom's trajectory and its internal degrees of freedom under the influence of the medium-altered field. As with the previous parties it is only the averaged effect and not the microscopic details of the atom's internal degree of freedom which we need for the description of the force.

In our last tier of coarse-graining we trace over $\vec{Q}_f$ the oscillator's internal dof to obtain the oscillator-reduced density matrix which characterizes the dynamics of the atom's trajectory determined by its interaction with all remaining parties

$$\rho_Z(z_f, z'_f; t_f) = \int_{CTP}^{\vec{z}_f, \vec{z}'_f} \mathcal{D}\vec{z}\; e^{i(S_Z[\vec{z}] - S_Z[\vec{z}'])} \mathcal{F}_Z[\vec{z}, \vec{z}']. \tag{24}$$

The environmental influences are now packaged in the oscillator-reduced influence functional $\mathcal{F}_Z[\vec{z}, \vec{z}']$ with their back-action accounted for in a self-consistent manner. The coarse-grainings are summarized in Table 1 at each tier the influence of all lower tiers on the remaining degrees of freedom is packaged in the influence functional.

We proceed by evaluating $\mathcal{F}_Z[\vec{z}, \vec{z}']$ perturbatively to lowest order in the atom-field coupling. In the fully dynamical case an exact calculation is not possible because the back-action of the field on the dynamics of the atom's dipole moment enters as a third time derivative and includes multiple reflections between the dielectric medium and the atom. At leading order in powers of the atom-field coupling, an expression for the force can be found which neglects the radiation reaction to the atom's dipole moment and the effects of multiple reflections.

Table 1 Layers of coarse-graining and physics remaining at each tier

Tier	C-Grain	Infunc'l	Detailed physics	Remainder
0			atom 2 + field + matter 2	$\vec{z}, \vec{Q}, A^\mu, \vec{P}, \vec{X}_\nu$
1	Tr_X	$\mathcal{F}_X$	atom + field + polarization	$\vec{z}, \vec{Q}, A^\mu, \vec{P}$
2	Tr_P	$\mathcal{F}_M$	atom + field	$\vec{z}, \vec{Q}, A^\mu$
3	Tr_A	$\mathcal{F}_\xi$	atom's ext & int dof	$\vec{z}, \vec{Q}$
4	Tr_Q	$\mathcal{F}_Z$	atom's motion	$\vec{z}$

To ease the notational burden we define

$$\langle ... \rangle_o = \int_{-\infty}^{\infty} d\vec{Q}_f \int_{CTP}^{\vec{Q}_f, \vec{Q}_f} \mathcal{D}\vec{Q}\, e^{i(S_Q[\vec{Q}] - S_Q[\vec{Q}'])}(...) \tag{25}$$

which represents the noninteracting time-dependent expectation value with respect to the oscillator's initial state. With this simplification the oscillator-reduced influence functional can be compactly expressed in terms of quantum and stochastic expectation values

$$e^{iS_{IF}[\vec{z}, \vec{z}']} \stackrel{def}{=} \mathcal{F}_Z[\vec{z}, \vec{z}'] = \left\langle\left\langle\left\langle e^{iS_{IF}^E} \right\rangle_o\right\rangle_{\xi_M}\right\rangle_{\xi_X} \tag{26}$$

which introduces the influence action, S_{IF}. With it we get the coarse-grained effective action, $S_{CGEA}[\vec{z}, \vec{z}'] \equiv S_Z[\vec{z}] - S_Z[\vec{z}'] + S_{IF}[\vec{z}, \vec{z}']$,

At the saddle point of (24) we get the semiclassical equation of motion for the atom's trajectory:

$$\left.\frac{\delta}{\delta z^{k-}(\tau)} S_{CGEA}[\vec{z}, \vec{z}']\right|_{z^{k-}=0} = 0 \Rightarrow M\ddot{z}_k(\tau) + \partial_k V[\vec{z}(\tau)] = f_k(\tau), \tag{27}$$

whereas the influence force, $f_k(\tau)$, is given by

$$\left.\frac{\delta}{\delta z^{k-}(\tau)} S_{IF}[\vec{z}, \vec{z}']\right|_{z^{k-}=0} = f_k(\tau), \tag{28}$$

which is the sought-after Casimir-Polder force.

5 Quantum Hierarchical System: Correlation Noise

In this section we implement what was described in Sect. 3, aiming at deriving the correlation noise from the higher order correlation functions in the Schwinger-Dyson hierarchy of an interacting quantum field. We first summarize the conceptual framework schematically, drawing the classical parallel of stochastic Boltzmann equation [17]. We then define our goal and present the methodology to reach that goal. Two subtopics used or related here are: The application of two-particle-irreducible (2PI) effective actions to quantum kinetic field theory [55], which is the starting point toward the construction of a master effective action [56]. For decoherent correlation histories, see [57, 58]. Further details of the subject matter covered here are in [59, 60]. (See also related work in, e.g., [61, 62].)

Finding an effective stochastic formulation to describe an interacting quantum system has a much broader scope and significance. Stochastic inflation [63] and stochastic gravity [64] are two noticeable themes in gravity and cosmology.

Conceptual Framework

(1) Classical: Boltzmann equation is the lowest order BBGKY hierarchy.
(2) BBGKY hierarchy governing the correlation functions $\Rightarrow$ Schwinger-Dyson hierarchy of equations (SDE) of interacting quantum field theory.
(3) Higher order correlation functions represented as a noise—the correlation noise. This is the key link.
(4) The inclusion of this noise is necessary by demand of the fluctuation-dissipation relation. This gives rise to a stochastic Schwinger-Dyson hierarchy, a QFT generalization of the stochastic BBGKY hierarchy, the lowest order being the stochastic Boltzmann equation.

Explanation: Points (1) and (2) As described earlier, in classical kinetic theory the full BBGKY hierarchy gives complete information of the closed system, say, of a molecular gas. It is upon (A) the truncation of the hierarchy and (B) the imposition of causal factorization conditions—the combination of these two procedures we call "slaving", the molecular chaos assumption being a familiar example—that the equations for the low order correlation functions (such as the Boltzmann equation for the one particle distribution function) acquires dissipative behavior.

Points (3) and (4). While the lower- order correlation functions constitute the system of interest, the slaved higher order correlation functions play the role of an environment of this effectively open system, their fluctuations being the source of noise in the stochastic Boltzmann equations. This stochastic generalization of the Boltzmann equation gives rise to a Boltzmann-Langevin equation and its field theoretical parallel is the stochastic Schwinger-Dyson equation.

Our goal is to obtain a self-contained dynamics for the propagators. What this means is that the propagators or correlation functions should carry in them the effect of their interaction with the higher correlation functions as embodied in the BBGKY or Dyson-Schwinger hierarchy. The procedure which makes this possible is slaving. (For Boltzmann, the imposition of the molecular chaos assumption). The necessary consequence is the appearance of dissipative behavior in the dynamics of the correlation functions, and, owing to the existence of a dissipation-fluctuation relation, the necessary existence of noise as well.

Pathway We start with classical kinetic theory and show a derivation of the stochastic Boltzmann equation. In the next subsection we shall sketch the main steps towards deriving a Boltzmann-Langevin for interacting QFT. The key link is correlation noise. The final stop is decoherence brought about by this noise, which results from the coarse-graining of the higher correlation functions. Again, for readers wary of too much technical details, here is a useful CG scheme for the rest of this section: After a glance of the classical stochastic Boltzmann Eqs. (48) with (50) and (46), go to Sect. 5.2 for the quantum field equivalent. All the formal essentials are there.

5.1 Boltzmann Equation for a Classical Relativistic Gas

Consider a dilute gas of relativistic classical particles. The system is described by its one particle distribution function $f(X, k)$, where X is a position variable, and k is a momentum variable. Momentum is assumed to lie on a mass shell $k^2 + M^2 = 0$. (We use the MTW convention, with signature –+++ for the background metric) and have positive energy $k^0 > 0$. In other words, given a spatial element $d\Sigma^\mu = n^\mu d\Sigma$ and a momentum space element d^4k, the number of particles with momentum k lying within that phase space volume element is

$$dn = -4\pi f(X, k)\,\theta\left(k^0\right)\delta\left(k^2 + M^2\right)k^\mu n_\mu\, d\Sigma\frac{d^4k}{(2\pi)^4} \tag{29}$$

The dynamics of the distribution function is given by the Boltzmann equation, in a notation adapted to our later needs:

$$k^\mu\frac{\partial}{\partial X^\mu}f(k) = I_{col}(X, k) \tag{30}$$

$$I_{col} = \frac{\lambda^2}{4}(2\pi)^3\int\left[\prod_{i=1}^{3}\frac{d^4p_i}{(2\pi)^4}\theta\left(p_i^0\right)\delta\left(p_i^2 + M^2\right)\right]\left[(2\pi)^4\,\delta\left(p_1 + p_2 - p_3 - k\right)\right]\mathbf{I} \tag{31}$$

$$\mathbf{I} = \{[1 + f(p_3)][1 + f(k)]f(p_1)f(p_2) - [1 + f(p_1)][1 + f(p_2)]f(p_3)f(k)\} \tag{32}$$

The entropy flux is given by

$$S^\mu(X) = \tag{33}$$
$$4\pi\int\frac{d^4p}{(2\pi)^4}\theta\left(p^0\right)\delta\left(p^2 + M^2\right)p^\mu\{[1 + f(p)]\ln[1 + f(p)] - f(p)\ln f(p)\}$$

while the entropy itself S is (minus) the integral of the flux over a Cauchy surface. Now consider a small deviation from the equilibrium distribution

$$f = f_{\rm eq} + \delta f \tag{34}$$

$$f_{\rm eq} = \frac{1}{e^{\beta p^0} - 1} \tag{35}$$

corresponding to the same particle and energy fluxes

$$\int \frac{d^4 p}{(2\pi)^4} \theta\left(p^0\right) \delta\left(p^2 + M^2\right) p^\mu \delta f(p) = 0 \tag{36}$$

$$\int \frac{d^4 p}{(2\pi)^4} \theta\left(p^0\right) \delta\left(p^2 + M^2\right) p^\mu p^0 \delta f(p) = 0 \tag{37}$$

Then the variation in entropy becomes

$$\delta S = -2\pi \int d^3 X \int \frac{d^4 p}{(2\pi)^4} \theta\left(p^0\right) \delta\left(p^2 + M^2\right) p^0 \frac{1}{\left[1 + f_{\text{eq}}(p)\right] f_{\text{eq}}(p)} (\delta f)^2 \tag{38}$$

In the classical theory, the distribution function is concentrated on the positive frequency mass shell. Therefore, it is convenient to label momenta just by its spatial components $\vec{p}$, the temporal component being necessarily $\omega_p = \sqrt{M^2 + \vec{p}^2} > 0$. In the same way, it is simplest to regard the distribution function as a function of the three momentum $\vec{p}$ alone, according to the rule

$$f^{(3)}(X, \vec{p}) = f\left[X, \left(\omega_p, \vec{p}\right)\right] \tag{39}$$

where f represents the distribution function as a function on four dimensional momentum space, and $f^{(3)}$ its restriction to three dimensional mass shell. With this understood, we shall henceforth drop the superscript, using the same symbol f for both functions, since only the distribution function on mass shell enters into our discussion. The variation of the entropy now reads

$$\delta S = -\frac{1}{2} \int d^3 X \int \frac{d^3 p}{(2\pi)^3} \frac{1}{\left[1 + f_{\text{eq}}(p)\right] f_{\text{eq}}(p)} (\delta f)^2 . \tag{40}$$

From Einstein's formula, we conclude that, in equilibrium, the distribution function is subject to Gaussian fluctuations, with equal time mean square value

$$\left\langle \delta f\left(t, \vec{X}, \vec{p}\right) \delta f\left(t, \vec{Y}, \vec{q}\right) \right\rangle = (2\pi)^3 \delta\left(\vec{X} - \vec{Y}\right) \delta\left(\vec{p} - \vec{q}\right) \left[1 + f_{\text{eq}}(p)\right] f_{\text{eq}}(p) \tag{41}$$

Note that this fluctuation formula is quite independent of the processes which sustain equilibrium; in particular, it holds equally for a free and an interacting gas, since it contains no coupling constants.

5.1.1 Stochastic Source to Sustain Fluctuations

In the interacting case, however, a stochastic source is necessary to sustain these fluctuations. Following the discussion of the fluctuatio-dissipation relation above, we compute these sources by writing the dissipative part of the equations of motion in terms of the thermodynamic forces

$$F\left(X,\vec{p}\right)=\frac{1}{\left[1+f_{\rm eq}\left(p\right)\right]f_{\rm eq}\left(p\right)}\frac{\delta f\left(X,\vec{p}\right)}{\left(2\pi\right)^{3}} \tag{42}$$

To obtain an equation of motion for $f\left(X,\vec{p}\right)$ multiply both sides of the Boltzmann equation (30) by $\theta\left(k^{0}\right)\delta\left(k^{2}+M^{2}\right)$ and integrate over k^{0} to get

$$\frac{\partial f}{\partial t}+\frac{\vec{k}}{\omega_{k}}\vec{\nabla}f=\frac{1}{\omega_{k}}I_{col} \tag{43}$$

Upon variation we get

$$\frac{\partial(\delta f)}{\partial t}+\frac{\vec{k}}{\omega_{k}}\vec{\nabla}(\delta f)=\frac{1}{\omega_{k}}\delta I_{col} \tag{44}$$

When we write δI_{col} in terms of the thermodynamic forces, we find local terms proportional to $F\left(k\right)$ as well as nonlocal terms where F is evaluated separately. We shall keep only the former, as it is usually done in deriving the "collision time approximation" to the Boltzmann equation (related to the Krook-Bhatnager-Gross kinetic equation), thus we write

$$\delta I_{col}\left(k\right)\sim-\omega_{k}\nu^{2}(X,\vec{k})F(X,\vec{k}) \tag{45}$$

where

$$\begin{aligned}\nu^{2}(X,\vec{k})=&\frac{\lambda^{2}}{4\omega_{k}}\left(2\pi\right)^{6}\int\left[\prod_{i=1}^{3}\frac{d^{4}p_{i}}{\left(2\pi\right)^{4}}\theta\left(p_{i}^{0}\right)\delta\left(p_{i}^{2}-M^{2}\right)\right]\\&\times\left[\left(2\pi\right)^{4}\delta\left(p_{1}+p_{2}-p_{3}-k\right)\right]I_{+},\end{aligned} \tag{46}$$

with $k^{0}=\omega_{k}$, and

$$I_{+}=\left[1+f_{\rm eq}\left(p_{1}\right)\right]\left[1+f_{\rm eq}\left(p_{2}\right)\right]f_{\rm eq}\left(p_{3}\right)f_{\rm eq}\left(k\right) \tag{47}$$

This linearized form of the Boltzmann equation provides a quick estimate of the relevant relaxation time. Let us assume the high temperature limit, where $f\sim T/M$, and the integrals in Eq. (46) are restricted to the range $p\leq M$. Then simple dimensional analysis yields the estimate $\tau\sim M/\lambda^{2}T^{2}$ for the relaxation time appropriate to long wavelength modes .

In [59] it is explained why the fluctuation-dissipation theorem (FDT) demands that a stochastic source j be present in the Boltzmann equation (30) (and its linearized form, Eq. (44)) which should assume the Langevin form:

$$\frac{\partial f}{\partial t}+\frac{\vec{k}}{\omega_{k}}\vec{\nabla}f=\frac{1}{\omega_{k}}I_{col}+j(X,\vec{k}) \tag{48}$$

Then

$$\langle j(X,\vec{p})\, j(Y,\vec{q})\rangle = -\left\{\frac{1}{\omega_p}\frac{\delta I_{col}(X,\vec{p})}{\delta F(Y,\vec{q})} + \frac{1}{\omega_q}\frac{\delta I_{col}(Y,\vec{q})}{\delta F(X,\vec{p})}\right\} \tag{49}$$

From Eqs. (45), (46) and (47) we find the noise auto-correlation

$$\left\langle j(X,\vec{k})j(Y,\vec{p})\right\rangle = 2\delta^{(4)}(X-Y)\,\delta\left(\vec{k}-\vec{p}\right)\nu^2(X,\vec{k}) \tag{50}$$

where ν^2 is given in Eq. (46). Equations (50) and (46) are the solution to our problem, that is, they describe the fluctuations in the Boltzmann equation, required by consistency with the FDT. Observe that, unlike Eq. (41), the mean square value of the stochastic force vanishes for a free gas.

In this discussion, of course, we accepted the Boltzmann equation as given without tracing its origin. We now want to see how the noises in Eq. (50) originate from a deeper level, that related to the higher correlation functions, which we call the correlation noises. We now turn to the statistical mechanics of interacting quantum fields.

5.2 Schwinger-Dyson Hierarchy for Interacting Quantum Fields

Consider a scalar field theory ϕ^a (we use DeWitt's condensed notation where the index a denotes both a spacetime point and one or the other branch of the closed-time-path (CTP) [54]) whose action is $S = S\left[\phi^1\right] - S^*\left[\phi^2\right]$. We shall use the two particle irreducible (2PI) representation [55, 65] where the (two-point) correlation function stands as an independent variable apart from the mean field. Thus there will be an additional separate source K_{ab} driving $\phi^a\phi^b$ over the usual $J_a\phi^a$ term in the 1PI representation. Our strategy is to seek a description of the field in terms of a new object $\mathbf{G}^{ab}$, namely, a stochastic correlation function whose expectation value over the noise average gives the usual two point functions reduces to G^{ab}, while its fluctuations reproduce the quantum fluctuations in the binary products of field operators. From the generating functional

$$Z[K_{ab}] = e^{iW[K_{ab}]} = \int D\phi^a\; e^{i\left\{S+\frac{1}{2}K_{ab}\phi^a\phi^b\right\}} \tag{51}$$

we have

$$G^{ab} = \left\langle\phi^a\phi^b\right\rangle = 2\left.\frac{\delta W}{\delta K_{ab}}\right|_{K=0} \tag{52}$$

and

$$\left.\frac{\delta^2 W}{\delta K_{ab}\delta K_{cd}}\right|_{K=0} = \frac{i}{4}\left\{\left\langle\phi^a\phi^b\phi^c\phi^d\right\rangle - \left\langle\phi^a\phi^b\right\rangle\left\langle\phi^c\phi^d\right\rangle\right\} \tag{53}$$

This suggests viewing the stochastic kernel $\mathbf{G}^{ab}$ as a Gaussian process defined (formally) by the relationships

$$\left\langle\mathbf{G}^{ab}\right\rangle - \left\langle\phi^a\phi^b\right\rangle; \qquad \left\langle\mathbf{G}^{ab}\mathbf{G}^{cd}\right\rangle = \left\langle\phi^a\phi^b\phi^c\phi^d\right\rangle \tag{54}$$

Or else, calling

$$\mathbf{G}^{ab} = G^{ab} + \Delta^{ab} \tag{55}$$

where

$$\left\langle\Delta^{ab}\right\rangle = 0; \qquad \left\langle\Delta^{ab}\Delta^{cd}\right\rangle = -4i\left.\frac{\delta^2 W}{\delta K_{ab}\delta K_{cd}}\right|_{K=0}. \tag{56}$$

5.2.1 Correlation Dynamics from 2PI Effective Action

We can define the process Δ^{ab} also in terms of a stochastic equation of motion. The Legendre transform of W is the two-particle-irreducible (2PI) effective action (EA)

$$\Gamma\left[G^{ab}\right] = W\left[K^*_{ab}\right] - \frac{1}{2}K^*_{ab}\mathbf{G}^{ab}; \qquad K^*_{ab} = -2\frac{\delta\Gamma}{\delta\mathbf{G}^{ab}} \tag{57}$$

We have the identities

$$\frac{\delta\Gamma}{\delta G^{ab}} = 0; \qquad \frac{\delta^2 W}{\delta K_{ab}\delta K_{cd}} = \frac{-1}{4}\left[\frac{\delta^2\Gamma}{\delta\mathbf{G}^{ab}\delta\mathbf{G}^{cd}}\right]^{-1} \tag{58}$$

the first of which is just the Schwinger-Dyson equation for the propagators; we therefore propose the following equations of motion for $ss\mathbf{G}^{ab}$

$$\frac{\delta\Gamma}{\delta\mathbf{G}^{ab}} = -\frac{1}{2}\kappa_{ab} \tag{59}$$

where κ_{ab} is a stochastic nonlocal Gaussian source defined by

$$\left\langle\kappa_{ab}\right\rangle = 0; \qquad \left\langle\kappa_{ab}\kappa_{cd}\right\rangle = 4i\left[\frac{\delta^2\Gamma}{\delta G^{ab}\delta G^{cd}}\right]^{\dagger} \tag{60}$$

If we linearize Eq. (59) around G^{ab}, then the correlation Eq. (60) for κ_{ab} implies Eq. (56) for Δ^{ab}. Consistent with our recipe of handling G^{ab} as if it were real we should treat κ_{ab} also as if it were a real source.

It is well known that the noiseless Eq. (59) can be used as a basis for the derivation of transport equations in the near equilibrium limit. Indeed, for a $\lambda\phi^4$ type theory, the resulting equation is simply the Boltzmann equation for a distribution function f defined from the Wigner transform of G^{ab}. It can be shown [59] that the full stochastic equation (59) leads, in the same limit, to a Boltzmann-Langevin equation, thus providing the microscopic basis for this equation in a manifestly relativistic quantum field theory.[3]

5.2.2 Dissipative and Stochastic Dynamics

Let us first examine some consequences of Eq. (60). For a free field theory, we can compute the 2PI EA explicitly

$$\Gamma\left[G^{ab}\right] = \frac{-i}{2}\ln\left[Det\mathbf{G}\right] - \frac{1}{2}c_{ab}\left(-\Box + m^2\right)\mathbf{G}^{ab}(x,x) \tag{61}$$

where c_{ab} is the 2×2 CTP metric tensor (see, e.g., [65]). We immediately find

$$\frac{\delta^2\Gamma}{\delta G^{ab}\delta G^{cd}} = \frac{i}{2}\left(G^{-1}\right)_{ac}\left(G^{-1}\right)_{db} \tag{62}$$

therefore

$$\left\langle \Delta^{ab}\Delta^{cd}\right\rangle = i\left[\frac{\delta^2\Gamma}{\delta G^{ab}\delta G^{cd}}\right]^{-1} = G^{ac}G^{db} + G^{da}G^{bc} \tag{63}$$

an eminently sensible result. Observe that the stochastic source does not vanish in this case, rather

$$\langle \kappa_{ab}\kappa_{cd}\rangle = G^{-1}_{ac}G^{-1}_{db} + G^{-1}_{da}G^{-1}_{bc} \tag{64}$$

[3] Constructing a stochastic representation of quantum field theory has caught increasing attention. Here are some noteworthy technical points about stochastic equations for the physical propagators, as exemplified in [60], which nicely completes the quest in [59]. In the 1PI theory, the ϕ^- field vanishes identically on-shell. However, in the stochastic approach we assign a nontrivial source ξ_-^{1PI} to it. It is by eliminating this auxiliary field that we recover the usual approach, with a single stochastic source $\tilde{\xi}$ whose self-correlation is given by the noise kernel. Similarly, in the quantum field theory problem the correlator $G^{--} = \langle\varphi^-\varphi^-\rangle$ vanishes identically, as a result of path ordering. However, in the stochastic approach, we consider it as an auxiliary field and couple a source to it. The authors of [59] failed to recognize the violation of the constraint $G^{--} = 0$, but guessed the right form for the noise self-correlation by introducing a sign change in their expression arising from adding a hermitian conjugation to the second derivative of the 2PI-EA. This was shown to be unnecessary by Calzetta [60] who provided a satisfactory explanation for this sign change: It is due to the elimination of the auxiliary field G^{--}, keeping only the physical degrees of freedom.

However

$$\left(G^{-1}\right)_{ac} \sim -ic_{ac}\left(-\Box + m^2\right) \tag{65}$$

does vanish on mass—shell. Therefore, when we take the kinetic theory limit, we shall find that for a free theory, there are no on—shell fluctuations of the distribution function. For an interacting theory this is no longer the case.

The physical reason for this different behavior is that the evolution of the distribution function for an interacting theory is dissipative, and therefore basic statistical mechanics considerations call for the presence of fluctuations. Indeed it is this kind of consideration which led us to think about a Boltzmann-Langevin equation in the first place.

Conceptually, as mentioned earlier, the correct approach is to view the two point functions as an effectively open system separated from, yet interacting with, the hierarchy of higher correlation functions obeying the set of Schwinger-Dyson equations. The averaged effect of its interaction with an environment of slaved higher irreducible correlations brings about dissipation and the attending fluctuations give rise to the correlation noise. This is the conceptual basis of our program laid out in [56], expounded in [59] and completed in [60].

5.3 *Master Effective Action*

We motivated the existence of a fluctuation term in the Boltzmann equation for an interacting quantum field with the 2PI effective action which include the contribution of both the one-particle distribution function and the two-point correlation function as independent variables, on the same footing. The generalization of this 2PIEA from n = 2 to $n \to \infty$ is called the master effective action [56], a functional of the whole string of Green functions of the interacting field theory whose variation generates the Dyson-Schwinger hierarchy. To complete our journey we give here a formal construction of the master effective action. Further details of nPI techniques can be found in [66, 67].

We consider then a scalar field theory whose action

$$S[\Phi] = \frac{1}{2}S_2\Phi^2 + S_{int}[\Phi] \tag{66}$$

decomposes into a free part and an interaction part

$$S_{int}[\Phi] = \sum_{n=3}^{\infty} \frac{1}{n!} S_n \Phi^n \tag{67}$$

All fields are to be defined on a closed-time-path and DeWitt's condensed notation is adopted. We also use the shorthand

$$K_n \Phi^n \equiv \int d^d x_1 ... d^d x_n \; K_{na^1...a^n}(x_1, ... x_n) \Phi^{a^1}(x_1) ... \Phi^{a^n}(x_n), \tag{68}$$

where the kernel K is assumed to be totally symmetric.

The 'source action' is given by

$$J[\Phi] = J_1 \Phi + \frac{1}{2} J_2 \Phi^2 + J_{int}[\Phi], \tag{69}$$

where $J_{int}[\Phi]$ contains the higher order sources

$$J_{int}[\Phi] = \sum_{n=3}^{\infty} \frac{1}{n!} J_n \Phi^n. \tag{70}$$

We define the generating functional

$$Z[\{J_n\}] = e^{iW[\{J_n\}]} = \int D\Phi \; e^{iS_t[\Phi, \{J_n\}]}, \tag{71}$$

where

$$S_t[\Phi, \{J_n\}] = J_1 \Phi + \frac{1}{2}(S_2 + J_2)\Phi^2 + S_{int}[\Phi] + J_{int}[\Phi]. \tag{72}$$

We shall also call

$$S_{int}[\Phi] + J_{int}[\Phi] = S_I. \tag{73}$$

As it is well known, the Taylor expansion of Z with respect to J_1 generates the expectation values of path—ordered products of fields

$$\frac{\delta^n Z}{\delta J_{1a^1}(x_1) ... \delta J_{1a^n}(x_n)} = \langle P\{\Phi^{a^1}(x_1) ... \Phi^{a^n}(x_n)\}\rangle \equiv F_n^{a^1...a^n}(x_1, ... x_n) \tag{74}$$

while the Taylor expansion of W generates the 'connected' Green functions (e.g., the 'linked cluster theorem' of Haag)

$$\frac{\delta^n W}{\delta J_{1a^1}(x_1) ... \delta J_{1a^n}(x_n)} = \langle P\{\Phi^{a^1}(x_1) ... \Phi^{a^n}(x_n)\}\rangle_{\text{connected}} \equiv C_n^{a^1...a^n}(x_1, ... x_n). \tag{75}$$

Comparing these last two equations, we find the rule connecting the F's with the C's. First, we must decompose the ordered index set $(i_1, ... i_n)$ $(i_k = (x_k, a^k))$ into all possible clusters P_n. A cluster is a partition of $(i_1, ... i_n)$ into N_{P_n} ordered subsets $p = (j_1, ... j_r)$. Then

$$F_n^{i_1 \dots i_n} = \sum_{P_n} \prod_p C_r^{j_1 \dots j_r}. \tag{76}$$

Now from the obvious identity

$$\frac{\delta Z}{\delta J_{n i_1 \dots i_n}} \equiv \frac{1}{n!} \frac{\delta^n Z}{\delta J_{i_1} \dots \delta J_{i_n}} \tag{77}$$

we obtain the chain of equations

$$\frac{\delta W}{\delta J_{n i_1 \dots i_n}} \equiv \frac{1}{n!} \sum_{P_n} \prod_p C_r^{j_1 \dots j_r}. \tag{78}$$

We can invert these equations to express the sources as functionals of the connected Green functions, and define the master effective action (MEA) as the full Legendre transform of the connected generating functional

$$\Gamma_\infty[\{C_r\}] = W[\{J_n\}] - \sum_n \frac{1}{n!} J_n \sum_{P_n} \prod_p C_r. \tag{79}$$

The physical case corresponds to the absence of external sources, whereby

$$\frac{\delta \Gamma_\infty[\{C_r\}]}{\delta C_s} = 0. \tag{80}$$

This hierarchy of equations is equivalent to the Dyson-Schwinger series.

In summary, in the hierarchical representation of a closed quantum system a closed-time-path master ($n = \infty PI$) effective action is constructed, whose functional variation generates the Schwinger-Dyson equations for the hierarchy of correlation functions. When the hierarchy is truncated, one obtains the ordinary closed system of correlation functions up to a certain order, which obey a set of time-reversal invariant equations of motion. But when the effect of the higher order correlation functions in the form of a correlation noise is included in one's consideration, the dynamics of the lower order correlations shows dissipative features. These two procedures: *'truncation'* versus *'slaving'* (or causal factorization—Boltzmann's molecular chaos assumption being a famous example) produce very different physical results. Truncation admits unitary evolution in the dynamical variables of both the lower sector and the higher sector of correlation functions, each can be regarded as a separate closed system; whereas slaving imparts dissipative dynamics in the lower-order correlation functions, as is well-known in the field-theory derivation of the Boltzmann equation by means of the 2PI effective action [55]. Furthermore, the existence of a fluctuation-dissipation relation for such an effectively open system mandates the existence of a stochastic term in the dynamical equation, representing the quantum fluctuations—the correlation noise—contributed by the higher order correlation functions. This is the origin of the stochastic Schwinger-Dyson equations, in parallel to the classical Boltzmann-Langevin equation, which depicts both the dissipative and stochastic dynamics of correlation functions in interacting quantum field theory.

5.4 Decoherence by Correlation Noise and Correlation Entropy

Having seen how correlation noise can be defined in a closed quantum system and how it imparts a stochastic component in the dissipative dynamics of the correlation functions, it is easy to see how decoherence comes about in a closed quantum system in a hierarchical representation. We mentioned earlier as the special role of the hydrodynamic variables which, thanks to the conservation laws they obey, habitually decohere and enable the emergence of relatively stable quasi-classical domains. Hierarchical representation offers the next best thing—the correlation noise captures the effect of the higher correlations which acts like an environment to the lower correlations which induces decoherence. One can calculate the decoherence time scales by examining the multi-time scales of this environment which is inherited from the higher correlations, or by directly solving the Boltzmann-Langevin equation.

A related physical quantity of fundamental importance is entropy. In the Boltzmann-BBGKY/Dyson-Schwinger framework Calzetta and Hu [68] proposed the definition of a *correlation entropy* of the nth order for an interacting quantum field. This is useful for addressing the thermalization of closed quantum systems and related quantum thermodynamic/kinetic theory issues. Using a CTP-2PI effective action under a next-to-leading-order large N approximation they proved an H theorem for the correlation entropy of a quantum-mechanical O(N) model.

Philosophically, the choice of hydrodynamical variables in the description of a closed quantum system of complex interactions is the key to the power and simplicity of a *hydrodynamics theory*, while a hierarchical representation of a closed quantum system focusing on the nth order correlations, whose dynamics is captured by the nPI effective action, evokes the perspective of *kinetic theory*. Moving from a hydrodynamic depiction to a hierarchical representation enables one to zoom in from the macroscopic to the mescoscopic domains of a closed system.

Acknowledgements The two key ideas expounded here: the correlation noise, is contained in prior work with Esteban Calzetta begun more than three decades ago, which he satisfactorily completed, while that of hierarchical coarse-graining, with Ryan Behunin, a decade ago and ongoing. I'm thankful for their invaluable contributions which deepened my understanding of these fundamental issues of physics.

References

1. H.D. Zeh, On the interpretation of measurement in quantum theory. Found. Phys. **1**(1), 69–76 (1970). H.D. Zeh, Toward a quantum theory of observation. Found. Phys. **3**(1), 109–116 (1973). H.D. Zeh, *The Direction of Time* (Springer, Berlin, 1989)
2. E. Joos, H.D. Zeh, The emergence of classical properties through interaction with the environment. Zeitschrift für Physik B Condensed Matter **59**(2), 223–243 (1985). D. Giulini, C. Kiefer, H.D. Zeh, Symmetries, super selection rules, and decoherence. Phys. Lett. A **199**(5–6), 291–298 (1995). H.D. Zeh, Roots and Fruits of Decoherence Seminaire Poincaré **2**, 1–19 (2005). arXiv:quant-ph/0512078v2
3. H.P. Breuer, F. Petruccione, *The Theory of Open Quantum Systems* (Oxford University Press, 2002). A. Rivas, S.F. Huelga, *Open Quantum Systems*, vol. 13 (Springer, Berlin, 2012)
4. E.A. Calzetta, B.-L.B. Hu, *Nonequilibrium Quantum Field Theory* (Cambridge University Press, 2008)
5. M. Kardar, R. Golestanian, The friction of vacuum, and other fluctuation-induced forces. Rev. Modern Phys. **71**(4), 1233 (1999). P.W. Milonni, *An Introduction to Quantum Optics and Quantum Fluctuations* (Oxford University Press, 2019)
6. H.B. Casimir, D. Polder, The influence of retardation on the London-van der Waals forces. Phys. Rev. **73**(4), 360 (1948)
7. E. Joos, H.D. Zeh, C. Kiefer, D.J. Giulini, J. Kupsch, I.O. Stamatescu, *Decoherence and the Appearance of a Classical World in Quantum Theory* (Springer Science & Business Media, 2013). M.A. Schlosshauer, *Decoherence: and the Quantum-to-Classical Transition* (Springer Science & Business Media, 2007)
8. M. Gell-Mann, J.B. Hartle, Adaptive coarse graining, environment, strong decoherence, and quasiclassical realms. Phys. Rev. A **89**(5), 052125 (2014)
9. J.P. Paz, S. Sinha, Decoherence and back reaction: the origin of the semiclassical Einstein equations. Phys. Rev. D 44(4), 1038 (1991). C. Kiefer, *The semiclassical approximation to quantum gravity, in Canonical Gravity: from Classical to Quantum* (Springer, Berlin, 1994), pp. 170–212
10. R. Zwanzig, *Nonequilibrium Statistical Mechanics* (Oxford University Press, 2001)
11. S.L. Adler, *Quantum Theory as an Emergent Phenomenon: the Statistical Mechanics of Matrix Models as the Precursor Of Quantum Field Theory* (Cambridge University Press, 2004)
12. G. 't Hooft, Emergent quantum mechanics and emergent symmetries, in *AIP Conference Proceedings American Institute of Physics*, vol. 957, no. 1 (2007), pp. 154–163
13. R.C. Balescu, *Equilibrium and Non-Equilibrium Statistical Mechanics (Wiley, 1975)* (H. Spohn, Large Scale Dynamics of Interacting Particles (Springer, 1991)
14. S. Chandrasekhar, Stochastic problems in physics and astronomy. Rev. Modern Phys. **15**(1), 1 (1943). N.G. Van Kampen, *Stochastic Processes in Physics and Chemistry*, vol. 1. (Elsevier, 1992)
15. M. Gell-Mann, J.B. Hartle, Classical equations for quantum systems. Phys. Rev. D **47**(8), 3345 (1993)
16. O. Kübler, H.D. Zeh, Dynamics of quantum correlations. Ann. Phys. **76**(2), 405–418 (1973)
17. M. Bixon, R. Zwanzig, Boltzmann-Langevin equation and hydrodynamic fluctuations. Phys. Rev. **187**(1), 267 (1969). R.F. Fox, G.E. Uhlenbeck, Contributions to nonequilibrium thermodynamics. II. Fluctuation theory for the Boltzmann equation. Phys. Fluids **13**(12), 2881–2890 (1970)
18. M.R. Gallis, G.N. Fleming, Environmental and spontaneous localization. Phys. Rev. A **42**(1), 38 (1990)
19. L. Diosi, Quantum master equation of a particle in a gas environment. EPL (Europhysics Letters) **30**(2), 63 (1995)
20. K. Hornberger, *Introduction to Decoherence Theory, in Entanglement and Decoherence* (Springer, Berlin, 2009), pp. 221–276
21. P.J. Dodd, J.J. Halliwell, Decoherence and records for the case of a scattering environment. Phys. Rev. D **67**(10), 105018 (2003)

22. J. Polonyi, Dissipation and decoherence by a homogeneous ideal gas. Phys. Rev. A **92**(4), 042111 (2015)
23. C.J. Riedel, W.H. Zurek, Quantum Darwinism in an everyday environment: Huge redundancy in scattered photons. Phys. Rev. Lett. **105**(2), 020404 (2010)
24. W.H. Zurek, Quantum Darwinism. Nat. Phys. **5**, 181–8 (2009). C.J. Riedel, W.H. Zurek, M. Zwolak, The rise and fall of redundancy in decoherence and quantum Darwinism. New J. Phys. **14**(8), 083010 (2012)
25. W. Marshall, C. Simon, R. Penrose, D. Bouwmeester, Towards quantum superpositions of a mirror. Phys. Rev. Lett. **91**(13), 130401 (2003)
26. M. Arndt, O. Nairz, J. Vos-Andreae, C. Keller, G. Van der Zouw, A. Zeilinger, Wave-particle duality of C^{60} molecules. Nature **401**(6754), 680–682 (1999)
27. C.H. Chou, B.L. Hu, T. Yu, Quantum Brownian motion of a macroscopic object in a general environment. Phys. A Stat. Mech. Appl. **387**(2–3), 432–444 (2008)
28. B.L. Hu, J.P. Paz, Y. Zhang, Quantum Brownian motion in a general environment: exact master equation with nonlocal dissipation and colored noise. Phys. Rev. D **45**(8), 2843 (1992)
29. T.A. Brun, L. Mlodinow, Decoherence by coupling to internal vibrational modes. Phys. Rev. A **94**(5), 052123 (2016)
30. M. Hillery, L. Mlodinow, V. Buek, Quantum interference with molecules: the role of internal states. Phys. Rev. A **71**(6), 062103 (2005)
31. J.C. Flores, Decoherence from internal degrees of freedom for clusters of mesoparticles: a hierarchy of master equations. J. Phys. A Math. Gen. **31**, 8623 (1988)
32. H.K. Park, S.W. Kim, Decoherence from chaotic internal dynamics in two coupled δ-function-kicked rotors. Phys. Rev. A **67**(6), 060102 (2003)
33. W.H. Zurek, Pointer basis of quantum apparatus: Into what mixture does the wave packet collapse? Phys. Rev. D **24**, 1516–1525 (1981). W.H. Zurek, Environment-induced superselection rules. Phys. Rev. D **26**(8), 1862 (1982). W.H. Zurek, Decoherence, einselection, and the quantum origins of the classical. Rev. Modern Phys. **75**(3), 715 (2003)
34. R.B. Griffiths, Consistent histories and the interpretation of quantum mechanics. J. Stat. Phys. **36**(1–2), 219–272 (1984). R.B. Griffiths, *Consistent Quantum Theory* (Cambridge University Press, 2003)
35. R. Omnes, Consistent interpretations of quantum mechanics. Rev. Modern Phys. **64**(2), 339 (1992). R. Omnes, *The Interpretation of Quantum Mechanics*, vol. 102 (Princeton University Press, 2018)
36. M. Gell-Mann, J.B. Hartle, Complexity, entropy and the physics of information. SFI Stud. Sci. Complex. **8**, 425 (1990). M. Gell-Mann, J.B. Hartle, in *Proceedings of the 4th Drexel Symposium on Quantum Non-Integrability: The Quantum-Classical Correspondence* edited by D.-H. Feng, B.-L. Hu (International Press, Boston, 1995)
37. J.B. Hartle, The quantum mechanics of closed systems, in *Directions in General Relativity*, vol. 1, ed. by B.-L. Hu, M.P. Ryan, C.V. Vishveshwara (Cambridge University Press, Cambridge, 1993), pp. 104–124. J.B. Hartle, The quasiclassical realms of this quantum universe. Found. Phys. **41**(6), 982–1006 (2011)
38. D.A. Lidar, I.L. Chuang, K.B. Whaley, Decoherence-free subspaces for quantum computation. Phys. Rev. Lett. **81**(12), 2594 (1998)
39. J.B. Hartle, R. Laflamme, D. Marolf, Conservation laws in the quantum mechanics of closed systems. Phys. Rev. D **51**(12), 7007 (1995)
40. J.J. Halliwell, Decoherent histories and hydrodynamic equations. Phys. Rev. D **58**(10), 105015 (1998)
41. J.J. Halliwell, Decoherent histories and the emergent classicality of local densities. Phys. Rev. Lett. **83**(13), 2481 (1999)
42. E.A. Calzetta, B.L. Hu, Influence action and decoherence of hydrodynamic modes. Phys. Rev. D **59**(6), 065018 (1999)
43. J.J. Halliwell, Decoherence of histories and hydrodynamic equations for a linear oscillator chain. Phys. Rev. D **68**(2), 025018 (2003)

44. T.A. Brun, J.B. Hartle, Classical dynamics of the quantum harmonic chain. Phys. Rev. D **60**(12), 123503 (1999)
45. E. Calzetta, B.L. Hu, Quantum fluctuations, decoherence of the mean field, and structure formation in the early universe. Phys. Rev. D **52**(12), 6770 (1995)
46. E. Calzetta, B.L. Hu, Stochastic behavior of effective field theories across the threshold. Phys. Rev. D **55**(6), 3536 (1997)
47. B.L. Hu, Fluctuation, dissipation and irreversibility in cosmology, in *The Physical Origin of Time-Asymmetry*, edited by J. J. Halliwell, J. Perez-Mercader, W.H. Zurek (Cambridge University, Cambridge, 1993), arxiv:gr-qc/9302021
48. M. Gell-Mann, J.B. Hartle, Quasiclassical coarse graining and thermodynamic entropy. Phys. Rev. A **76**(2), 022104 (2007)
49. E.M. Lifshitz, Sov. Phys. JETP 2(1), 73. 27. I.E. Dzyaloshinskii, E.M. Lifshitz, L.P. Pitaevskii, Adv. Phys. **10**(38), 165 (1961). S.Y. Buhmann, *Dispersion Forces I: Macroscopic Quantum Electrodynamics and Ground-State Casimir, Casimir-Polder and van der Waals Forces*, vol. 247 (Springer, 2013)
50. S. Shresta, B.L. Hu, Moving atom-field interaction: quantum motional decoherence and relaxation. Phys. Rev. A **68**(1), 012110 (2003)
51. J.B. Pendry, Shearing the vacuum-quantum friction. J. Phys. Condens. Matter, **9**(47), 10301 (1997). J.B. Pendry, Quantum friction- fact or fiction? New J. Phys. **12**(3), 033028 (2010). S. Buhmann, *Dispersion Forces II: Many-Body Effects, Excited Atoms, Finite Temperature and Quantum Friction*, vol. 248 (Springer, 2013). F. Intravaia, R.O. Behunin, D.A. Dalvit, Quantum friction and fluctuation theorems. Phys. Rev. A **89**(5), 050101 (2014)
52. R.O. Behunin, B.L. Hu, Nonequilibrium forces between atoms and dielectrics mediated by a quantum field. Phys. Rev. A **84**(1), 012902 (2011)
53. R.O. Behunin, B.L. Hu, Nonequilibrium forces between neutral atoms mediated by a quantum field. Phys. Rev. A **82**(2), 022507 (2010)
54. J. Schwinger, Brownian motion of a quantum oscillator. J. Math. Phys. **2**(3), 407–432 (1961). L.V. Keldysh, Diagram technique for nonequilibrium processes. Sov. Phys. JETP **20**(4), 1018–1026 (1965)
55. E. Calzetta, B.L. Hu, Nonequilibrium quantum fields: closed-time-path effective action, Wigner function, and Boltzmann equation. Phys. Rev. D **37**(10), 2878 (1988)
56. E. Calzetta, B.L. Hu, *Correlations, decoherence, dissipation, and noise in quantum field theory*, in *Discourses in the Mathematics and its Applications* No. 4, edited by S. A. Fulling (Texas A & M University Press, College Station, 1995) Arxiv:hep-th/9501040
57. E. Calzetta, B.L. Hu, Decoherence of correlation histories. Dir. Gen. Relat. **2**, 38–65 (1993)
58. C. Anastopoulos, Coarse grainings and irreversibility in quantum field theory. Phys. Rev. D **56**(2), 1009 (1997)
59. E. Calzetta, B.L. Hu, Stochastic dynamics of correlations in quantum field theory: from the Schwinger-Dyson to Boltzmann-Langevin equation. Phys. Rev. D **61**(2), 025012 (1999)
60. E. Calzetta, Fourth-order full quantum correlations from a Langevin-Schwinger-Dyson equation. J. Phys. A Math. Theor. **42**(26), 265401 (2009)
61. P.G. Reinhard, E. Suraud, Stochastic TDHF and the Boltzman-Langevin equation. Ann. Phys. **216**(1), 98–121 (1992)
62. C. Greiner, S. Leupold, Stochastic interpretation of Kadanoff-Baym equations (1998). arXiv:hep-ph/9809296
63. A.A. Starobinsky, Stochastic de Sitter (inflationary) stage in the early universe. Lect. Notes Phys. **246**, 107–126 (1986). V. Vennin, A.A. Starobinsky, Correlation functions in stochastic inflation. Eur. Phys. J. C **75**(9), 413 (2015)
64. B.L. Hu, E. Verdaguer, Stochastic gravity: theory and applications. Living Rev. Relat. **11**(1), 3 (2008). B.L. Hu, E. Verdaguer, Semiclassical and stochastic gravity: quantum field effects on curved spacetime (Cambridge University Press, 2020)
65. S.A. Ramsey, B.L. Hu, O (N) quantum fields in curved spacetime. Phys. Rev. D **56**(2), 661 (1997)

66. J. Berges, N-particle irreducible effective action techniques for gauge theories. Phys. Rev. D **70**(10), 105010 (2004)
67. M.E. Carrington, Techniques for calculations with nPI effective actions, in *EPJ Web of Conferences*, vol. 95 (EDP Sciences, 2015), p. 04013
68. E.A. Calzetta, B.L. Hu, Correlation entropy of an interacting quantum field and H theorem for the O (N) model. Phys. Rev. D **68**(6), 065027 (2003)

Decoherence and the Puzzle of Quantum Brownian Motion in a Gas

Lajos Diósi

1 Introduction

The long time of ignorance after Zeh's publications [1, 2] in 1970–71 on environmental decoherence got broken by Wigner. He summarized his own revelation and Zeh's discovery [3] as follows: *This writer's earlier belief that the physical apparatus' role can always be described by quantum mechanics [...] implied that the "collapse of the wave function" takes place only when the observation is made by a living being—a being clearly outside of the scope of our quantum mechanics. The arguments which convinced me that quantum mechanics' validity has narrower limitation, that it is not applicable to the description of the detailed behavior of macroscopic bodies, is due to D. Zeh. (1970) [...]. The point is that a macroscopic body's inner structure, i.e. its wave function, is influenced by its environment in a rather shot time even if it is in intergalactic space. Hence it cannot be an isolated system [...].* Wigner raised the question: *Can an equation for the time-change of the state of the apparently not-isolated system be proposed?*

Nowadays, after decades, the answer is part of the theory of open quantum systems. But in 1983, it was a novelty that Wigner showed a master equation for the massive object's density matrix, modeling the decoherence of its rotational motion:

$$\frac{d\hat{\rho}}{dt} = -\frac{i}{\hbar}[\hat{H},\hat{\rho}] - \sum_{\ell m} \varepsilon_\ell \left[\hat{L}_{\ell m}, [\hat{L}_{\ell m}, \hat{\rho}]\right] , \tag{1}$$

where $\hat{H}$ is the Hamiltonian of the macroscopic object and $\hat{L}_{\ell m}$ are the multipole operators of its angular momentum; strengths of their decoherence are given by the parameters ε_ℓ.

In 1985, Joos and Zeh (JZ) found [4] that decoherence of the center-of-mass position $\hat{\mathbf{x}}$ would be more typical and, what is important, its derivation is simple.

L. Diósi (✉)
Wigner Research Centre for Physics, 1525, P.O.Box 49, Budapest 114, Hungary
e-mail: diosi.lajos@wigner.hu

Eötvös Loránd University, Budapest, Hungary

C. Kiefer (ed.), *From Quantum to Classical*, Fundamental Theories of Physics 204,
https://doi.org/10.1007/978-3-030-88781-0_10

2 Equation of Positional Decoherence

Consider a macroscopic object, e.g. a dust particle, of mass M under the influence of incoming plane waves of particles, e.g. molecules, of mass $m \ll M$ that are scattered independently by the macroscopic object. JZ took the following unitary transition per single collisions, valid if $M \to \infty$:

$$|\mathbf{p}_i\rangle \otimes |\mathbf{k}_i\rangle \Rightarrow |\mathbf{p}_i\rangle \otimes |\mathbf{k}_i\rangle + \frac{i}{2\pi k_i} \int d\mathbf{k}_f f(\mathbf{k}_f, \mathbf{k}_i) \delta(k_f - k_i) \left|\mathbf{p}_f\right\rangle \otimes \left|\mathbf{k}_f\right\rangle, \quad (2)$$

$\mathbf{p}_{i/f}, \mathbf{k}_{i/f}$ are the initial/final momenta of the object and the particle, respectively, where $\mathbf{p}_f = \mathbf{p}_i + \mathbf{k}_i - \mathbf{k}_f$ ensures momentum conservation, and f is the standard scattering amplitude. The authors pointed out that repeated scatterings of the incoming particles contribute to gradual localization of the object, i.e., the off-diagonal terms of the positional density matrix $\rho(\mathbf{x}', \mathbf{x})$ become damped. If the distribution $\rho^{\mathcal{E}}(\mathbf{k}_i)$ of the incoming environmental particles is isotropic then the collisions contribute to the following master equation:

$$\frac{d\hat{\rho}}{dt} = -\frac{i}{\hbar}[\hat{H}, \hat{\rho}] - \Lambda[\hat{\mathbf{x}}, [\hat{\mathbf{x}}, \rho]]; \quad (3)$$

valid if the coherent extension of the object's position is much smaller than the wavelength of the particles:

$$|\mathbf{x}' - \mathbf{x}| \ll \hbar/k. \quad (4)$$

JZ determined the parameter Λ of localization rate:

$$\Lambda = \frac{1}{\hbar^2} \times \text{incoming flux of particles} \times k^2 \sigma_{\text{eff}}. \quad (5)$$

They calculated the effective cross section σ_{eff} from the differential cross section $|f|^2$.
The JZ master equation (3) is paradigmatic in decoherence theory. It describes the gradual damping of the off-diagonal elements of the positional density matrix $\rho(\mathbf{x}', \mathbf{x})$ which is called positional decoherence on one hand and yields localization of the coherent extension of the object on the other. In 1990 Gallis and Fleming [5] revisited the considerations of JZ and refined their derivation of positional decoherence and its rate Λ. It is not clear when, lately, was the localization rate Λ related to the classical diffusion coefficient for the first time. But the research moved to that direction.

3 Quantum Brownian Motion in a Gas

The JZ master equation (3) has an alternative interpretation, independent of and older than the concept of decoherence. It corresponds to momentum diffusion of the Brownian object, with the coefficient D_p of momentum diffusion:

$$\Lambda = \frac{D_p}{\hbar^2} \,. \tag{6}$$

As a beneficial consequence, to obtain and understand the dynamics of decoherence, also to complete the JZ master equation (3) by a term of friction, we could have used the standard quantum theory of Brownian motion in a gas. Just this standard theory did not exist at the time. And it has since remained problematic despite efforts of a community of researchers including myself. We all were motivated by our foundational interest in the quantum behavior of macroscopic objects under the influence of their uncontrollable environments. The efforts [6–10] started in 1995 and culminated in the Vacchini-Hornberger review [10]. These autors say ... *the seminal paper on decoherence by Joos and Zeh [...], seeking to explain the absence of quantum delocalization in a dust particle by the scattering of photons and air molecules, derived and studied what the authors called a Boltzmann-type master equation. Two decades later, the long quest for the characterization of the phenomenon of collisional decoherence has now reached a mature theoretical description, permitting its quantitative experimental confirmation.* Let me outline the story, in my—selective and certainly subjective— interpretation.

In 1995 [6], without mentioning my foundational motivations, I asked the question: what is the quantum Brownian dynamics of the dust in a dilute gas at thermal equilibrum? First I solved the classical problem by the linear variant of the classical Boltzmann-equation where the molecule-molecule collision term is just replaced by the dust-molecule collision term. Unlike the classical case, the derivation of the quantum linear Boltzmann equation (QLBE) was not straightforward. Quantum mechanically, a single collision corresponds to the following unitary transition, generalizing (2) for finite M:

$$|\mathbf{p}_i\rangle \otimes |\mathbf{k}_i\rangle \Rightarrow |\mathbf{p}_i\rangle \otimes |\mathbf{k}_i\rangle + \frac{i}{2\pi k_i^*} \int d\mathbf{k}_f^* f(\mathbf{k}_f^*, \mathbf{k}_i^*) \delta(k_f^* - k_i^*) \left|\mathbf{p}_f\right\rangle \otimes \left|\mathbf{k}_f\right\rangle \,, \tag{7}$$

where $\mathbf{k}_{i/f}^*$ are the initial/final momenta of the particle, respectively, in the center-of-mass frame. As before, $\mathbf{p}_f = \mathbf{p}_i + \mathbf{k}_i - \mathbf{k}_f$ ensures momentum conservation, total energy conservation is ensured by the delta-function. When imposing the distribution $\rho^{\mathcal{E}}(\mathbf{k}_i)$ of the gas molecule momenta, I had to introduce a heuristic *maneouvre of square-root* (MSqR); otherwise the correct mathematical structure [11, 12] of the desired quantum master equation wouldn't have been achieved. The MSqR was equivalent to a deliberate adding off-diagonal elements to the standard diagonal density matrix $\rho^{\mathcal{E}}(\mathbf{k}, \mathbf{k}') \propto \rho^{\mathcal{E}}(\mathbf{k})\delta(\mathbf{k} - \mathbf{k}')$ of the ideal gas molecules. The choice was

$$\rho^{\mathcal{E}}(\mathbf{k}, \mathbf{k}') = \sqrt{\rho^{\mathcal{E}}(\mathbf{k})}\sqrt{\rho^{\mathcal{E}}(\mathbf{k}')} \,. \tag{8}$$

This MSqR and some other simple assumptions led to the first QLBE of Brownian motion in a gas. In the diffusion limit it yields the quantum Fokker-Planck equation (QFPE):

$$\frac{d\hat{\rho}}{dt} = -\frac{i}{\hbar}[\hat{H}, \hat{\rho}] - \frac{D_p}{\hbar^2}[\hat{\mathbf{x}}, [\hat{\mathbf{x}}, \hat{\rho}]] - i\frac{\eta}{2\hbar}[\hat{\mathbf{x}}, \{\hat{\mathbf{p}}, \hat{\rho}\}] - \frac{D_x}{\hbar^2}[\hat{\mathbf{p}}, [\hat{\mathbf{p}}, \hat{\rho}]] \tag{9}$$

The coefficients of momentum diffusion $D_p = \eta M k_{\mathrm{B}} T$ and friction η correspond to those in the classical Fokker-Planck equation:

$$\frac{d\rho}{dt} = \{H, \rho\}_{\mathrm{Poisson}} + D_p \left(\frac{\partial}{\partial \mathbf{p}}\right)^2 \rho + \eta \frac{\partial}{\partial \mathbf{p}} \mathbf{p}\rho \,. \tag{10}$$

However, the quantum version (9) contains a strange term of *position diffusion* which would be nonsense classically. Position diffusion of the Brownian object is a pure quantum effect, the celebrated GKLS theorem [11, 12] puts the following lower bound on the coefficient of position diffusion:

$$\frac{D_x}{\hbar^2} \geq \frac{\eta^2}{4D_p} = \frac{\eta}{4Mk_{\mathrm{B}}T} \,, \tag{11}$$

which was satisfied in [6] by construction. Hornberger [8], applying the MSqR, found an ambiguity—nicely elucidated later in [10]—and derived an alternative QLBE. His was more natural than mine, in particular because his QLBE had the minimum possible value of D_x [cf. (11)], i.e., the minimum rate of the strange position diffusion.

The context of quantum Brownian motion theory and the related results — achieved by physicists mostly working on quantum foundations otherwise—were summarized in 2009 [10] by Vacchini and Hornberger. The QLBE of Hornberger [8] seemed to be the true and ultimate quantum version of the classical linear Boltzmann equation. But soon, an elementary argument of decoherence popped up and questioned it together with all previous versions, including mine.

4 Complete Momentum Decoherence (CMD)

To understand the overlooked dramatic phenomenon indicated by the title above, we only need the momentum and energy conservation of collision in one dimension first:

$$p_i + k_i = p_f + k_f \,, \tag{12}$$

$$\frac{p_i^2}{2M} + \frac{k_i^2}{2m} = \frac{p_f^2}{2M} + \frac{k_f^2}{2m} \,. \tag{13}$$

We express the final momentum of the mass M in the following form:

$$p_f = \mu_+ k_i + \mu_- k_f \ , \tag{14}$$

where $\mu_\pm = (M/m \pm 1)/2$. Observe that the initial momentum p_i of the mass M canceled! Its final momentum depends on the initial k_i and final k_f of the scattered mass m! This fact yields a crisis quantum mechanically. Observe that the reduced post-collision state of the mass M remains the same if we measure the post-collision momentum k_f of the other mass. Assume we do so and measure k_f. Then the above expression of p_f means that we measure the final momentum p_f of the mass M as well and, as a consequence, momentum superpositions for the mass M can exist no more after a single collision! *Any single collision causes complete momentum decoherence* of the mass M.

This trivial fact of CMD surfaced in 2009 [13] and in 2010 [14] in the general case of the three-dimensional collision (7) where the expression (14) survives for the components of $\mathbf{p}_f$, $\mathbf{k}_i$, $\mathbf{k}_f$ parallel to the momentum transfer $\mathbf{k}_f - \mathbf{k}_i$ only:

$$p_f^\| = \mu_+ k_i^\| + \mu_- k_f^\| \ . \tag{15}$$

CMD in all three components of $\mathbf{p}_f$ requests just three collisions in a row. CMD is obviously unphysical. It would, in particular, suggest a divergent coefficient $D_x = \infty$ of position diffusion in the QFPE (9).

The obligate question follows: how did the derivations of QLBEs from 1995 over fifteen years got finite D_x against the trivial CMD which imposes $D_x = \infty$. How did they regularize the divergent position diffusion?

5 Collision and Methods Revisited

We go back to the type of elementary considerations of JZ, this time taking the exact collision kinematics like (7) into the account, instead of the approximate (2). To detect CMD and the role of the MSqR in its regularization, it is sufficient to simplify the derivations from three to one dimension. First, let us find the counterpart of (7) in one dimension. Assume, again for simplicity, the repulsive hard-wall potential between the dust and a molecule so that we can ignore that they tunnel through each other. Then the unitary transition (7) in a collision reduces to:

$$|p\rangle \otimes |k\rangle \Rightarrow |p + 2k^*\rangle \otimes |k - 2k^*\rangle \ . \tag{16}$$

[For brevity, we stop indicating that all momenta are the initial ones.] Remember the center-of-mass initial momentum $k^* = (Mk - mp)/(M + m)$. If the initial state of the dust is a superposition of momentum eigenstates, the transition of an off-diagonal element of the density matrix reads:

$$|p\rangle\langle p'| \otimes |k\rangle\langle k| \Rightarrow |p+2k^*\rangle\langle p'+2k^*| \otimes |k-2k^*\rangle\langle k-2k^{*\prime}| \,, \tag{17}$$

where $k^{*\prime} = (Mk - mp')(M+m)$. Take the partial trace of both sides, yielding

$$|p\rangle\langle p'| \Rightarrow 0 \,, \tag{18}$$

because the two post-collision states of the molecule, scattered on $|p\rangle$ and $|p'\rangle$, respectively, became orthogonal:

$$\langle k_i - 2k^{*\prime} | k - 2k_i^*\rangle = 0 \,. \tag{19}$$

This proves CMD analytically and confirms the previous measurement theoretical argument: *Any single collision causes CMD* of the dust. This cannot happen in the reality since it would completely delocalize the wave function. It is now obvious that the mentioned two post-collision states of the molecule should overlap!

Our derivations [6, 8–10] of QLBE's created this overlap formally via the MSqR (8), without any awareness or reference to the above physical background. We assumed an environmental ideal gas, i.e., a mixture of plane waves of thermal distribution $\rho^{\mathcal{E}}(k) \propto \exp(-k^2/mk_{\mathrm{B}}T)$ of temperature T. But at a certain later stage towards the QLBE, we took the MSqR (8) and postulated the following density matrix:

$$\rho^{\mathcal{E}}(k,k') = \sqrt{\rho^{\mathcal{E}}(k)}\sqrt{\rho^{\mathcal{E}}(k')} \,, \tag{20}$$

which represents a single normalized central real Gaussian wave function $\psi_k^{\mathcal{E}} \propto \exp(-k^2/2mk_{\mathrm{B}}T)$, i.e., a central real Gaussian wave packet standing at the origin:

$$\psi^{\mathcal{E}}(x) \propto \exp\left(-\frac{mk_{\mathrm{B}}Tx^2}{2\hbar^2}\right) . \tag{21}$$

This single pure state served as an effective *substitute* of the single molecule mixed state in the ideal gas. The translation invariance was lost obviously. Nonetheless it became restored since all plane wave components $|k\rangle$ were assumed to collide with the dust:

$$\int dk \psi_k^{\mathcal{E}} |p\rangle \otimes |k\rangle \Rightarrow \int dk \psi_k^{\mathcal{E}} |p+2k^*\rangle \otimes |k-2k^*\rangle \,. \tag{22}$$

It is of course hard to interpret this assumption but it was implicit in all derivations and, most importantly, restored the translation invariance of the resulting QLBE.

Due to the MSqR (20) the two post-collision states of the molecule are, unlike in (19), no more orthogonal; they overlap, and the effect of CMD disappears, gives its role to finite momentum decoherence . Detailed calculations, omitted here, yield the QFPE (9) with the standard momentum diffusion $D_p = \eta M k_{\mathrm{B}} T$ and the finite coefficient of position diffusion (momentum decoherence) D_x saturating the constraint (11). In all historic QLBEs [6, 8–10] it is the MSqR that removes the divergence of D_x.

Are these D_x's physical? In view of the meaning of the MSqR that a standing wave packet substitutes the ideal gas single-molecule density matrix, a finite D_x may well be an artifact of the MSqR, as suggested by [13, 14].

6 Farwell MSqR

What other, more physical mechanism could explain the finite physical momentum decoherence (position diffusion) if it is tractable at all via the independent collisions. Should one improve on the single molecule density matrix of the ideal gas by taking molecule-molecule interactions into account? Unfortunately, one should not. The diagonal form $\rho(\mathbf{p}, \mathbf{p}') \propto \delta(\mathbf{p} - \mathbf{p}')$ remains because the due translation invariance of the gas equilibrium state. To mitigate CMD, playing with the quantum state $\rho^{\mathcal{E}}(\mathbf{k}, \mathbf{k}') \propto \rho^{\mathcal{E}}(\mathbf{k})\delta(\mathbf{k} - \mathbf{k}')$ of the particles is useless. We play with the collision.

When MSqR turned out to be a kind of unphysical elimination of CMD, the following consideration arose. CMD assumes idealized quantum scatterings that means, e.g., infinite intercollision time $\tau = \infty$. If one takes the finite τ valid even in dilute gas then energy conservation (13) in single collision becomes unsharp and CMD becomes relaxed. This was certainly a more justified mechanism to mitigate CMD than the MSqR had been, I thought in [13], and got a finite coefficient D_x of momentum decoherence (position diffusion):

$$D_x = \frac{1}{3}\left(\frac{\tau^2}{M}\right) D_p. \tag{23}$$

Hornberger and Vacchini [15] claimed that the CMD issue was nonexistent in their QLBE [8–10] which contains the ultimate physics of quantum Brownian motion—I disagreed [16]—as long as binary independent collisions are considered between the dust and the molecules. Also Kamleitner and Cresser [14] blamed the idealization of the scattering process for CMD and introduced a nonzero collision (interaction) time instead of the idealized zero. Apparently, no consensus has since been reached as to the value of D_x neither to the very existence of momentum decoherence (position diffusion).

This issue is not yet too burning since the effect is not testable currently. The experimental significance of a non-zero D_x was anticipated long time ago [17], a possible test was mentioned tangentially [18], a fundamental experiment [19] used and confirmed the QLBE prediction for momentum diffusion only.

7 Epilogue

Many times, questioning the conservative and confirmed wisdom respecting quantum mechanics turns out to be unproductive. Zeh's criticism was different and changed our abstraction and practice about coherence in quantum theory. I only wished to illustrate how Zeh's work, apart from its impact on foundations, opened the Pandora's box of a standard unsolved problem independent of foundations. What is our theory of a quantum Brownian particle in a gas? Theory ran into a puzzle that—I'm afraid—has remained unsolved so far.

Acknowledgements Support from the FQXi (minigrant 2021) and from the NRDIO (Grant No. KKP133827) is acknowledged.

References

1. H.D. Zeh, Found. Phys. **1**, 69 (1970)
2. H.D. Zeh, in *Foundations of Quantum Mechanics*. ed. by B. d'Espagnat (Academic, New York, 1971), p. 263
3. E. Wigner, in *Quantum Optics, Experimental Gravitation, and Measurement Theory*, ed. by P. Meystre, M.O. Scully (Plenum, New York, 1983), p. 43
4. E. Joos, H.D. Zeh, Z. Phys. B 59, 223 (1985)
5. M.R. Gallis, G.N. Fleming, Phys. Rev. A **42**, 38 (1990)
6. L. Diósi, Europhys. Lett. **30**, 63 (1995)
7. B. Vacchini, Phys. Rev. E **63**, 066115 (2001)
8. K. Hornberger, Phys. Rev. Lett. **97**, 060601 (2006)
9. K. Hornberger, B. Vacchini, Phys. Rev. A **77**, 022112 (2008)
10. B. Vacchini, K. Hornberger, Phys. Rep. **478**, 71 (2009)
11. V. Gorini, A. Kossakowski, E.C.G. Sudarshan, J. Math. Phys. **17**, 821 (1976)
12. G. Lindblad, Commun. Math. Phys. **48**, 119 (1976)
13. L. Diósi, Phys. Rev. A **80**, 064104 (2009)
14. I. Kemleitner, J. Cresser, Phys. Rev. A **81**, 012107 (2010)
15. K. Hornberger, B. Vacchini, Phys. Rev. A **82**, 036101 (2010)
16. L. Diósi, Phys. Rev. A **82**, 036102 (2010)
17. K. Jacobs, I. Tittonen, H.M. Wiseman, S. Schiller, Phys. Rev. A **60**, 538 (1999)
18. J.Z. Bernád, L. Diósi, T. Geszti, Phys. Rev. Lett. **97**, 250404 (2006)
19. K. Hornberger, S. Uttenthaler, B. Brezger, L. Hackermüller, M. Arndt, A. Zeilinger, Phys. Rev. Lett. **90**, 160401 (2003)

On the Statistical Viewpoint Concerning the Second Law of Thermodynamics—A Reminder on the Ehrenfests' Urn Model

Domenico Giulini

1 Prologue

This contribution is dedicated to the memory of H. Dieter Zeh, with whom I had many discussions over a period of time that easily spans a quarter of a century. These discussions were usually controversial, sometimes *very* controversial, so that often we could only agree do disagree. From these discussions I learned a lot!

I came to know Zeh—the university teacher—in my earlier student days. In June 1984 a friend gave me Zeh's "Die Physik der Zeitrichtung" as a birthday present [14]. Below the preface my friend wrote in his own hand: "I am sure this book will find your undivided approval", and it did! Not so much because I felt competent enough to judge the content, but because I felt a degree of veracity behind it that appealed to me very much.

The book of which I received a copy was the first edition, based on notes of lectures Zeh had given since 1979. When the first edition finally came out it quickly circulated amongst the younger generation of students, even outside the physics department (in fact, my friend was in the department of chemistry and knew about the book before I did). For many of us Zeh represented the serious and uncompromising urge for real "understanding" that many of us hoped to find at the university and, in particular, in the department of theoretical physics. This hope was not always fulfilled but Zeh was definitely someone proving that our hopes had not been in vain.[1]

I should add that our trust was based not so much on the fact that Zeh talked about the "big issues", like "arrow of time", or "interpretation of quantum mechanics", which clearly fascinate all beginning student alike, but rather on the fact that he

[1] To give a contrasting example, I remember from my first lecture on quantum field theory, in which after the scheme of canonical quantisation was introduced and the interaction picture mentioned, the professor said: "There is a theorem due to Rudolf Haag, according to which the interaction picture does not exist; but we shall henceforth ignore that!" How should a serious beginning student deal with such a casually tossed comment?.

D. Giulini (✉)
Institute for Theoretical Physics, Leibniz University Hannover,
Appelstrasse 2, 30167 Hannover, Germany
e-mail: giulini@itp.uni-hannover.de

C. Kiefer (ed.), *From Quantum to Classical*, Fundamental Theories of Physics 204,
https://doi.org/10.1007/978-3-030-88781-0_11

touched upon these issues while at the same time striving for full clarity as regards the "small issues". The only lecture I took by Zeh was on analytical mechanics. From that I remember his explanation of the Euler angles, which he gave by means of meticulous blackboard drawing that he had prepared before the lecture, showing four systems of orthonormal frames in different colours with the relevant rotation axes and angles. His comment was that he did not understand the corresponding drawing in the standard textbook then widely used, so he developed everything from scratch again.

Many years later Zeh gave me the third english edition [15] as a present. That edition contains a new Appendix on a simple numerical toy model, the so called *ring model*, taken from [8, Chap. 3], which is meant to illustrate the concept of a "Zwanzig projection", that plays a most central role in Zeh's book, as one can already see from the book's bibliography, that contains 43 entries for that term (the only entry under "Z"). Zeh also mentions the Ehrenfest's *urn-model* as a popular and widely known one to illustrate conceptual points connected with statistical statements in physics. I remember discussing that with Zeh and that I praised the Ehrenfests' model for its ability to illustrate basic but non-trivial concepts from statistical mechanics by means of *exact* mathematical expressions. It was then that I worked out some details of that model, just for my own understanding, illustrating—as Zeh used to say—the "fact-like" rather than "law-like" nature of entropy increase. This is what I wish to present here.

Following the third, there were two more editions of Zeh's classic text, the last (fifth) in Springer's *Froniers Collection* [16], that I reviewed in [6].

2 Introduction

The point that the likely statistical increase of entropy does *not* as such prefer a direction of time has been first made by [4] in connection with their *urn model* [3]; see also their general review [5] (english translation [2]). It has again been emphasised by [13].[2] This insight is not new and should be a commonplace, though that is far from true according to my experience.

Without going into any analytical details, Zeh said it very clearly at the beginning of paragraph 3 of [16, p. 40] on the thermodynamical arrow of time:

> While statistical considerations are indeed essential for the understanding of thermodynamical concepts, statistics as a method of counting has nothing a priori to do with dynamics. Therefore, it cannot by itself explain dynamically 'irreversible' processes – characterised by $\{dS/dt\}_{\text{int}} > 0$. This requires *additional* assumptions, which often remain unnoticed, since they appear 'natural' to our prejudiced way of thinking in terms of causes (exclusively in the past). These hidden assumptions have therefore to be carefully investigated in order to reveal the true origin of the thermodynamical arrow."[3]

---

[2] The conclusions Weizsäcker drew from this insight are, however, problematic; see, [9].

[3] In the first (german) edition and in the following english editions up to, and including, the third, the "has nothing a priori to do with dynamics" reads instead: "has nothing a priori to do with the

Zeh continues by comparing the four possibilities of processes

$$\text{probable / improbale state} \quad \xrightarrow{t} \quad \text{probable / improbable state}$$

pointing out in particular, that there are as much processes from improbable to probable than from probable to improbable states. Hence, an asymmetry in their number of occurrences must be connected with an *additional*, symmetry breaking assumption of an improbable state at one end of the time axis. Zeh calls that prescribed state the "initial" one, which at this point may be read as an implicit definition of "initial", namely as that end at which the condition is put, for otherwise there is so far no objective difference between the two ends of the time axis.

In this contribution I will employ the celebrated urn model to exemplify these points. This model has once even been called "probably one of the most instructive models in the whole of physics" [8, p. 73]. The rather simple analytical features of this model help to guide one's own expectations and reduce the danger of possible misconceptions. For the issue to be discussed here, namely that of the "likely increase" of entropy is a time symmetric statement, possible misconceptions have to do with a failure to appreciate the fact that the probabilities are conditioned in time, and that their interpretation, namely as being either "retarded" or "advanced", are in itself indistinguishable unless a time orientation has already been established.

I will proceed as follows: In the next section I will try to put the qualitative statement just made into more precise words. In the following section this is then quantitatively analysed in terms of the urn model, where probabilities can actually be calculated in an explicit and elementary way. The remaining sections then discuss Boltzmann entropy, Gibbs entropy and H-theorem, and the thermodynamic limit and deterministic dynamics. Some elementary notions from probability theory that we made freely use of are collected in a short appendix.

3 The Basic Statements

In this section we start by making some of the previous statements more precise. We think of an idealised system, whose state may only change at sharp, discrete times. This allows us to speak unambiguously about "next" and "previous" points in time. Now we make the following

Assumption. At time t_i the system is in a state $z(t_i)$ of *non-maximal* entropy.

The statistical 2nd law now makes the following statement about conditioned probabilities (the condition, which is just this assumption, will not be repeated):

---

physical concept of time or its direction" [15, p. 37], or "...jedoch hat die Statistik als mathematische Disziplin nichts mit der physikalischen Zeit zu tun und vermag daher den Zeitpfeil auch nicht zu begründen" [14, p. 23].

Statement 1 The probability, that the state $z(t_i)$ will develop *in the future* to a state $z(t_{i+1})$ of larger entropy, is larger than the probability for a development into a state of smaller entropy. □

Statement 2 The probability, that the state $z(t_i)$ has developed in the *past* from a state $z(t_{i-1})$ of larger entropy, is larger than the probability of a development from a state of smaller entropy. □

Consequence 1 The likely increase of entropy in the future state-development $z(t_i) \mapsto z(t_{i+1})$ does not imply a likely decrease for the (fictitious) past development $z(t_i) \mapsto z(t_{i-1})$, but rather also a likely increase. □

Consequence 2 The most likely development $z(t_{i-1}) \mapsto z(t_i)$ is that of decreasing entropy. Somewhat ironically, one may say that it is more likely for the state $z(t_i)$ to come about through the improbable development from a more probable state $z(t_{i-1})$ than through the probable development from an improbable state. □

To properly understand the last consequence, recall that our **condition** is placed on $z(t_i)$, that is at time t_i. For $z(t_i) \mapsto z(t_{i+1})$ this means a *retarded* or *initial* condition, for $z(t_{i-1}) \mapsto z(t_i)$, however, an *advanced* or *final* condition. It is this change of condition which makes this behaviour of entropy possible.

Consequence 3 The mere (likely) increase of entropy does not provide an orientation of time. It does *not* serve to define a 'thermodynamic arrow of time'. Rather, an orientation is usually given by considering a definite time-interval (usually of finite length) and imposing a low-entropy condition at one of the two *ends* of the interval. Without further structural elements that would serve to distinguish the two ends, the apparently existing *two* possibilities to place the low-entropy conditions are, in fact, identical. An apparent distinction is sometimes introduced by stating that the condition at one end is to be understood as *initial*. But at this level this merely defines *initial* to be used for that very end at which the condition is placed. □

4 The Urn-Model

This model was introduced in [3, 4] and quickly entered textbooks and other pedagogical oriented discussions. More detailed mathematical discussion of it are contained in [10], reprinted in [12, pp. 349–357], and [7, 8]. Reference [10] also reports on actual experiments done in order to determine the Boltzmann H-curve for this model.

Think of two urns, U_0 and U_1, among which one distributes N numbered balls. For exact equipartition to be possible we assume N to be even. A *microstate* is given by the individual numbers (names) of balls contained in U_1. (The complementary set of numbers then label the balls in U_0.) To formalise this, we associate a two-valued quantity $x_i \in \{0, 1\}$, $i \in \{1, \dots, N\}$, to each ball, where $x_i = 0$ ($x_i = 1$) stands for the i'th ball being in U_0 (U_1). This identifies the set of microstates, which we will call Γ (it corresponds to phase space), with $\Gamma = \{0, 1\}^N$, a discrete space of of 2^N elements. It can be further identified with the set of all functions $\{1, \cdots, N\} \to \{0, 1\}$, $i \mapsto x_i$. Mathematically speaking, the space Γ carries a natural measure, μ_Γ, given by associating to each subset $\Lambda \subset \Gamma$ its cardinality: $\mu_\Gamma(\Lambda) = |\Lambda|$. We now make the physical assumption, that the probability measure (normalized measure) $\nu_\Gamma := 2^{-N}\mu_\Gamma$ gives the correct *physical* probabilities. Note that this is a statement about the dynamics, which here my be expressed by saying, that in the course of the dynamics of the system, all microstates are reached equally often on time average.

Physical observables correspond to functions $\Gamma \to \mathbb{R}$. We call the set of such functions $\mathcal{O}$. Conversely, it is generally impossible to associate a physically realisable observable to any element in $\mathcal{O}$. Let $\{O_1, \ldots, O_n\} =: O_{\text{re}} \subset \mathcal{O}$ be the physically realisable ones,[4] which we can combine into a single n-component observable $O_{\text{re}} \in \mathcal{O}^n$. If $O_{\text{re}} : \Gamma \to \mathbb{R}^n$ is injective, the state is determined by the value of O_{re}. In case of thermodynamical systems it is essential to be far away from injectivity, in the sense that a given value $\alpha \in \mathbb{R}^n$ should have a sufficiently large pre-image $O_{\text{re}}^{-1}(\alpha) \subset \Gamma$. The coarse-grained of macroscopic state space is then given by the image $\Omega \subset \mathbb{R}^n$ of the realized observables O_{re}. To every macrostate $\alpha \in \Omega$ corresponds a set of microstates: $\Gamma_\alpha := O_{\text{re}}^{-1}(\alpha) \subset \Gamma$. The latter form a partition of Γ: $\Gamma_\alpha \cap \Gamma_\beta = \emptyset$ if $\alpha \neq \beta$ and $\bigcup_{\alpha \in \Omega} \Gamma_\alpha = \Gamma$.

The realised observable for the urn-model is given by the number of balls in U_1, that is, $O_{\text{re}} = \sum_{i=1}^N x_i$. Its range is the set $\Omega = \{0, 1, \ldots, N\}$ of macrostates, which contains $N + 1$ elements. The macrostates are denoted by z. To z there corresponds the set Γ_z of $\binom{N}{z}$ microstates. The probability measure ν_Γ induces so-called 'a-priori-probabilities' for macrostates z:

$$W_{\text{ap}}(z) = \nu_\Gamma(\Gamma_z) = 2^{-N} \binom{N}{z}. \tag{1}$$

Let $X : \Omega \to \mathbb{R}$ be the random variable $z \mapsto X(z) = z$. Its expectation value, denoted by E, and its standard deviation, denoted by S, with respect to the a-priori-distribution (1) are given by

$$E(X, \text{ap}) = \frac{N}{2}, \tag{2}$$

$$S(X, \text{ap}) = \frac{\sqrt{N}}{2}. \tag{3}$$

[4] The subscript 're' can be read as abbreviation for 'realised' or 'relevant'.

This follows from

$$
\begin{aligned}
E(X;\mathrm{ap}) &= 2^{-N}\sum_{z=1}^{N} z\binom{N}{z} = 2^{-N}N\sum_{m=0}^{N-1}\binom{N-1}{m} = \frac{N}{2}\,,\\
E(X^2-X;\mathrm{ap}) &= 2^{-N}\sum_{z=2}^{N} z(z-1)\binom{N}{z}\\
&= 2^{-N}N(N-1)\sum_{m=0}^{N-2}\binom{N-2}{m} = \frac{N(N-1)}{4}\,,\\
S^2(X;\mathrm{ap}) &= E(X^2-X;\mathrm{ap}) + E(X;\mathrm{ap}) - E^2(X;\mathrm{ap})\,.
\end{aligned}
\tag{4}
$$

The system has a Markovian random evolution, which is defined as follows: At every time t_i, where $i = \{0, 1, 2, \cdots\}$ with $t_j > t_i$ for $j > i$, a random generator picks a number n in the interval $1 \leq n \leq N$. Subsequently the ball with number n changes the urn. There are two possibilities: The ball with number n has been in urn U_0 so that the change of macrostate is given by $z \to z + 1$. Alternatively, the ball has been in U_1 and the change of macrostate is given by $z \to z - 1$. The conditional probabilities, $W(z \pm 1; t_{i+1}|z; t_i)$, that given the state z at time t_i the evolution will yield the state $z \pm 1$ at time t_{i+1}, are given by

$$W(z+1; t_{i+1}|z; t_i) = \frac{N-z}{N} =: W_{\mathrm{ret}}(z+1|z), \tag{5}$$

$$W(z-1; t_{i+1}|z; t_i) = \frac{z}{N} =: W_{\mathrm{ret}}(z-1|z). \tag{6}$$

Since these are independent of time, we can suppress the arguments t_i. We just have to keep in mind that the left entry, $z \pm 1$, is one time step *after* the time of z, that is, the probabilities are *past-conditioned* or *retarded*. We indicate this by writing W_{ret}.

Let $W(z; t_i)$ denote some chosen absolute probability for the state to be z at time t_i and $W_i : z \to W(z; t_i)$ the probability distribution at time t_i. The dynamics described above will now induce a dynamical law, $W_i \to W_{i+1}$, on such distributions, given by

$$
\begin{aligned}
W(z; t_{i+1}) &= W(z; t_{i+1}|z+1; t_i)\, W(z+1; t_i)\\
&\quad + W(z; t_{i+1}|z-1; t_i)\, W(z-1; t_i) &\qquad (7)\\
&= \frac{z+1}{N}\, W(z+1; t_i) + \frac{N-z+1}{N}\, W(z-1; t_i), &\qquad (8)
\end{aligned}
$$

whose Markovian character is obvious. To be sure, W_i, $i > 0$, will depend on the initial distribution W_0. This dependence will be essential if W_0 is far from equilibrium and the number of time steps i not much larger than the number N of balls. Conversely, one expects that for W_i will approach an equilibrium distribution W_{stat} for $i \gg N$, where W_{stat} is independent of W_0. Its uniqueness is shown by

Theorem 1 *A distribution W_{stat} which is stationary under (8) is uniquely given by W_{ap} in (1).*

Proof We show, that W_{stat} can be uniquely determined from (8). To this end, we assume a time independent distribution W_{stat} and write (8) in the form

$$W_{stat}(z+1) = \frac{N}{z+1} W_{stat}(z) - \frac{N-z+1}{z+1} W_{stat}(z-1). \tag{9}$$

Since $W_{stat}(-1) = 0$ we have for $z = 0$ that $W_{stat}(1) = N W_{stat}(0)$, hence recursively $W_{stat}(2) = \frac{1}{2} N(N-1) W_{stat}(0)$ and $W_{stat}(3) = \frac{1}{6} N(N-1)(N-2) W_{stat}(0)$. By induction we get the general formula $W_{stat}(z) = \binom{N}{z} W_{stat}(0)$. Indeed, inserting this expression for z and $z-1$ into the right hand side of (7), we obtain

$$\begin{aligned} W_{stat}(z+1) &= \left[\frac{N}{z+1}\binom{N}{z} - \frac{N-z+1}{z+1}\binom{N}{z-1}\right] W_{stat}(0) \\ &= (N-z)\frac{N(N-1)\cdots(N-z+1)}{(z+1)!} W_{stat}(0) \\ &= \binom{N}{z+1} W_{stat}(0). \end{aligned} \tag{10}$$

The value of $W_{stat}(0)$ is finally determined by the normalization condition:

$$1 = \sum_{z=0}^{n} W_{stat}(z) = W_{stat}(0) \sum_{z=0}^{N} \binom{N}{z} = W_{stat}(0)\, 2^N \Rightarrow W_{stat}(0) = 2^{-N}. \tag{11}$$

□

4.1 *Future-Conditioned Probabilities and Bayes' Rule*

Consider a probability space and a set of events, $\{A_1, \ldots, A_n\}$, which is (1) complete, i.e. $A_1 \cup \cdots \cup A_n = \mathbf{1}$ (here **1** denotes the certain event), and (2) mutually exclusive, i.e. $i \neq j \Rightarrow A_i \cap A_j = \mathbf{0}$ (here **0** denotes the impossible event). The probability of an event B then obeys the well known rule $W(B) = \sum_{k=1}^{n} W(B|A_k) W(A_k)$. This is just what we used in (7). Now, *Bayes' rule*, which we here regard as an independent assumption,[5] will now allow us to deduce the inversely conditioned probabilities:

$$W(A_k|B) = \frac{W(B|A_k) W(A_k)}{\sum_{i=1}^{n} W(B|A_i) W(A_i)}. \tag{12}$$

[5] Therefore we avoid to call it Bayes' theorem.

We now identify the A_i with the $N+1$ events $(z'; t_i)$ at the fixed time t_i, where $z' = 0, \ldots, N$, and A_k with the special event $(z \pm 1; t_i)$. Further we identify the event B with $(z; t_{i+1})$, i.e. with the occurrence of z at the *later* time t_{i+1}. Then we obtain:

$$W(z \pm 1; t_i \mid z; t_{i+1}) = \frac{W(z; t_{i+1}|z \pm 1; t_i)W(z \pm 1; t_i)}{\sum_{z'=0}^{N} W(z; t_{i+1}|z'; t_i)W(z'; t_i)} \tag{13}$$

$$= \frac{W(z; t_{i+1}|z \pm 1; t_i)W(z \pm 1; t_i)}{W(z; t_{i+1})}. \tag{14}$$

Hence, given W_i, a formal application of Bayes' rule allows us to express the future conditioned ('advanced') probabilities in terms of the past conditioned ('retarded') ones. In our case we think of the latter ones as given by (5)–(6). Hence we obtain the conditioned probability for $(z \pm 1; t_i)$, given that at the later time t_{i+1} the state will z occur:

$$W(z+1; t_i|z; t_{i+1}) = \frac{W(z+1; t_i)}{W(z+1; t_i) + \frac{N-z+1}{z+1} W(z-1; t_i)}, \tag{15}$$

$$W(z-1; t_i|z; t_{i+1}) = \frac{W(z-1; t_i)}{W(z-1; t_i) + \frac{z+1}{N-z+1} W(z+1; t_i)}. \tag{16}$$

4.2 Flow Equilibrium

The condition for having flow equilibrium for the pair of times t_i, t_{i+1} reads

$$W(z \pm 1; t_{i+1}|z; t_i)W(z; t_i) = W(z; t_{i+1}|z \pm 1; t_i)W(z \pm 1; t_i). \tag{17}$$

It already implies $W_i = W_{\text{ap}}$, since (5)–(6) give[6] $W(z+1; t_i) = \frac{N-z}{z+1} W(z; t_i)$ which leads to $W(z; t_i) = \binom{N}{z} W(0; t_i)$. Since $1 = \sum_z W(z; t_i)$ we have $W(0; t_i) = 2^{-N}$. Using Theorem 1, we conclude that flow equilibrium at t_i implies $W_j = W_{\text{ap}}$ for $j \geq i$.

---

[6] Without using (5)–(6) one gets

$$\begin{aligned} W(z \pm 1; t_{i+1}|z; t_i)W(z; t_i) &= W(z; t_{i+1}|z \pm 1; t_i)W(z \pm 1; t_i) \\ &= W(z \pm 1; t_i|z; t_{i+1})W(z; t_{i+1}), \end{aligned} \tag{18}$$

where the last equality is the identity $W(a|b)W(b) = W(b|a)W(a)$. The local (in time) condition of flow equilibrium is therefore equivalent to (cf. 20)

$$\frac{W(z \pm 1; t_{i+1}|z; t_i)}{W(z \pm 1; t_i|z; t_{i+1})} = \frac{W(z; t_{i+1})}{W(z; t_i)}. \tag{19}$$

4.3 Time-Reversal Invariance

To be distinguished from flow equilibrium is time-reversal invariance. The latter is given by the following equality of past- and future-conditioned probabilities:

$$W(z \pm 1; t_{i+1} | z; t_i) = W(z \pm 1; t_i | z; t_{i+1}) \tag{20}$$

$$\overset{(14)}{=} W(z; t_{i+1} | z \pm 1; t_i) \frac{W(z \pm 1; t_i)}{W(z; t_{i+1})}, \tag{21}$$

$$\overset{(5,6)}{\Longleftrightarrow} \quad W(z; t_{i+1}) = \frac{z+1}{N-z} W(z+1; t_i) \tag{22}$$

$$= \frac{N-z+1}{z} W(z-1; t_i). \tag{23}$$

It is interesting to note that the condition of time-reversal invariance is weaker that that of flow equilibrium. The former is implied by, but does not itself imply, the equilibrium distribution. Let us explain this in more detail: Equations (22)–(23) imply (8), since $\frac{N-z}{N} \times (22) + \frac{z}{N} \times (23) = (8)$. Hence (22)–(23) are stable under time evolution (8). Conversely, (22)–(23) is implied by (8) and the following equation, expressing the equality of the right hand sides of (22) and (23):

$$W(z+1; t_i) = \frac{N-z}{z+1} \frac{N-z+1}{z} W(z-1; t_i). \tag{24}$$

Indeed, eliminating $W(z+1; t_i)$ in (8) using (24), one gets

$$W(z; t_{i+1}) = \frac{N-z+1}{z} W(z-1; t_i) \overset{(18)}{=} \frac{z+1}{N-z} W(z+1; t_i), \tag{25}$$

hence (22)–(23). Time-reversal invariance for future times is therefore equivalent to the 'constraint' (24) for the initial condition. It allows for a one-parameter family of solutions, since it determines W_i for given $p := W(0; t_i)$ and $q := W(1; t_i)$. Indeed, in analogy to the proof of Theorem 1 one gets $W_i(z) = p\binom{N}{z}$ for z even and $W_i(z) = \frac{q}{N}\binom{N}{z}$ for z odd. Since $\sum_{z=\text{even}}\binom{N}{z} = \sum_{z=\text{odd}}\binom{N}{z} = 2^{N-1}$, the normalization condition leads to $1 = 2^{N-1}(p + \frac{q}{N}) \Rightarrow q = N(2^{-(N-1)} - p)$. This shows that $p \in [0, 2^{-(N-1)}]$ faithfully parameterizes all distributions obeying (24). One should note that solutions to (24) are closed under convex sums. In this way one sees, that the obtained distributions are the convex sum $W_i = pW^e + (1-p)W^o$ of the 'even' distribution, $W^e(z) = (1 - (-1)^{z-1})2^{-N}\binom{N}{z}$ and 'odd' distribution, $W^o(z) = (1 - (-1)^z)2^{-N}\binom{N}{z}$. Solutions to (24) form a closed interval within the simplex Δ^N, which connects the point W^e in the $\frac{N}{2}$–sub-simplex $\Delta^{13\ldots N-1}$ with the point W^o on the $(\frac{N}{2}+1)$–sub-simplex $\Delta^{24\cdots N}$. If we call this interval Δ^*, we have

Theorem 2 *The set $\Delta^* \subset \mathcal{W}$ is invariant under time evolution. The future development using $W(z; t_{i+1}|z'; t_i)$ and the past development using $W(z; t_i|z'; t_{i+1})$ coincide.*[7]

It is of central importance to note that the past development is, mathematically speaking, *not* the inverse operation to the future development. The reason being precisely that such a change in the direction of development is linked with a change from retarded to advanced conditionings in the probabilities.

5 General Consequences

In the following we want to restrict to the equilibrium condition. In this case the future-conditioned probabilities are independent of the t_i and we can write $W(z \pm 1; t_i|z; t_{i+1}) =: W_{\text{av}}(z \pm 1|z)$. Hence we have:

$$W_{\text{ret}}(z+1|z) = W_{\text{av}}(z+1|z) = \frac{N-z}{N}, \tag{28}$$

$$W_{\text{ret}}(z-1|z) = W_{\text{av}}(z-1|z) = \frac{z}{N}, \tag{29}$$

from which statements 1 and 2 made in the Introduction follow. Indeed, let $z = z(t_i) > N/2$, then the probabilities that at time t_{i-1} or t_{i+1} the state was or will be $z-1$ is, in both cases, given by $\frac{z}{N}$. The probability for the state $z+1$ at time t_{i-1} or t_{i+1} is $\frac{N-z}{N}$. Now, every change of state in the direction of the equilibrium distribution leads to an increase in entropy (see below). Hence the probability of having a higher entropy at t_{i-1} or t_{i+1} is $\frac{z}{N-z}$ times that of having a lower entropy. If $z = z(t_i) < N/2$ we have to use the inverse of that.

5.1 Boltzmann Entropy

Boltzmann Entropy S_B is a function $S_B : \Omega \to \mathbb{R}$. We stress that since Ω is defined only *after* a choice of coarse graining (i.e. a choice of O_{re}) has been made, Boltzmann

---

[7] Explicitly one can see the preservation of (24) under time evolution (8) as follows: Given that the initial distribution W_i satisfies (24), the development (8) is equivalent to (22)–(23). Hence

$$W(z-1; t_i) = \frac{z}{N-z+1} W(z; t_{i+1}) \tag{26}$$

$$W(z+1; t_i) = \frac{z+2}{N-z-1} W(z+2; t_{i+1}), \tag{27}$$

which allows to rewrite (24) for W_i into (24) for W_{i+1}.

Entropy, too, must be understood as relative to that choice.[8] The value $S_B(z)$ in the macrostate z is defined by $S_B(z) := \ln \mu_\Gamma(\Gamma_z)$. For the urn model this corresponds to the logarithm of microstates that correspond to the macrostate z. In what follows it will sometimes be more convenient to label the macrostate not by $z \in [0, N]$, but rather by a parameter $\sigma \in [-1, 1]$ of range independent of N. Let the latter be defined by $z = \frac{N}{2}(1+\sigma)$. If we assume that $N, z, (N-z) \gg 1$ and approximate $\ln N! = N \ln N - N + O(\ln N)$ (Stirling formula), we obtain the following expression for the Boltzmann entropy:

$$S_B(z) = N \ln N - z \ln z - (N-z)\ln(N-z), \tag{30}$$

$$S_B(\sigma) = -\frac{N}{2}\left[\ln\frac{1-\sigma^2}{4} + \sigma \ln\frac{1+\sigma}{1-\sigma}\right]. \tag{31}$$

It obeys $S_B(\sigma) = S_B(-\sigma) = S_B(|\sigma|)$, which just corresponds to the invariance of the first expression under $z \mapsto N - z$. Considered as function of $|\sigma|$, $S_B : [0, 1] \to [\ln 2^N, 0]$ is strictly monotonically decreasing. That $S_B(\sigma = 1) = 0$ is best seen in the limit $z \to N$ of (30). Despite Stirling's approximation this value is, in fact, exact, as one easily infers from the fact that $z = N$ just corresponds to a single microstate. In contrast, the given value at $\sigma = 0$ is only approximately valid.

---

[8] This apparently non-objective character of entropy is often complained about. But this criticism is based on a misconception, since the term *thermodynamical system* is not defined without a choice for O_{re}. This is no different in phenomenological thermodynamics, where the choice of 'work degrees of freedom', $\{y^i\}$, (the 'relevant' or 'controlled' degrees of freedom) is part of the definition of 'system'. Only after they have been specified can one *define* the differential one-form of heat, called ω, as the difference between the differential of total energy, dE, and the differential one-form of reversible work, called $\alpha := f_i dy^i$; hence $\omega := dE - \alpha$. Note that neither ω nor α are exact. In particular, $\omega \neq dQ$ for some function of state Q. In contrast to E, which is a function of *states*, ω and α are each a function of *processes*, which means that given a curve γ on the manifold of (equilibrium) states, ω and α can be evaluated on (i.e. integrated along) γ. But it is meaningless to ask for the 'value' of heat and work on states. The value of heat associated to a process depends on the choice of α, which in turn depends on the choice of 'relevant' $\{y^i\}$. Roughly speaking, heat is the amount of energy not transmitted in the channels (degrees of freedom) controlled by the $\{y^i\}$. This dependence of heat on the $\{y^i\}$ is directly inherited by entropy S, through $T\,dS = \omega$, where T (temperature) and S (entropy) are functions of state. They exist if and only if ω has an integrating factor (here $1/T$), which is the case if and only if $\omega \wedge d\omega = 0$, or in differential-geometric terminology, if the kernel distribution of ω is integrable. This integrability is, in turn, equivalent to the statement that in any neighbourhood of a given state there is another state that cannot be connected to the given one by a process (curve) on which the value of ω vanishes. To require that this latter be the case is just Carathéodory's *principle of adiabatic inaccessibility* [1], which allows to deduce the existence of S whose dependence on $\{y^i\}$ is now obvious.

5.2 Consequences 1 and 2

The quantitative form of Consequences 1 and 2 are given by the solution to the following exercises: Let the state at time t_i be $z = z(t_i)$. Calculate the conditioned probabilities for

(i) $z(t_i)$ being a local maximum,
(ii) $z(t_i)$ being a local minimum,
(iii) $z(t_i)$ lying on a segment of positive slope,
(iv) $z(t_i)$ lying on a segment of negative slope.

Let the corresponding probabilities be $W_{\text{max}}(z)$, $W_{\text{min}}(z)$, $W_\uparrow(z)$, and $W_\downarrow(z)$, respectively. These are each given by the product of one past and one future conditioned probability. This being a result of the Markovian character of the dynamics, i.e. that for given (z, t_i) the dynamical evolution $(z; t_i) \to (z \pm 1; t_{i+1})$ is independent of $z(t_{i-1})$. Using (28)–(29) we obtain:

$$W_{\text{max}}(z) = W_{\text{av}}(z-1|z)W_{\text{ret}}(z-1|z) = \left(\frac{z}{N}\right)^2, \tag{32}$$

$$W_{\text{min}}(z) = W_{\text{av}}(z+1|z)W_{\text{ret}}(z+1|z) = \left(1-\frac{z}{N}\right)^2, \tag{33}$$

$$W_\uparrow(z) = W_{\text{av}}(z-1|z)W_{\text{ret}}(z+1|z) = \frac{z}{N}\left(1-\frac{z}{N}\right), \tag{34}$$

$$W_\downarrow(z) = W_{\text{av}}(z+1|z)W_{\text{ret}}(z-1|z) = \frac{z}{N}\left(1-\frac{z}{N}\right). \tag{35}$$

For $z/N > \frac{1}{2}$ ($z/N < \frac{1}{2}$) the probability W_{max} (W_{min}) dominates the other ones. Expressed in terms of σ the ratios of probabilities are given by the simple expressions:

$$W_{\text{max}}(\sigma) : W_{\text{min}}(\sigma) : W_\uparrow(\sigma) : W_\downarrow(\sigma) = \frac{1+\sigma}{1-\sigma} : \frac{1-\sigma}{1+\sigma} : 1 : 1. \tag{36}$$

In the limiting case of infinitely many t_i we get that the state z is $z^2/(N^2 - z^2) = (1+\sigma)^2/2(1-\sigma)$ times more often a maximum than any other of the remaining three possibilities.

We also note an expression for the expected recurrence time, $T(z)$, for the state z.[9] It is derived in [7] (there formula (66)). If the draws from the urns have constant time separation Δt one has

$$T(z) = \frac{\Delta t}{W_{\text{ap}}(z)}, \tag{37}$$

[9] Note that we talk about recurrence in the space Ω of macrostates ('coarse grained' states), not in the space Γ of microstates.

and hence a connection between mean recurrence time and entropy:

$$S(z) = \ln\left[\frac{2^N \Delta t}{T(z)}\right]. \tag{38}$$

Kac [7] also shows the recurrence theorem, which for discrete state spaces asserts the recurrence of each state with certainty. More precisely: let $W'(z'; t_{i+n}|z; t_i)$ be the probability that for given state z at time t_i the state z' occurs at time t_{i+n} for the *first* time after t_i (this distinguishes W' from W), then $\sum_{n=1}^{\infty} W'(z; t_{i+n}|z; t_i) = 1$.

5.3 Coarse Grained Gibbs Entropy and the H-Theorem

We recall that the Gibbs entropy S_G lives on the space of probability distributions (i.e. normed measures) on Γ and is hence independent of the choice of $\mathcal{O}_{\text{re}}$. In contrast, the coarse grained Gibbs entropy, S_G^{cg}, lives on the probability distributions on Ω, $S_G^{cg} : \mathcal{W} \to \mathbb{R}$, and therefore depends on $\mathcal{O}_{\text{re}}$. Since the former does serve, after all, as a $\mathcal{O}_{\text{re}}$ independent definition of entropy (even though, thermodynamically speaking, not a very useful one), we distinguish the latter explicitly by the superscript 'cg'. If at all, it is S_G^{cg} and not S_G that thermodynamically can we be compared to S_B. The function S_G^{cg} is given by

$$S_G^{\text{cg}}(W) = -\sum_{z=0}^{N} W(z) \cdot \ln\left[\frac{W(z)}{W_{\text{stat}}(z)}\right]. \tag{39}$$

The structure of this expression is highlighted by means of the generalized H-theorem, which we explain below.[10] Since the two entropies S_B and S_G^{cg} are defined on different spaces, Ω and $\mathcal{W}$, it is not immediately clear how to compare them. To do this, we would have to agree on what value of S_G^{cg} we should compare with $S_B(z)$, i.e. what argument $W \in \mathcal{W}$ should correspond to $z \in \Omega$. A natural candidate is the distribution centered at z, that is, $W(z') = \delta_z(z')$, which is 1 for $z' = z$ and zero otherwise. From (39) we then obtain

$$S_G^{\text{cg}}(\delta_z) = S_B(z) - N \ln 2\,. \tag{40}$$

Let us now turn to the generalized H-theorem. Let $\Phi : \mathbb{R} \to \mathbb{R}$ be a convex function. Then for any finite family $m := \{x_1, \ldots, x_n\}$ of not necessarily pairwise distinct points in $\mathbb{R}$ we have the following inequality $\Phi(\sum_i \alpha_i x_i) \leq \sum_i \alpha_i \Phi(x_i)\, \forall\, \alpha_i \in \mathbb{R}_{\geq 0}$

[10] Usually this expression is called the *relative* entropy [of W relative to W_{stat}]. As [absolute] entropy of W one then understands the expression $-\sum_z W(z) \ln W(z)$. The H-theorem would be valid for the latter only if the constant distribution (in our case $W(z) = 1/(N+1)$) is an equilibrium distribution, which is not true for the urn model.

with $\sum_i \alpha_i = 1$, where equality holds iff there is no index pair i, j, such that $x_i \neq x_j$ and $\alpha_i \cdot \alpha_j \neq 0$. In the latter case the convex sum is called trivial. We now define a function $H : \mathcal{W} \times \mathcal{W} \to \mathbb{R}$ through

$$H(W, W') := \sum_{z=0}^{N} W'(z)\Phi\left[\frac{W(z)}{W'(z)}\right]. \tag{41}$$

Consider a time evolution $W_i \mapsto W_{i+1}$, $W_{i+1}(z) := \sum_{z'} W(z|z')W_i(z')$, where clearly $W(z|z') \geq 0$ and $\sum_z W(z|z') = 1$. We also assume that no row of the matrix $W(z|z')$ just contains zeros (which would mean that the state labelled by the corresponding row number is impossible to reach). We call such time evolutions and the corresponding matrices *non-degenerate*. In what follows those distributions $W \in \mathcal{W}$ for which $W(z) > 0\,\forall z$, i.e. from the interior $\mathring{\mathcal{W}} \subset \mathcal{W}$, will play a special role. We call them *generic*. The condition on $W(z|z')$ to be non-degenerate then ensures that the evolution leaves the set of generic distributions invariant. After these preparations we formulate

Theorem 3 (generalized H-theorem) *Let W_i' be generic and the time evolution non-degenerate; then $H(W_{i+1}, W_{i+1}') \leq H(W_i, W_i')$.*

Proof (Adaptation of the proof of theorem 3 in [11] for the discrete case.) We define a new matrix $V(z, |z') := [W_{i+1}'(z)]^{-1}W(z|z')W_i'(z')$, which generates the time evolution for $W_i(z)/W_i'(z)$ and obeys $\sum_{z'} V(z|z') = 1$. It follows:

$$H(W_{i+1}, W_{i+1}') = \sum_{z=0}^{N} W_{i+1}'(z)\,\Phi\left[\frac{W_{i+1}(z)}{\mathcal{W}_{i+1}'(z)}\right] \tag{42}$$

$$= \sum_{z=1}^{N} W_{i+1}'(z)\,\Phi\left[\sum_{z'=0}^{N} V(z|z')\frac{W_i(z')}{W_i'(z')}\right] \tag{43}$$

$$\leq \sum_{z'=0}^{N}\sum_{z=0}^{N} W_{i+1}'(z)V(z|z')\,\Phi\left[\frac{W_i(z')}{W_i'(z')}\right] \tag{44}$$

$$= \sum_{z'=0}^{N} W_i'(z')\,\Phi\left[\frac{W_i(z')}{W_i'(z')}\right] \tag{45}$$

$$= H(W_i, W_i'). \tag{46}$$

Equality in (44) holds, iff the convex sum in the square brackets of (43) is trivial. □

Picking a stationary distribution for W', which in our case is the unique distribution W_{stat}, then H is a function of just one argument which does not increase in time. Taking in addition the special convex function $\Phi(x) = x\ln(x)$, then we obtain with $S_G^{cg} := -H$ the above mentioned entropy formula.

Let from now on Φ be as just mentioned. Then we have, due to $\ln(x) \geq 1 - x^{-1}$, with equality iff $x = 1$:

$$H(W, W') = \sum_{z=0}^{N} W(z) \ln\left[\frac{W(z)}{W'(z)}\right] \geq \sum_{z=0}^{N} (W(z) - W'(z)) = 0, \tag{47}$$

$$= 0 \Leftrightarrow W(z) = W'(z) \quad \forall z. \tag{48}$$

Let us denote by a *distance function* on a set M any function $d : M \times M \to \mathbb{R}_{\geq 0}$, such that $d(x, y) = d(y, x)$ and $d(x, y) = 0 \Leftrightarrow x = y$. (This is more general than a *metric*, which in addition must satisfy the triangle inequality.) A map $\tau : M \to M$ is called non-expanding with respect to d, iff $d(\tau(x), \tau(y)) \leq d(x, y) \, \forall x, y \in M$. We have

Theorem 4 $D : \overset{\circ}{\mathcal{W}} \times \overset{\circ}{\mathcal{W}} \to \mathbb{R}$, $D(W, W') := H(W, W') + H(W', W)$ *is a distance function with respect to which every proper non-degenerate time evolution is non-expanding.*

Proof Symmetry is clear and (48) immediately implies $D(W, W') \geq 0$ with equality iff $W = W'$, as follows from the separate positivity of each summand. Likewise (46) holds for each summand, so that no distance increases. □

6 Thermodynamic Limit and Deterministic Dynamics

In this section we wish to show how to get a deterministic evolution for random variables in the limit $N \to \infty$. To this end we first consider the discrete, future directed time evolution of the expectation value of the random variable $X(z) = z$. We have

$$E(X, t_{i+1}) = \sum_{z'=0}^{N} z' W_{i+1}(z') = \sum_{z'=0}^{N} \sum_{z=0}^{N} z' W_{\text{ret}}(z'|z) W_i(z) \tag{49}$$

$$= \sum_{z=0}^{N} \left[(z+1)\frac{N-z}{N} + (z-1)\frac{z}{N}\right] W_i(z)$$

$$= 1 + \left(1 - \frac{2}{N}\right) E(X, t_i). \tag{50}$$

In the same way we get

$$
\begin{aligned}
E(X^2, t_{i+1}) &= \sum_{z=0}^{N}\left[(z+1)^2\frac{N-z}{N} + (z-1)\frac{z}{N}\right] W_i(z) \\
&= 1 + 2E(X, t_i) + (1 - 4/N)\, E(X^2, t_i)\,, \qquad (51) \\
V(X, t_{i+1}) &= E(X^2, t_{i+1}) - E^2(X, t_{i+1}) \\
&= (1 - 4/N)\, V(X, t_i) + \frac{4}{N} E(X, t_i) - \frac{4}{N^2} E^2(X, t_i)\,. \qquad (52)
\end{aligned}
$$

By the evolution being 'future directed' one means that W_{ret} and not W_{av} are used in the evolution equations, as explicitly shown in (49). In this case one also speaks of 'forward-directed evolution'.

In order to carry out the limit $N \to \infty$ we use the new random variable $\Sigma : \Omega \to \sigma$, where $\sigma = \frac{2z}{N} - 1$ as above; hence $X = \frac{N}{2}(1 + \Sigma)$. Simple replacement yields

$$
E(\Sigma, t_{i+1}) = (1 - 2/N)\, E(\Sigma, t_i)\,, \qquad (53)
$$

$$
V(\Sigma, t_{i+1}) = (1 - 4/N)\, V(\Sigma, t_i) + \frac{4}{N^2}\left(1 - E^2(\Sigma, t_i)\right)\,. \qquad (54)
$$

In order to have a seizable fraction of balls moved within a macroscopic time span τ, we have to appropriately decrease the time steps $\Delta t := t_{i+1} - t_i$ with growing N, e.g. like $\Delta t = \frac{2}{N}\tau$, where τ is some positive real constant. Its meaning is to be the time span, in which $N/2$ balls change urns. Now we can take the limit $N \to \infty$ of (53) and (54),

$$
\frac{d}{dt}E(\Sigma, t) = -\frac{1}{\tau}E(\Sigma, t) \Longrightarrow E(\Sigma, t) = E_0 \exp\left(\frac{-(t - t_1)}{\tau}\right), \qquad (55)
$$

$$
\frac{d}{dt}V(\Sigma, t) = -\frac{2}{\tau}V(\Sigma, t) \Longrightarrow V(\Sigma, t) = V_0 \exp\left(\frac{-2(t - t_2)}{\tau}\right), \qquad (56)
$$

where E_0, V_0, t_1, t_2 are independent constants. These equations tell us, that (1) the expectation value approaches the equilibrium value $\Sigma = 0$ exponentially fast in the future, and (2) it does so with exponentially decaying standard deviation. The half mean time of both quantities is the time for $N/2$ draws.

According to the discussions in previous sections it is now clear, that in case of equilibrium identical formulae would have emerged if W_{av} instead of W_{ret} had been used, for then $W_{\text{av}} = W_{\text{ret}}$. Most importantly to note is, that the backward evolution is *not* obtained by taking the forward evolution and replacing in it $t \mapsto -t$. The origin of this difference is the fact already emphasized before (following Theorem 2), that $W_{\text{av}}(z; z')$ is not the inverse matrix to $W_{\text{ret}}(z; z')$, but rather the matrix computed according to Bayes' rule.

7 Appendix

In this Appendix we collect some elementary notions of probability theory, adapted to our specific example.

The space of elementary events[11] is $\Omega = \{0, 1, \ldots, N\}$. By

$$\mathcal{X} := \{X : \Omega \to \mathbb{R}\}, \tag{57}$$

$$\mathcal{W} := \{W : \Omega \to \mathbb{R}_{\geq 0} \mid \sum_{z\in\Omega} W(z) = 1\}, \tag{58}$$

we denote the sets of random variables and probability distributions respectively, where $\mathcal{W} \subset \mathcal{X}$. The map $\mathcal{X} \to \mathbb{R}^{N+1}$, $X \mapsto (X(0), X(1), \cdots, X(N))$ defines a bijection which allows us to identify $\mathcal{X}$ with $\mathbb{R}^{N+1}$. This identifies $\mathcal{W}$ with the N-simplex

$$\Delta^N := \{(W(0), \cdots, W(N)) \in \mathbb{R}^{N+1} \mid W(z) \geq 0, \sum_z W(z) = 1\} \subset \mathbb{R}^{N+1}. \tag{59}$$

Its boundary, $\partial\Delta^N$, is the union of all $(N-K)$-simplices:

$$\Delta^{i_1 \cdots i_K} := \{(W(0), \ldots, W(N)) \in \Delta^N \mid 0 = W(i_1) = \cdots = W(i_K)\}, \tag{60}$$

for all K. Its interior is $\mathring{\mathcal{W}} := \mathcal{W} - \partial\mathcal{W}$, so that $W \in \mathring{\mathcal{W}} \Leftrightarrow W(z) \neq 0 \forall z$.

Expectation value E, variance V, and standard deviation S are functions $\mathcal{X} \times \mathcal{W} \to \mathbb{R}$, defined as follows:

$$E : \mathcal{X} \times \mathcal{W} \to \mathbb{R}, \quad E(X, W) := \sum_{z\in\Omega} X(z)W(z), \tag{61}$$

$$V : \mathcal{X} \times \mathcal{W} \to \mathbb{R}_{\geq 0}, \quad V(X, W) := E((X - \langle X \rangle)^2, W) = E(X^2, W) - E^2(X, W), \tag{62}$$

$$S : \mathcal{X} \times \mathcal{W} \to \mathbb{R}_{\geq 0}, \quad S(X, W) := \sqrt{V(X, W)}, \tag{63}$$

where in (62) $\langle X \rangle$ simply denotes the constant function $\langle X \rangle : z \mapsto E(X, W)$, and $E^2(X, W) := [E(X, W)]^2$. In the main text we also write $E(X, s)$ if the symbol s uniquely labels a point in $\mathcal{W}$, like $s = \text{ap}$ for the a priori distribution (1), or $E(X, t_i)$ for the distribution W_i at time t_i.

[11] 'Elementary' is merely to be understood as mathematical standard terminology, not in any physical sense. For example, in the urn model, Ω is obtained after coarse graining from the space of physically 'elementary' events.

References

1. C. Carathéodory, Untersuchungen über die Grundlagen der Thermodynamik. Math. Ann. **67**, 355–386 (1909). https://doi.org/10.1007/BF01450409
2. P. Ehrenfest, T. Ehrenfest, *The Conceptual Foundations of the Statistical Approach in Mechanics* (Dover Publications, Inc., New York, 1990). Translation of the 1912 german original by Michael J. Moravcsik
3. P. Ehrenfest, T. Ehrenfest, Über eine Aufgabe aus der Wahrscheinlichkeitsrechnung, die mit der kinetischen Deutung der Entropievermehrung zusammenhängt. Mathematisch-Naturwissenschaftliche Blätter **3**(11,12) (1906)
4. P. Ehrenfest, T. Ehrenfest, Über zwei bekannte Einwände gegen das Boltzmannsche H-Theorem. Phys. Zeitsch. **9**, 311–314 (1907)
5. P. Ehrenfest, T. Ehrenfest, Begriffliche Grundlagen der statistischen Auffassung in der Mechanik, in *Encyklopädie der Mathematischen Wissenschaften mit Einschluss ihrer Anwendungen. Vierter Band: Mechanik*, ed. by F. Klein, C. Müller, vol. IV:2:II (B.G. Teubner, Leipzig, 1912)
6. D. Giulini, Book review: the physical basis of the direction of time (fifth edition) by H.D. Zeh. Class. Quantum Grav. **25**(20), 209003 (2 pp) (2008). https://doi.org/10.1088/0264-9381/25/20/209003
7. M. Kac, Random walk and the theory of Brownian motion. Am. Math. Mon. **54**(7), 369–391 (1947). https://doi.org/10.2307/2304386
8. M. Kac, *Probability and Related Topics in Physical Sciences* (Interscience Publishers, London, 1959). Proceedings of the Summer Seminar in Applied Mathematics by the American Mathematical Society at Boulder, Colorado, June 23 to July 19, 1957
9. C. Kiefer, Weizsäckers Zeitbegriff aus heutiger Sicht. Acta Historica Leopoldina **63**, 177–186 (2014)
10. F. Kohlrausch, E. Schrödinger, Das Ehrenfestsche Modell der H-Kurve. Phys. Zeitsch. **27**, 306–313 (1926)
11. R. Kubo, H-theorems for markoffian processes, in *Perspectives in Statistical Physics (M.S. Green Memorial Volume)*. ed. by H.J. Raveché (North-Holland Publishing Company, Amsterdam, 1981), pp. 101–110
12. E. Schrödinger, *Gesammelte Abhandlungen, Band 1: Beiträge zur statistischen Mechanik* (Verlag der Österreichischen Akademie der Wissenschaften und Friedrich Vieweg & Sohn, Wien und Braunschweig/Wiesbaden, 1984)
13. C.F. von Weizsäcker, Der zweite Hauptsatz und der Unterschied von Vergangenheit und Zukunft. Ann. der Phys. **428**(3–4), 275–283 (1939). https://doi.org/10.1002/andp.19394280309. Reprinted in C.F.v.W. book "Die Einheit der Natur", Carl Hanser Verlag, München (1971), chapter II 2, pp. 172–182
14. H.D. Zeh, *Die Physik der Zeitrichtung*. Lecture Notes in Physics, vol. 200, 1st edn. (Springer, Berlin, 1984)
15. H.D. Zeh, *The Physical Basis of the Direction of Time*, 3rd edn. (Springer, Berlin, 1999)
16. H.D. Zeh, *The Physical Basis of the Direction of Time*. The Frontiers Collection. 5th edn. (Springer, Berlin, 2007)

Explanation, the Progress of Physical Theories and Computer Simulations

Ion-Olimpiu Stamatescu

Abbreviations

CM	Classical mechanics
CSM	Classical statistical mechanics
CS	Complex systems theory
ED	Classical electrodynamics
GRT	General relativity theory
QED	Quantum electrodynamics
QFT	Quantum field theory
LFT	Lattice field theory
QG	Quantum gravity (in spe)
QM	Quantum mechanics (in the text used sometimes as paradigm for quantum theories in general)
SM	Standard model: SME (of elementary particles), SMC (of cosmology)
SRT	Special relativity theory
SST	Solid state theory
Th	Thermodynamics

The ideas developed here have been to no small extent influenced by a long interaction with Heinz-Dieter Zeh and the other members of a FEST[1] project, Domenico Giulini, Erich Joos, Claus Kiefer and Joachim Kupsch. This project led among others

[1] Forschungsstätte der Evangelischen Studiengemeinschaft (Protestant Institute for Interdisciplinary Research), Heidelberg.

[2] Joos et al. [32].

I.-O. Stamatescu (✉)
Institut für Theoretische Physik, Universität Heidelberg, 69120 Heidelberg, Germany
e-mail: stamates@thphys.uni-heidelberg.de

Forschungsstätte der Evangelischen Studiengemeinschaft, 69118 Heidelberg, Germany

C. Kiefer (ed.), *From Quantum to Classical*, Fundamental Theories of Physics 204,
https://doi.org/10.1007/978-3-030-88781-0_12

to the book "Decoherence and the appearance of a classical world in quantum theory" (1st edition 1996 and 2nd edition 2003, Springer, Heidelberg, New York[2]) and to a Workshop we have organised at ZiF ("Decoherence, Theoretical, Experimental and Conceptual Problems"[3]) and I am happy to mention here the high intellectual and human capital I drew from the interaction in this group. A brief account of these considerations also entered a talk at the Conference "Mechanistic Explanation, Computability and Complex Systems" of the *International Academy for the Philosophy of the Sciences*, Dortmund, 2016 and extend ideas I have published previously.[4] This essay is dedicated to my daughters.

I want to thank Claus Kiefer for a careful reading of the manuscript and for his important comments and good suggestions. I also want to thank him generally for his engagement and work in bringing forth and preparing this volume. Our group mentioned above went apart many years ago but I still have a good and thankful memory of our discussions. I am indebted to Lukas Barth, a very talented young physicist with both solid mathematical and philosophical interest, for reading the manuscript and for his useful comments. I am indebted to Erhard Seiler for his invaluable help in achieving consistence and improving precision and understandability of this essay and for the time he invested in that. I take this opportunity also to thank my friend and partner in physical projects Erhard for the many discussions and the collaboration we had over many years and his occasionally successful efforts to make me appreciate the depths and abysses of mathematical physics. And finally I want to thank my colleague and long time friend Michael Schmidt in Heidelberg for many discussions and common projects, for his deep physical understanding, and not the least for his personal support since our first meeting 1969.

Both Heinz-Dieter Zeh and Joachim Kupsch passed away in the recent years which was a sad event for many of us.

1 Introduction

I propose a discussion concerning the features of explanation in modern physics and also the role of computers therein. My approach is pragmatic and I apologize for unconventional use of philosophical concepts. This approach appears supported by the discussion provided by well known physicists in articles or books—see also the excellent overview by *Erhard Scheibe*.[5] I think this perspective is important for a genuine interaction between physics and philosophy of physics.

Nature is the object of physics and physics the object of philosophy of physics. Therefore it is not only understandable but in fact compulsory that the evolution of physical knowledge should keep opening new horizons to the philosophical understanding, in the same way the former is driven by the study of nature itself. This

[3] Blanchard et al. (Eds.) [9].

[4] I.-O. *Stamatescu* in Ferrari and Stamatescu (Eds.) [22] and in Seiler and Stamatescu (Eds.) [48].

[5] Scheibe [44].

happened in the seventeenth century with the Copernican–Galilean revolution, in the nineteenth century with the establishing of classical physics, in the twentieth century with the advent of quantum physics—with corresponding changes in our world view and new perspectives in the philosophy, cf. e.g. the discussion on causality in *E. Cassirer* "Zur modernen Physik".[6] But this also suggests that physics at any stage cannot be fully analysable according to strict philosophical schemes since it must bear the seed of change implanted in it by nature. With Einstein: "The scientist … cannot afford to carry his striving for epistemological systematics [too] far. He accepts gratefully the epistemological conceptual analysis; but the external conditions, which are set for him by the facts of experience, do not permit him to let himself be too much restricted in the construction of his conceptual world by the adherence to an epistemological system."[7]

The philosophy of explanation is a far developed field with many streams—between a "realistic" line supported, say, by Salmon[8] and an "epistemic" (anti-realistic) line represented by Hempel,[9] branching in a wide landscape, and also including computational,[10] cognitive,[11] communicative, and other models. The Theory of Explanation is a very interesting topic in the frame of the philosophy of science. It aims at analysing the structure itself of what we want to call explanation and the consistency, the a priori basis and the achievements of the various models. It typically tries to reach a general understanding of explanation over the borders of particular sciences and to abstract in this way its essential character. This will not be, however, our approach here as hinted at in the above remarks. First, we shall concentrate our discussion on physics (and I hope—but do not claim—that we can obtain here paradigmatic insights). Secondly, we shall not start from one of the many philosophical explanation models and try to apply it to physical theories. I want instead to start from physical theories, their features and their progress and try to see in what sense and how can we then speak of explanation. This should also give us the means to follow the development of physics, which I assert to be an essential aspect in a discussion of explanation. This is a bottom-up approach as compared to the top-down one of the philosophical discussion. I do not intend to define an "interface" between these approaches, but there may be some hints to take from each other. So, for instance, the coherence question from the philosophical discussion which is not immediate in the bottom-up proceeding, or the wholeness character of explanation implied by the closed systems of concepts of the physical theories which seems not a major point of view in the top-down approach.

---

[6] Cassirer [15].

[7] A. Einstein, "Reply to criticisms", in Cushing [18], p. 357. This is not "against philosophy", this just means that physics needs its free space to do its job.

[8] Salmon [43].

[9] Hempel [28].

[10] See, e.g. Thagard [50].

[11] See, e.g. Churchland [16].

My assertion is that the physical theories provide explanations for the physical world and not mere ad-hoc instruments for classifying the observations. In the following I shall try to make this statement concrete.

I shall proceed from the poignant discussion given by Duhem.[12] After renouncing (with him) metaphysical models I shall take as basis for my approach his identified aim of physical theories as promoting "natural classifications" and recognise the latter as "explanations".[13] This of course is not Duhem's concept of explanation which he associates with providing a metaphysical model which he wants to exorcize from physics. The basis of my discussion, sketched here and made explicit in the next sections is summarized in the following two statements.

I want to consider explanation as that which is produced in (provided by) physical theories and try to discuss the features of the former in terms of the properties of the latter.[14] Notice that I shall not speak of the particular explanation a certain law provides for a particular phenomenon, unless this represents a valid specialization of the theory or of a closed approximation to it. The interesting object to consider in connection to explanation is the theory as a "consistent system of concepts" (Duhem thesis).

I contend that realistic points of view understood as the claim that physical concepts—or physical symbols, Sect. 2.1*—have a referent in the real world are perfectly defensible.* Although in the evolution of the physical knowledge both the symbols and the "objects" they are supposed to refer to change, a steady referential correlation exists, is well defined and follows the evolution of the theories. Correspondingly, explanation is to be seen as a process which can itself be understood from the evolution of the theories.

Thereby I shall make a number of qualified omissions. I shall not present a critical discussion of the relation between experimental results and observations on the one hand, and what we shall call "empirical facts" on the other hand—neither at the level of controlling the former (which we assume fulfilled by standard procedures[15]) nor at the more subtle one of the interpretation of our perceptions.[16] We shall not discuss further procedures in translating empirical facts in mathematical objects, such as idealization and generalization, besides mentioning their role in explanation. We shall not go into details of the physical theories and their progress beyond the aspects most relevant for the discussion intended here. I shall also not enter the philosophical

[12] Duhem, "La theorie physique, son objet et sa structure" (1908), english translation [20].

[13] A discussion of the understanding of "natural classification" as assumed in the following will be given in Sect. 2.3. For the moment I shall only refer to Duhem [20], Chapt. 2 with the suggestion that the order provided by an accomplished theory tends to be "the reflex of an ontological order".

[14] There are of course also other notions of explanation as related to the context, to other understanding of causation or to particular phenomena—cf. for instance P. Weingarten, M. Massimi in proceedings of the AIPS Conference 2016, "Mechanistic Explanation" (to appear). For a general discussion see Woodward [55].

[15] This implies of course theory and mathematics and can be very complex—the "cosmic distance ladder", for instance, belongs to the empirical part of cosmology but it implies both observation and theoretical models.

[16] H. von Helmholtz, "Die Tatsachen in der Wahrnehmung" (1878) in [27].

discussion of the many models concerning the foundations of physical knowledge and the character of the research. Instead I shall follow Helmholtz, view realism as a good hypothesis[17] and try to see what conditions does modern physics impose on it. I would consider myself as an adherent of a "critical scientific realism", but instead of trying to find a good name for this I shall just let it hopefully become clear in the following.

The most important omission, however, is that I shall only consider established physical theories, going up to the Standard Model of elementary particles (SME, a QFT) and of cosmology (SMC). The latter are called "models" since they do not provide a (sought for) full unification for the fundamental phenomena, not even for the electro-weak and strong interactions. They are seen therefore as steps in an advancing search for unification which eventually should include all interactions especially also gravity. At present we only have a classical theory for the latter which relates gravity with the structure of space–time (General Relativity, GRT). The cosmological standard model SMC is based on GRT but introduces concepts from the SME without achieving a unification of the two theoretical schemes.

In spite of them being considered as models SME and SMC are to a certain degree established and provide a basis for advancing and testing any further developments (beyond searching for solutions in the framework of the SMs, such as dark matter). Some further developments, however, are either limited to the description of a number of fundamental paradigmatic phenomena such as black holes which, however, could not be bound in an overall, consistent theory of Quantum Gravity (QG). Or they belong to a field of partly competing ideas and models such as string theory, AdS/CFT, asymptotically safe gravity or loop gravity which also do not yet build up a unique, overriding and established theoretical scheme. Therefore I found it difficult to recognise and select an overriding symbolic structure, although there are very many exciting developments which also involve new mathematics. It might be that explanation starts now exploring new directions with a new hypothetical basis of metaphysics and principles, however I could not provide interesting speculations beyond those already populating the field. It is not even clear in some cases that my approach of seeing in theories the achievement of a "natural classification" can be meaningful unless we take a platonic perspective.[18]

I shall also not discuss other *Ansätze* particularly concerning QM for which a deterministic metaphysics is postulated. Among the furthest studied are the de Broglie-Bohm model[19] which uses a non-local process and 't Hooft's classical cellular automaton model.[20] Likewise I shall not discuss unifying *Ansätze* based on an universal quantum theory of matter and mind (*Geist*), such as Görnitz and

---

17 H. von Helmholtz [27]. In modern terms his argument can be described as IBE (inference to the best explanation)....

18 Tegmark [49] and related papers. This perspective can appear as the logical completion of a certain form of "structural realism" which appeals to mathematical physicists such as Poincaré and Weyl and which cuts short the question why, say, equations and the objects they both define and refer to should have different ontological status.

19 See Bohm and Hiley [12], also Bell [5].

20 For a comprehensive account see 't Hooft [51].

Görnitz [25] which display a monist entirety picture of everything in a development connecting to the "Ur-model" of C.F. [52]. All these competing approaches introduce various metaphysics. As interesting as these developments and those mentioned before can be, such discussions need a different frame than the one provided by this essay.

In the first two sections of the first part I shall briefly discuss some relevant features of our theories and their progress, before approaching in Sect. 3 the connection to explanation. In Sects. 2.4–2.6 I shall go into some detail concerning aspects of this connection and how this is reflected in our view of explanation. Section 2.7 is a brief recall of the so called decoherence, both because it enters some of the arguments in the main discussion and because it was the basis of the common work in the project with Heinz-Dieter Zeh mentioned at the beginning. Section 2.8 addresses the role of mathematics in theoretical physics and Sect. 2.9 closes the discussion.

As I try to convey in the first part an important aspect is the mathematical structure of physical theories.[21] This aspect follows from a basic interaction between physics and mathematics. Physics uses mathematical schemes, but many developments in mathematics are prompted by physical questions. One essential feature of the role of mathematics concerns the fundamental quantitative aspect of physics. This introduces computation and computability as indispensable components of this structure and computers as their major basis. This will be the subject of the second part of this essay. We shall see there a sketch of the various uses of computers in physics and of the further developments concerning both use and definition of computing. Finally I shall discuss how the participation of computers in explanation may be viewed.

2 Physical Theories and Explanation

2.1 Physical Symbols and Physical Theories

Symbols and physical knowledge; The character of physical symbols; Theories as closed conceptual systems

For a discussion of explanation in physics I propose to see symbols as the basic elements in building physical knowledge. I prefer the notion of symbols to that of sign and of concept: the symbol points at some thing (like a sign) but also bears[22] an inner structure (like a concept) allowing it to connect to other symbols. This is rather akin to what Leibniz says about ideas and the capability to build ideas as fundamental

[21] In fact this interaction reaches also the meta-theoretical level when comparing different theories, cf Barth [4].

[22] Or indicates a structure which can be deployed when asked for.

for thinking: ... there must be something in me which not only leads to the thing but it also expresses it."[23]

A succinct modern history of symbol is given in the article "Sources for the history of the concept of symbol from Leibniz to Cassirer" by Massimo Ferrari.[24] Particularly intuitive for a physicist is the proposal of Heinrich Hertz" We construct internal appearances or symbols of external objects, and we make them such, that what results by thought-necessity from such symbols will always be a symbol of that what follows by nature-necessity from the symbolised objects ...".[25] This frequently is referred to as a constructivist testimony whereby it is overlooked that one speaks here of "symbols of *external objects*" (Hertz) —which the symbols "*express*" (Leibniz). In the following I shall stay at the above pragmatic level and sometimes also use the word physical concept meant however as symbol.

Of course one can ask which are, say, the "things" mentioned above, what is their status, etc. A discussion of this and other aspects of the use of symbols in physics is provided in the mentioned volume *Symbol and physical knowledge*[26] which is dedicated to this approach and collects contributions from philosophers and physicists.

Physical symbols appear and act in the process of evolving physical knowledge, they however "solidify" in the frame of theories, when the process finds its realisations in "closed systems of concepts (or symbols)".[27] It is therefore useful to primarily refer to physical theories. For definiteness we shall mostly stay within the so called "fundamental physics",[28] that is the theories of fundamental phenomena where some of the problems to be discussed appear most clearly.

First let us notice that the physical symbols are bound in a network of connections with two main valencies: On the one hand, they have to support the abstract, *mathematical structure* of a theory as expressed, e.g. in the equations of the latter. Therefore they need to be defined as mathematical quantities. On the other hand, they must support the *interpretation* of that abstract scheme down to phenomena and therefore must act in logical chains of symbols aiming at experiment and observation. We thus must define measurable quantities and measurement or observation rules. A rude illustration is given in Fig. 1.

---

[23] "... es muss also etwas in mir geben, das nicht nur zu der Sache führt, sondern sie auch ausdrückt." see Leibniz, Quid sit idea", in C.I. Gerhardt (Ed), vol. VII, p. 263 sq. [24].

[24] In Ferrari and Stamatescu (Eds.) (2002) [22].

[25] "Wir machen uns innere Scheinbilder oder Symbole der äußeren Gegenstände, und zwar machen wir sie von solcher Art, daß die denknotwendigen Folgen der Bilder stets Bilder seien von den naturnotwendigen Folgen der abgebildeten Gegenstände ..." [30]. Notice that this parallels Leibniz' conception, which is also worth mentioning: "Dass eine Idee von Dingen in uns ist, heißt deshalb nichts anderes, als dass Gott, Urheber gleichermaßen der Dinge wie des Geistes, diese Fähigkeit des Denkens dem Geist eingeprägt hat, damit ⟨derselbe⟩ aus seinen Tätigkeiten dasjenige ableiten kann, was vollkommen demjenigen entspricht, was aus den Dingen folgt." in C.I. Gerhardt (Ed), vol. VII, p. 263 sq. [24].

[26] Articles in Ferrari and Stamatescu (Eds.) [22].

[27] W. Heisenberg, in Scheibe and Süssman (1973), p. 140 [45].

[28] For an actual assessment see Seiler and Stamatescu (Eds.) (2007) [22].

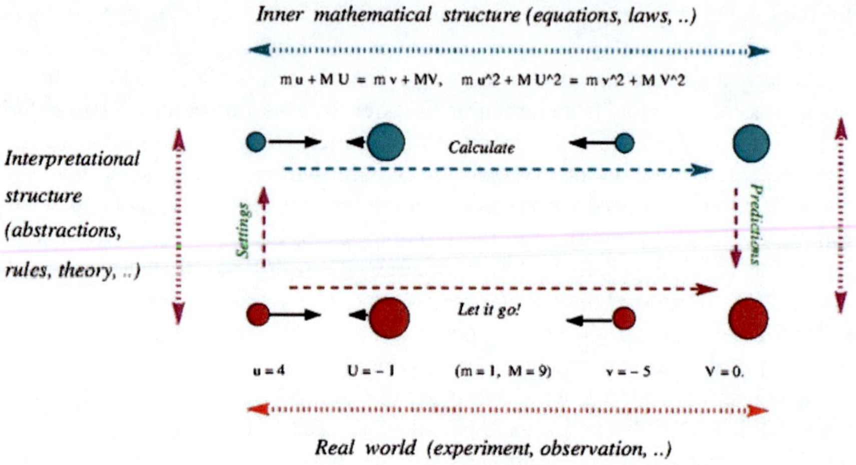

Fig. 1 Abstract collision with a slow lorry in an extended Hertz scheme. © by the author

Now, the evolution of physical knowledge as we can follow it, say, since Galilei and Newton (and if we discard some few dead ends like *caloric*, *phlogiston* or *ether*) appears to proceed by producing systems of symbols which are closed within a mathematical scheme and over a class of reproducible phenomena well defined by general criteria (e.g., scale, type of interactions, complexity). This evolution is not cumulative but progressive in that it produces hierarchies of theories with increasing capacity of reducing the multiplicity of the phenomena to few rules (laws). In such hierarchies the theories remain related in the sense that the "lower" theory can be approached in the frame of the "higher" one.

We have as *fundamental theories*: *Classical Mechanics* (CM), *Thermodynamics* (Th*), Special Relativity* (SRT) and *General Relativity* (GRT), C*lassical Electrodynamics* (ED), *Quantum Mechanics* (QM), *Quantum Electrodynamics* (QED) and the *Standard Models of Elementary Particles* (SME) and *of Cosmology* (SMC)- which involve particular cases of *Quantum Field Theories* (QFT).[29] Directed at the description of complexity is S*tatistical Mechanics* (CSM) and more generally the *Theory of Complex Systems* (CST). *Solid State Theory* (SST) uses QM, ED and CSM to build up a theoretical scheme defined by its domain of application. In the following I shall refer to QM as a paradigm of quantum theories, unless I shall explicitly consider one or other particular quantum theory.[30]

The line ED—QM—QED is a particular example of the evolution of theories and their symbolic structures. In all three we have, e.g., a symbol for the *electron* as a

---

[29] The SM are in fact a collection of partial QFTs without complete unification, GRT and partial models.

[30] A succinct presentation of the features and the general implications of quantum theory including most of the themes discussed here is Kiefer [34], see also Kiefer [33] for a brief introduction and discussion of the main quantum effects.

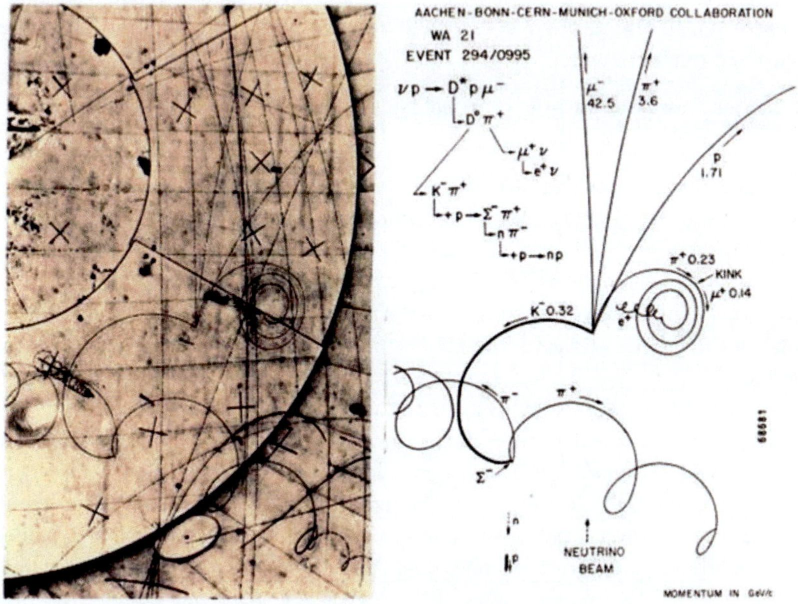

Fig. 2 Bubble chamber BEBC event, Dec. 1978, D-meson, © 1978–2017 CERN

fundamental object of the theory. It is typically indexed by *e*, but it clearly does not have identical properties in the different theories. We find:

e^{ED}: classical particle, identifiable, conserved, obeying SRT.
e^{QM}: quantum particle, not identifiable, conserved, obeying CM (Galilean symmetry).
e^{QED}: quantum particle of a quantum field obeying SRT, not identifiable, not conserved.[31]

In an absolute sense these symbols have no unique referent. However, from a pragmatic point of view in an empirical and a theoretical sense they directly address reality: they are well defined in the frame of each theory and point each time onto a reproducible empirical fact of a class of phenomena well defined by general criteria. They are compatible by connecting to each other, especially when a "higher" theory overrides a "lower" one, incorporating its empirical domain giving its limitations and developing its symbolic scheme.

This is nicely illustrated in Fig. 2 which represents a celebrated bubble chamber event (evidence of *charm*) at CERN. We see on the same picture production of particles (a QFT effect) which then produce traces in the medium (where also *decoherence*

[31] In fact e^{QED} is not really a fundamental object in the theory. The fundamental objects in QFT are the quantum fields, in which the fundamental relations of the theory (local interactions) are expressed. Particles are special manifestations of the fields. We also have unstable particles and resonances (e.g., Higgs boson!) and we observe a continuous transition between what we understand by particle and by just signal (enhancement) in a collision cross section. See also Falkenburg [21].

plays a role, a QM effect, see Sect. 2.7) and move according to *classical* ED in the superimposed magnetic field.

Based on these observations, the proposal is:

When conceptual systems close in coherent, successful theories over a class of phenomena (Heisenberg) they define a symbolic structure and symbols having a conditional claim for reality (referential properties), necessity, truth-carrying properties, relative to a both theoretically and empirically well circumscribed and reproducible class of phenomena (typically defined by the energy scale). The theory as a whole inherits the above properties. We shall speak therefore of a conditional character of the theories relative to a given step in the development of the physical knowledge.

Notice that the symbols show certain continuity properties in the theory developments and generalisation capabilities: e.g., in proceeding from QM to QED we use the symbols e^{ED} and e^{QM} as basis for developing the new symbol e^{QED}. We even use the symbols of QM particles as generalized to unstable particles observed as "resonances" in collision experiments. These symbols also seem to build up a ladder in their capabilities to describe the phenomena, therefore we may ask whether they might represent milestones on a track, let us call it **e** leading our search. Whether this track converges and onto what is a good question: it may completely change its character or be replaced by another concept. We don't know whether a *Final Theory* is possible and if it will entail an e^{FT} symbol. But at least for a while this track appears to be a good conception and represent an "element of reality" with its stages having well defined referential relations.

The referential relations, however, are more complex than apparently suggested, for instance in connection with the ED-QM-QED line above. The electron, say of QED, e^{QED} is a very complicated object which reveals its properties in the relations to the other symbols of the theory. These include its participation in the mathematical-conceptual structure of the theory as well as in relating to empirical observations. It is a basic effort to define its predicted properties as well as to envisage and perform the corresponding experiments. Such is, for instance the spin-magnetic moment of the electron which is by a so called gyromagnetic *g-factor* larger than its classical value. The relativistic, quantum mechanical Dirac equation gives $g = 2$ which already fullfils certain empirical bounds and the overriding quantum-field theory QED gives a correction $g\text{-}2 = 0.0023193048(8)$ agreeing to 13 digits with the refined experimental value 0.00231930430256(13)! The referential relation is dynamical, paralleling the evolution of the theories and of the empirical performance. This also emphasises the role of calculations.

Each of the mentioned fundamental theories is closed in its mathematical and interpretational scheme and also concerning its empirical reference domain.[32] It

[32] The notion "closed" as used here is rather sloppy and does not imply that the dynamics may not lead to "unphysical" situations, such as singularities and divergences. Sometimes these can be tamed by supplementary procedures, as in QFT, but they also can be of a more fundamental nature, as in CM, ED or GRT and signal the need for an overriding theory.

provides a good description of the phenomena in some restricted validity domain as a controlled approximation of an overriding theory, so for instance *Newtonian mechanics* (CM) as approximation to *Special Relativity* (SRT) or approximation scheme for the *General Relativity* (GRT)—the so called post-Newtonian approach. Moreover it contributes to the lines of theories' development by providing symbols and symbolic schemes which can be further evolved (Lagrangean, action, equations of motion,..). It is significant that all the theories mentioned above belong to the normal curriculum of the physics study, which stresses the special continuity of physical knowledge.

To summarise: Introducing symbols as basic elements for our discussion helps us in describing physical theories and the way they may be said to address reality. These symbols are bound in a network of relations carrying both the mathematical and the interpretational structure. Following the evolution of our physical understanding we need to qualify the referential features of these symbols and thus also the relation of the theories to physical reality. A brief account of this evolution is sketched in the next section.

2.2 *On the Progress of Theories*

Principles, symmetries, scales and thresholds

Some arguments oppose a view, that physical theories proceed by "accumulation", to one of sudden "jumps". This "myth" grossly simplifies the situation. Physical theories never develop just by accumulation and also never simply make jumps. *Ptolomaeus*' system developed by accumulation, but this was not a theory. The electrodynamics before Maxwell was not a theory either, but a collection of partial findings.[33] Special relativity was the work of a genius, but it was motivated by *Maxwell's Electrodynamics* and came in a context of very many ideas and results (*Lorentz transformations,* etc.). The myth resides on a misunderstanding: on the one hand, accumulation of facts and ideas is a necessary step *before* a theory can emerge. This accumulation is provided in partial findings and laws (Kepler, Coulomb, Faraday,..), in tests using the old theories (such as black-body radiation) or in unexpected empirical facts (radiation, beta-decay). On the other hand, once a theory is established we still have to find and test its consequences for all possible situations—such as describing the movement of a number of bodies in gravitational interaction, or the mass-spectrum of QCD. A theory is typically given by a limited number of relations and we must decode its inner structure to apply it to phenomena. This work is accumulation of information *about and from* the theory, and it will also help us to find its failures[34] and

[33] This accumulation was the ground on which Maxwell's ED could be established, but the latter is not just the sum of these partial laws. Accumulation and jumps are strongly intercorrelated, a modern example is the development of the SME.

[34] Theoretical (e.g., the divergent self energy of electrons) as well as empirical (e.g. the stability of matter) in ED.

proceed to a new level of physical understanding. It prepares the field on which new ideas can be born and based on new empirical findings ingenious work would realise a *jump* to a new theory. We found, for instance, that Newtonian gravitation cannot describe the internal movement of visible matter in galaxies. The main present work in this connection is testing hypotheses for dark (unvisible) matter or for modified gravitation laws which should show whether an extrapolation of old symbols may work or new ideas are called for. The evolution of theories is a process with interacting empirical and theoretical steps. An essential step is *promoting hypotheses.* Besides empirical information and theoretical indications we are helped here also by some general principles.

For Aristotle principles are the foundation of knowledge.[35] Ernst Cassirer speaks of principles as the matrices for building laws of nature.[36] In the present day physics principles are indeed used in establishing laws of nature and constructing theories. However, they may depend on the context in which they are introduced and may sometimes be replaced by a different concept. For instance, the shortest path principle of geometric optics (expressing the minimal action principle of Maupertuis) is not needed in wave optics, where using Huygens principle and the formation of wave fronts leads to the same and many more other results from interference effects in the propagation of the light waves.

Among the principles acting in establishing physical theories a primordial role is played by symmetries. Symmetries are statements about the invariance of physical laws under certain kinds of transformations. The most evident symmetries concern space–time, so for instance the *Galilean symmetry* of CM (translations, rotations and uniform motion) or the *Poincaré symmetry* of the so called *special relativistic* field theories (ED and QFT). Less intuitive are the so called internal symmetries, especially the so called *local gauge symmetries* which can be understood as a certain freedom in locally defining a field while ensuring comparability of fields at different points by transporting information of the local properties from point to point.[37] In establishing the *Standard Model* we found how such symmetries can be combined, and in searching for a so called *Grand Unified Theories* (GUT) how they can be derived from higher symmetries.

One interesting symmetry in the evolution of theories corresponds to *scale transformations.* To introduce these transformations we must first speak of "dimension" (as measured in meters m, seconds s, Joule J or eV). The existence of universal laws and fundamental physical constants which are pure numbers implies that we can translate the various units into each other. This brings in the so called *natural units system* in which there is only one "dimension", say the *length*, and all the others are

[35] Barnes [3], Aristotle, JB 53::Analytica posteriora I 2 71b.

[36] Cassirer [15], p. 189.

[37] This concept was hypothesized following the empirical observation of conservation laws (e.g. charge conservation) and other correlations between observations. It led to a powerful theoretical *constructive principle.* It is not a genuine symmetry but has rather the status of a *covariance* since physical information is normally carried by gauge invariant quantities.

powers or inverse powers of it. The *energy*, for instance, is an inverse length and typical units like GeV (about the rest energy of a proton $=1.6021773 \times 10^{-10}$ J) and fm ($= 10^{-15}$ m, about the "size" of a proton) are related by 1 GeV $\times$ 1 fm $\simeq$ 5.068. Scale may refer to length or to energy. Scale symmetry (invariance under scale transformations) would appear as scale independence of certain phenomena or theories not involving dimensional parameters (lengths or masses).

We have before differentiated theories by the scale of phenomena to which they apply, for instance CM—QM—QFT, or CM—SRT by increasing the *energy scale*. This however, was not precise enough.

Firstly, the domain of phenomena addressed by the higher theory includes that of the lower one. The latter domain can be efficiently approached in a sufficiently good approximation by the lower theory while the higher theory is valid "everywhere" in the enlarged domain.

Secondly, scale transformations are continuous transformations, while the chain of theories is discrete. In fact increasing the *energy scale* we encounter thresholds and the need to introduce new theoretical elements in order to obtain a closed system of concepts. With increasing velocities the *Galilean relativity* of CM became increasingly inadequate but the new theory (SRT) required a new, consistent mathematical scheme introducing a new symmetry (*Lorentz group*). The step from QM to QFT was called for empirically by the (expected and observed) opening of thresholds for particles creation at high energy and theoretically by the necessity of incorporating SRT in the theoretical schemes. But in order to account for this a new theoretical scheme was necessary: not only a new symmetry (SRT) but also a redefinition of the fundamental physical quantities (*quantum fields*) and of the complete mathematical, conceptual and interpretation scheme.[38]

It may be useful for the discussion of continuity to briefly follow the CM-QM-QFT chain. We can indeed reproduce CM effects in a genuine QM analysis (*decoherence*) but not the logical structure of CM. Conversely, the QM scheme is not *deducible* in the frame of CM. Even clearer is the case in the QM-QFT connection, where the change in the fundamental symmetry (from Galilean to SRT) forces a redefinition of the fundamental quantities for which the fundamental equations hold (from *particles* to *fields*).[39]

In a certain sense when a "higher" theory takes over the domain of a "lower" one one can say the lower theory was wrong and if we were clever enough we should from the beginning go for the higher one. But this is not so easy. Take CM-SRT: it would have been a undefined procedure to look beyond CM without any theoretical or empirical hints, and it would also be cumbersome to approach phenomena at low

[38] A relativistic QM cannot be established as a consistent theory see e.g. Wachter [53]. This makes clear that one cannot simply add a new hypothesis to an old theory but one must build up a new, closed theoretical scheme "incorporating" the former. In the case at hand it means going from particles to fields as fundamental with particles as a manifestation of these.

[39] We can, however, on the basis of a given theoretical scheme valid at a high energy scale derive *effective theories* at lower scales by the "renormalization group" procedure which allows to redefine the theory for adequate applications there.

velocity, easily described by CM due to the Galilean symmetry, with the full SRT arsenal. CM is a necessary step before and a useful one after SRT. Even worse is the case for the CM-QM pair. Here we need to determine a whole process (decoherence) besides taking limits in order to obtain a description for phenomena like scattering of particles or moving in an external field.

But there is more to it. The logical structure of each theory in a pair is evolved on the basis of a class of phenomena it is confronted with. The success of these theories means that *nature does behave that way there*. In my brand of a "critical realistic" view the symbols of these theories find their referents there. Therefore in the progress from lower to higher theories we just follow the sometimes complicated build up of nature itself. It is nature which suggests to us closed theoretical schemes since it apparently follows a hierarchical structure—which should then be mirrored in the progress of theories. And developing the lower theory is indispensable also because this offers symbols to be further developed.

Before proceeding, however, to relate theories and explanation we should say a word about identifying objects. Again, this discussion does not belong to the philosophy of science,[40] but to physics. Most symbols in a theory can be seen as "objects" and we may ask how are these objects identified (which implies both finding and constructing them, see below). Many of them are intuitively found as chunks of matter (e.g., the particles mentioned in Sect. 2.1) or of energy (e.g., the photons) but they may also be subject to metamorphoses, so is, for instance the "force" of classical physics and of QM replaced by the exchange of certain particles in QFT. As already suggested by these comments—and by the discussion of Sect. 2.1—identifying these objects is itself a process between experimental observations and theoretical hypotheses ending up in a consistent symbolic network which is the theory with its mathematical, interpretational and empirical structure. Here a role is also played by metaphysical conjectures and the "continuation" of objects from other (typically preceding) theories, metaphorical suggestions, models, etc.—but the last word is said when the complete scheme of the theory closes. The processes of object identification are included in the progress of theories.[41]

2.3 How Do We Understand Physical Theories and the Question of Explanation

Features, status and the development of a theory; What does a theory achieve; Explanation and natural classification; Explanation and physical theories

With the previous discussion in mind we can now proceed to make the connection to explanation and substantiate the assertion in the introduction. As already suggested there we consider explanation as provided by mature theories. A theory is a self

[40] For a brief overview see Rettler and Bailey [42].

[41] For a transcendental perspective see Bitbol et al. [8].

consistent, tight conceptual scheme with a mathematical inner structure and laws for relating its symbols to observation. These are fundamental aspects, since only then can we comply with the requirement, already voiced by Aristotle, that knowing means to know the grounds by which some thing is and also know that it cannot be otherwise.[42] A theory must claim necessity and universality with respect to a well defined class of reproducible phenomena, since only then can we learn, from agreement and from disagreement, and evolve our knowledge.

A theory may be built upon a number of (partly already established) partial rules and laws which it binds in a self consistent overriding scheme. The theory is more than the sum of these partial laws. Consider as example *Electrodynamics* (ED) where this can be observed in detail: it is preceded by the laws of electrostatics (*Coulomb*), currents (*Ampère*), induction (*Faraday*), ... but it is the genius of Maxwell to put forward the universal scheme of ED. This is now a coherent, tight scheme. It is significant, for instance, that this scheme introduced, for purely mathematical-theoretical reasons, a new law which could not be guessed from the partial ones (the *displacement current*). As a result a resetting of the space–time symmetries was promoted which finally led to Einstein's SRT. Another new result was the identification of light and electromagnetic waves.

The logic of the theory development is to follow the "how" (directly related to observation and based on induction) to a "why" (receded from observation and based on hypotheses). Thereby the multitude of observations typically gathered in empirical laws is *reduced* to few fundamental laws: From Kepler's *kinematic* laws describing the movements of the planets to Newton's gravitation theory which introduces the *dynamical* universal attraction law of gravity and from which Kepler's *kinematic* laws can be *derived*; from the laws of atomic spectra (Lyman, Balmer, Paschen) and of the black body radiation (Rayleigh- Jeans, Wien, Planck) to QM.

As argued in the previous sections we must and can refer to the theories closing as conceptual systems at each step with reference to a class of phenomena which give them a conditional (relative) character as discussed in Sect. 2.1.

With this in mind let us ask: What does a theory achieve? Duhem's claim, for instance, is that a physical theory does not provide explanation in the sense of revealing a metaphysical structure, however it also is not just simple classification. It provides instead.

A "natural classification" of laws and objects, and the better the theory becomes in describing and predicting phenomena, "the more we apprehend that [its] logical order ... is the reflex of an ontological order".[43]

If we renounce therefore of searching for metaphysical revelation it seems then natural to relate explanation to physical theories and see the task of explanation as

[42] "[wir meinen] etwas zu wissen, wenn wir glauben, sowohl die Ursache zu kennen, aufgrund derer ein Ding ist (und zu wissen, dass diese seine Ursache ist), als auch, dass es nicht anders sein kann" Barnes [3], Aristotle, JB 52: Analytica posteriora I 2 71b.

[43] Cf. Duhem [20], Ch. II, & 4. This and other statements in Ch. II and at other places suggest that the "instrumentalism" positioning of Duhem may need a more refined discussion, as also hinted at in Ariew [2], see also the article of Karl-Norbert Ihmig in Ferrari and Stamatescu (Eds.) [22].

putting into evidence *"natural classifications"* aiming at uncovering the *mechanisms of nature.*[44] The suggestion is therefore.

to consider explanation to be provided by physical theories and just as these to be committed to an evolutionary process and have at each stage a relative (or conditional) status.

Explanation in this view is therefore a process: we can "explain the world" at a certain level and can push our knowledge further with the theoretical progress, *resetting* the old explanation in the new frame.[45] And as with the evolution of theories, the older explanation is not lost: it is retrieved, e.g. as approximation and it has helped building the new explanation. Notice that the mentioned relativism is fundamental and not a question of perspective, since it follows from the direct relation to theories. Of course for one and the same theory there may be different assumptions about some kind of metaphysics or "interpretations" of its basis (QM is a good example: collapse, relative states, etc.) which will enter as alternative metaphysics. This is a relativism superimposed onto the basic processual character.

This basic relative or conditional character will also apply to our understanding of "natural classification" itself and also to other concepts such as "natural kinds".[46] In the pragmatic, realistic approach followed here we can think of natural kinds as the basic components of the theory together with their empirical referents. Is e of Sect. 2.1 a candidate for a"natural kind"? We can only speculate whether e is a relevant object in some final theory, if at all reachable. In the perspective of a process character for explanation we may rather consider the family $\{e^{\mathrm{ED}}, e^{\mathrm{QM}}, e^{\mathrm{QED}}..\}$ as a natural kind, searching for (and finding) its actual referent in each actually given theoretical and empirical environment.

2.4 Some Aspects of Explanation in Various Theoretical Schemes

Metaphysics; Principles; Theory and model; Fundamental and effective theories; Special theories

[44] Referred to by I. Kant in *The critique of practical reason, Ch.III*, also as "physico-mechanical connections in nature" in *The critique of pure reason, Appendix.*

[45] We can, for instance, derive *classical behaviour* as a certain QM effect *(Decoherence)*, see Joos et al. [32] and Sect. 2.7.

[46] This is of course an important discussion which, however, is beyond the aim of this essay. In ED, for instance, we should consider also the electromagnetic fields as "natural kinds" and possibly also the potentials because of their role, say, in quantisation. This would imply, however, redefining identity, classes, etc. taking into account gauge transformations and then adapt our discussion on their relative character considering the progress of theories. For a succinct overview of the philosophy of science discussion hereto see Bird and Tobin [7]. Here however we shall only retain the observation that natural kinds just as natural classification have the same conditional status as the theories and their symbols, see Sect. 2.1, and are also bound in the process of evolving the physical knowledge and should be empirically and theoretically well defined at each level.

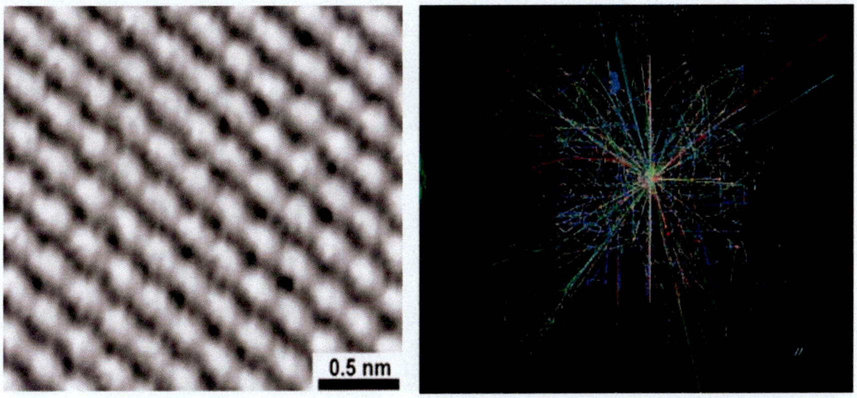

Fig. 3 **a** *Left*: scanning tunnel microscope picture of a graphite surface showing an ordered lattice of atoms (from commons.wikimedia.org, gemeinfrei). **b** *Right*: high energy LHC event in the ALICE detector, particle shower from a Proton-Pb collision (© 2012 CERN, ALICE)

Relating explanation to physical theories allows us to derive features of the former from those of the latter. One interesting question is, whether metaphysics completely disappears from explanation. If we understand metaphysics hypothetically this is not the case. Instead we find metaphysical settings as basic *assumptions* which can be very efficient in developing theories. An example is the atom hypothesis, which proved very fruitful in leading the development of physics—and one can even reply today to Mach's scepticism ("did you ever see one?") with pictures from field microscope—see Fig. 3 left. A more refined example is that of elementary particles[47]—see Fig. 3, right. Metaphysical settings can therefore enter explanation if we remember the relative and process character of the latter and take into account a hypothesis-character for the former.

A similar situation pertains to principles. Both they and metaphysical settings lead the progress of physical theories and are therefore in some sense "meta-theoretical". As already remarked, however, like metaphysical settings principles may also be related to a certain stage in the evolution of physical knowledge. This will also affect their role in explanation (see also Sect. 2.9).

Other interesting questions concern partial theories, models and effective theories as compared to what we called "fundamental" theories. Effective theories are full grown theories with only the amendment that they comprise a number of parameters and rules which are expected to be fixed at a higher level. As an example, the *standard model* SME is considered an effective theory to be derived from an envisaged more powerful theory with a more comprehensive symmetry at a higher energy scale. The overriding theory should both fix masses and other parameters of SME and provide a higher order of unification (GUT—*Grand Unified Theories*). Correspondingly, the explanation provided by the effective theory is relative in the sense discussed before,

[47] "Particle" is a powerful concept in promoting hypotheses, so for instance it dominates the search for "dark matter". See also Falkenburg [21].

but it additionally acknowledges incompleteness since it refers to another (higher) theory to fill the explanation gaps.

A certain subclass of phenomena can be successfully treated by setting up a theoretical scheme based on the particularities of that subclass of phenomena. Such is, for instance, the BCS-theory of *superconductivity*, which uses ED and QM in a *solid state theory* (SST) frame applied to the special case of the electron flow in some solids at very low temperatures. It identifies special effects due to the coupling between the movement of the electrons and the oscillations of the lattice of ions, introducing new symbols—the so called *Cooper pairs*. It thus provides an explanation in the framing of SST but spelled out for this concrete case. Notice that SST itself borrows basic concepts and laws from QM, ED and *statistical mechanics* (CSM) while also introducing new symbols and rules. The explanatory capacity of the SST implies therefore reference to other theories.

A different situation pertains to models. Many models typically do not represent self consistent theoretical schemes, neither mathematical nor interpretational. Their essential role is to suggest connections among symbols and test hypotheses in a simplified context. Their contribution to explanation is correspondingly sketchy and needs justification through the theories to which they lead or in the frame of which they are developed. The *Bohr atom model* is an example. It is theoretically inconsistent since it is set up in the frame of *classical electrodynamics* while introducing hypotheses incompatible with the classical movement of charges in an electric field—*stable orbits*. However, by suggesting (and testing) these hypotheses it directed the search leading to *quantum mechanics*.

The Landau model for phase transitions in *statistical mechanics* is another kind of example. It is derived from the fundamental mathematical object of CSM (the *partition function)* under some general conditions. It has universality character and can be applied to various concrete statistical mechanics models. It provides therefore elements for explanation of the phenomena due to complexity, independently of the particular interactions involved.

We also find models representing a self consistent approximation of a theory or a reduced account of a substructure thereof with the corresponding subclass of phenomena and aimed at emphasising the role and effect of some relevant connections in the theory, so for instance the non-linear σ-models or the chiral perturbation theory. Such models can be sometimes transported to other theoretical schemes.

Models have a very important role in the evolution of the theories, sometimes just because their explanatory program is less strong and is flexible enough to take into account analogies and metaphors which are important in building up hypotheses, or because they help selecting dominant theoretical concepts and rules.

2.5 Understanding and Explanation

Explanation and understanding; Intuition; Right concepts

In my pragmatic view a theory provides *explanation* for the phenomena in its field, and *explanation* as such is well defined. A different question pertains, however, *understanding*. In every day language we say, we explain some physical relations to somebody, but what we mean is making him *understand* them. Let us consider Maxwell's ED. Already the concept of field is not easy to understand. Faraday's concept of field is a very fruitful abstract symbol, his image of lines of force filling the space is intuitive but incorrect: there are two vectors in ED associated with each point in space, varying in a correlated way in time. There is only one vector in Newton's gravity. Intuition and imagination may help a bit but in the end we may need to accept a certain symbol with its embedding into the theoretical structure as fundamental and use imagination only as a vague suggestion to circumscribe it.[48]

Understanding and *explaining* are not identical, the former is more complex and less well defined. On the one hand it depends on explanation. So no understanding can come from *Ptolomaeus*' system and not even from *Kepler's* laws since they do not themselves ground theories and thus do not provide explanation. *Newton's* gravitational law (*universal attraction force*), however, does and can therefore bring understanding. On the other hand understanding may precede and generally will interact with explanation. So the concept (or symbol) of *stationarity* introduced in Bohr's model and illustrated by radiation free orbits will become a basic concept in QM. Understanding uses softer concepts than explanation while relying on the latter to give them solidity if required. These concepts are often generated in models preceding a theory (such as Bohr's model) and finalised in the theory or constructed in the frame of a theory (such as the *Cooper pairs* in the BCS-theory of superconductivity). Finally understanding may be context and perspective dependent and therefore not unique.

Being able to describe the simple effects in the theory may be a first step in providing understanding—so, for instance, the two-slit experiment in QM. However, one cannot go far enough this way: most interesting QM effects (*entanglement, decoherence*, etc.) cannot be reduced to intuitive notions and images. Instead, according to *Heisenberg*:

> We have understood a group of phenomena when we have found the right concepts for describing these phenomena.[49]

Finding the *right concepts* is a very important issue, *not only in understanding but in fact already in developing theories and in spelling out explanation.* For that we may need to accept abstract symbols as *"right concepts"*, appearing as nodes in the mathematical and interpretation scheme of the theory. Electromagnetic fields and potentials in ED may be such right concepts. They are not very intuitive (*anschaulich*) but show universality, capability for further developments and underlie a big class of

---

[48] In a brief section Feynman [23], II, 20–9 discusses the deficiency of the imagination in science while finding intellectual beauty in the wave equation due to the regularities and further developments it suggests. If we take this over to understanding it says that it is not the direct imagination which counts but the multitude of relations implied by a symbol.

[49] W. Heisenberg, "The concept of understanding in theoretical physics", in Blum [10] CIII, p. 335. Notice that "right concepts" and "natural kinds" need not be related.

effects in the theory. These qualities make them also important in the development of theories. The latter selects the adequate symbols for the higher theory, e.g. in going to QED the potentials, which in ED are helpful but not essential, prove to be *the* fundamental symbols for quantisation (local interaction).

Entanglement, superposition, operators, states, measurement are such *right concepts* in QM, some fully non-classical, other distinguished by new properties. There are in fact many further proposals hereto, sometimes as alternatives: the wave function collapse of the Copenhagen Interpretation, Everett's relative states, Griffiths' consistent histories, ETH events, decoherence, etc. in various combinations.

Notice that QM comes together with a fundamental break not only in the interpretation scheme but also in its mathematics. Classical physics relies on real numbers, complex numbers may simplify calculations (ED) but are not essential. On the contrary, the mathematics of QM is fundamentally based on complex numbers, starting with the Hilbert space as space of states. This is, e.g. evident in the proof of the *Bell theorem*[50] which also summarises the "picture" of QM showing that the statistics of the latter cannot result from local, hidden variables, which would lead to classical statistics. The difference is made clear by considering a certain correlated experiment and is enforced by the complex number character of the QM quantities.[51] Therefore it may be a bold suggestion to add as a further "right concept" for understanding quantum physics that of complex numbers (complex Hilbert space).

2.6 *Explanation and the Structure of Theories*

Observation and experiment; Hypotheses; Mathematical scheme

The building elements of the physical theories are also mirrored in explanation. So we find again in explanation a discussion of the experimental protocols and observational procedures providing the empirical basis of the theory, e.g. the observation of particles in QFT as *tracks in the detectors*, or the *cosmic distance ladder* in cosmology (which both involve theoretical knowledge at various levels). They define the way the symbols of the theory are related to phenomena and thus what explanation pertains to. This also involves the handling of data, interpolation, generalization, etc.

The next step in setting up a theory is the construction of a system of hypotheses concerning the physical quantities and their relations, which define the symbolic structure of the theory. We stressed that due primarily to the mathematical basis of the theory the symbolic structure is a tight scheme and the test of a hypothesis typically involves a bundle of other hypotheses. For explanation this means that we cannot arbitrarily delete or modify an isolated element in the theory and also that we

[50] Bell [5].

[51] See I.-O. Stamatescu, Appendix 4, in Joos et al. [32].

must always mention (or at least keep in mind) the complex of symbols and rules accompanying any particular explanation.[52]

Finally, since mathematics plays a key role in physical theories it will also constrain the explanation at various stages: In deriving post- and prediction for phenomena (*celestial mechanics, super-conductivity*); in the search for adequate concepts in the mathematical-theoretical structure of theories (*gravitational waves, quark-gluon plasma*); by proposing or deriving models *(quark model, Cooper pairs)*; in the development of the theories (*symmetries, renormalization).* And since physics is a quantitative science this also involves calculations and thus explanation also depends on calculability.

An intriguing aspect relates to the observation made at the end of the previous section. We became used in classical physics to accept increasingly abstract mathematical objects as symbols for the physical objects we want to refer to: vectors for velocity and force, tensors for electromagnetic fields or space–time curvature, etc. We fail to faithfully represent them intuitively (in "Anschauung") but we can give a scheme to "reconstruct" them from our simple intuitions based on real numbers—they remain somehow "familiar".

Not so in quantum physics. The symbols, for instance those for the "right concepts" are defined on the field of complex numbers. This is reflected in the measurement results which are real numbers but fulfil unusual correlations violating the so called Bell-type inequalities.[53]

All this is well and clearly represented as algebraic relations, diagrams, etc. However, it appears difficult to have some representation of the fundamental quantities and relations of quantum physics the same way we do in classical physics and realise a similar familiarity. This may be a reason why we hesitate to accept the "reality" of QM—besides the lack of well defined "conceptual interpretations". But in the view that theories provide explanations and if we do not consider QM as incomplete we may ask which is the "natural classification" QM refers to. Let us assume for a moment that QM offers us a natural classification, and thus leads us to reality, in the relative meaning advanced in Sect. 2.3 (see also Sect. 2.1). Why is this reality, even accepting its relative status and its abstract theoretical representation, so much less accessible to us intuitively than the classical one? We do not even seem aware of the fullness of quantum effects which direct our daily life (starting with the mere existence of the world—the stability of matter).

A simple answer (which might not be wrong) is that QM itself takes care of that. We live in a world of innumerable, uncontrollable interactions which steadily carry away QM phase information. QM *predicts* in this case that as a result the genuine QM effects are damped by *decoherence*[54] and the world *appears* as classical. In fact decoherence may also control the working of our brains and thus our understanding

[52] Cf. the note on relativistic QM (footnote 38).

[53] A metaphysics based on complex numbers is proposed, for instance, in the "Ur-model" of v. Weizsäcker [52], extrapolated to a universal *Ansatz* for everything in Görnitz and Görnitz [25].

[54] Joos et al. [32]. There exist of course macroscopic quantum effects unaffected by decoherence such as superconductivity, laser, transistors with many practical applications.

itself.[55] We always try to use classical concept in every description. We never had the chance to develop "quantum intuitions" based upon complex numbers even in an evidently fundamentally quantum world.[56] The only thing we can do at present is trying to keep in mind the typical effects and quantum laws as an underground world of beautiful beasts, which we only started to tame—see also the last section of this essay. And with increasing penetration of evident quantum effects (such as quantum information, quantum computing, etc.) in our daily life we may also be able to develop basic quantum insights and intuitions.

2.7 A Few Words About Decoherence

Since we mentioned decoherence in various contexts it may be useful to briefly introduce it here, the more so since there were Heinz-Dieter Zeh and Erich Joos who among the first correctly appreciated its importance, both practically and theoretically.[57] It is in fact a standard QM effect, thoroughly studied with its applications in Joos et al. [32], see also Zurek [33, 34]. It is also briefly reviewed in Wikipedia.

For a simple example consider a system *S* built up from two subsystems, *1* and *2*, which each can be in two states, **1** and **1′**, respectively **2**, **2′** like, say, an electron with spin up or down (in some basis). *Classically S* can have just four states and each of them is evidently just one of the pairs, **1 2, 1′ 2, 1 2′** and **2 2′** and in each of these states the two subsystems obviously have well defined states. Quantum mechanically, however, any complex linear superposition of possible states of a system is also a possible state. This ***superposition principle*** holds for *1* and *2* and also for the composite system *S*, therefore any linear superposition of the four pair states above is a possible state of *S*. But clearly such a *generic* state of *S* cannot be associated with a pairing of one state of *1* and one state of *2* from their superpositions.

Consider, e.g., the state $\mathbf{1\,2'} + \mathbf{1'\,2}$ of *S*, it is not possible to write it as $(a\mathbf{1} + a'\mathbf{1'})(b\mathbf{2} + b'\mathbf{2'})$, that is as a product of subsystems' states from the their superpositions (clearly one needs all coefficients *a, a', b, b'* to be non-zero, but then the product contains unwanted terms..).

[55] The hypothesis that quantum effects are at work in our brains, may be problematic since under the typical working conditions there (temperature, external influences) decoherence may be expected to destroy local quantum coherence. In an evolutionary perspective there seems to be no necessity for a quantum organ, see Hepp and Koch [29].

[56] There are macroscopic quantum effects which in a sense can belong to daily experience, such as superconductivity, or which by enhancement produce macroscopic effects, such as nuclear energy. And of course X-rays, LED, MRT, etc. They prove that the world *is* fundamentally quantum at all levels but the genuine quantum character is not direct enough to help us build quantum intuitions, and their interpretation remains abstract—the more so that QM itself does not offer us a clear and consistent interpretation for its most fundamental rules, e.g. measurement (albeit when using classical concepts …).

[57] There is very much literature on this subject—see, e.g., the bibliography in our mentioned book on decoherence, [32], Therefore I shall only mention some of the earliest studies hereto which also have a direct connection with our discussion here: Zeh [57], Joos and Zeh [31, 32], Zurek [58].

This is ***entanglement*:** at variance to the classical case there are states of S in which the subsystems do not posses well defined states for themselves.

What we observe (measure) in QM are *expectation values* of certain physical quantities O, typically called *observables* and the QM rule is to take their "projections" in the system's states, in the above case:

$$\begin{aligned}&\left(\mathbf{1}\,\mathbf{2}' + \mathbf{1}'\,\mathbf{2}|O|\mathbf{1}\,\mathbf{2}' + \mathbf{1}'\,\mathbf{2}\right)\\&= \left(\mathbf{1}\,\mathbf{2}'|O|\mathbf{1}\,\mathbf{2}'\right) + \left(\mathbf{1}\,\mathbf{2}'|O|\mathbf{1}'\,\mathbf{2}\right) + \left(\mathbf{1}'\,\mathbf{2}|O|\mathbf{1}\,\mathbf{2}'\right) + \left(\mathbf{1}'\,\mathbf{2}|O|\mathbf{1}'\,\mathbf{2}\right)\end{aligned}$$

Now assume the subsystem *1* to be at our disposal and the observable to refer only to it. Assume *2* to represent the environment of subsystem *1* and after interacting with it go beyond our reach—like, e.g., innumerable particles scattering on *1* and then escaping in all directions. We may then safely assume that these particles no longer care for what happens in our laboratory, or for each other. Their states, here indicated by **2** and **2'** at different edges of the world are therefore orthogonal:

$$\left(\mathbf{2}|\mathbf{2}'\right) = \left(\mathbf{2}'|\mathbf{2}\right) = 0 \text{ while } \left(\mathbf{2}|\mathbf{2}\right) = \left(\mathbf{2}'|\mathbf{2}'\right) = 1 \text{ (normalisation)}$$

and since our observable O is restricted to the laboratory, the result of our observation will simply be.

$$\begin{aligned}&\left(\mathbf{1}\,\mathbf{2}' + \mathbf{1}'\,\mathbf{2}|O|\mathbf{1}\,\mathbf{2}' + \mathbf{1}'\,\mathbf{2}\right)\\&= \left(\mathbf{1}|O|\mathbf{1}\right)\left(\mathbf{2}'|\mathbf{2}'\right) + \left(\mathbf{1}|O|\mathbf{1}'\right)\left(\mathbf{2}'|\mathbf{2}\right) + \left(\mathbf{1}'|O|\mathbf{1}\right)\left(\mathbf{2}'|\mathbf{2}\right) + \left(\mathbf{1}'|O|\mathbf{1}'\right)\left(\mathbf{2}'|\mathbf{2}\right)\\&= \left(\mathbf{1}|O|\mathbf{1}\right) + \left(\mathbf{1}'|O|\mathbf{1}'\right)\end{aligned}$$

This is ***decoherence*:** it destroys the effect of entanglement in a local observation. This implies *classical behaviour* since the result of our observation is described by classical statistics over well defined states of subsystem *1*. This is also effective in any process in which entanglement would be relevant.

Decoherence is a widely discussed topic. I took the risk to bore the reader in order to stress that this effect is obtained in a standard QM analysis independently of any "metaphysical interpretation"[58] of QM: Copenhagen collapse, Everett relative states, Consistentthistories, ETH events, etc. It involves but does not "solve" ("explain") any of the basic QM postulates (measurement, probabilities), however it is frequently used to ensure consistency of some hypotheses in one or other interpretation.[59] The essential aspect is, however, that it allows quantitative estimates[60] which is important in applications.

One important application concerns quantum computing. Here entanglement is the key factor and the functioning of a quantum computer relies on ensuring stable

---

[58] This is not the interpretational structure mentioned in Sect. 2.1 but the attempt to provide a metaphysics for QM.

[59] See, for instance, Zeh [57] and the discussion in Joos et al. [32].

[60] See e.g. Joos et al. [32].

entanglements for the time needed for a computation. This means, however, one needs to evaluate and limit the decoherence effects by good shielding, very low temperature, etc. The presently achieved coherence time is of the order of seconds for a few q-bits. Generally, any process based on QM coherence must take into account decoherence.

Decoherence enters our discussion of explanation in a number of ways.

- By showing that classical observations can be accounted for in QM, decoherence stresses the universality claim of the latter.
- The involvement of the decoherence arguments in "interpretation" proposals such as Everett's relative states or the coherent trajectories stresses its meta-theoretical role.
- Decoherence provides an explicit connection between CM and QM going beyond formal proofs such as the Ehrenfest theorems. It thus confirms the continuity in the progress of theories and at the same time qualifies it by fixing the theoretical and empirical conditions under which we can observe and how we must understand it.
- Since it is a quantitative effect it can be continuously tuned between quantum and classical observation, for instance by varying the vacuum or the temperature in an experimental set-up to observe QM interference. Thus we can understand both the empirical and the mathematical aspects of this effect and how it describes a continuous "transition" between quantum and classical phenomena.

2.8 On the Role of Mathematics in Physical Sciences

Foundation; Basic roles; Calculation

As a final point we should speak of the role of mathematics in physical theories, since this role can be taken one to one in the question of explanation in physics.

1960 Eugene Wigner wrote a beautiful small paper *"The Unreasonable Effectiveness of Mathematics in the Natural Sciences"*[61] Of course stressing the essential role of mathematics in physics is a constant issue in the historical development—..., Roger Bacon, Galileo Galilei, Pierre Duhem, Ernst Cassirer, ... such that it is taken as self evident. But the interesting question this paper addresses is: why (and how)? I shall not present this paper here and leave to an interested person the pleasure to read it himself (if he did not already do it). I shall only sketch a few of its main arguments.

It starts from some *basic observations*, such as:

[61] Wigner [54].

- Nature shows regularities and this implies that there are laws of nature and that we can obtain knowledge of them. This is the assumption—postulate (*Peirce*), *lawfulness* of nature-hypothesis (*Helmholtz*), general causality principle (*Cassirer*), ...—explicitly or implicitly made by most physicists.[62]
- It is possible to select restricted sets of phenomena for consideration within arbitrarily good approximation (approximate *separability*). Electromagnetic and gravitational effects for instance can be separately studied in corresponding experimental design. The legendary experiment attributed to Galilei needed separating the dominant effect (free fall) from disturbances (air friction). And we can give procedures to control the approximation.
- It is possible to use idealization, generalization and other procedures permitting to translate the experimental findings into physical quantities (symbols) adequate for the mathematical treatment and to compare the predictions of the theory to phenomena.

I just wanted to recall the conditions by which in Wigner's account mathematics enters physical theories. In fact an assumed deep compatibility between our thinking and *"the mechanisms of nature"* hinted at here has led to many discussions[63] including the suggestion that our theories just discover the Platonic algorithms which are at the basis of these mechanisms.

The paper then counts *basic roles and aspects* of mathematics in natural sciences:

- Mathematics is a *language* providing basic concepts in formulating theories (*e.g. state spaces, operators*), providing the rules for connecting these concepts into a theory (*equations, limiting procedures*) and of course for handling the empirical information and translate theoretical pre/post-dictions into measurable effects. Mathematics also provides the tools for connecting and developing theories (*symmetries, scale transformations*) and formulating and testing hypotheses.
- Since mathematical concepts are "transportable" mathematics allows the development of new theories by developing the mathematical-symbolic scheme of older theories (e.g., *in proceeding from CM to QM: Hamiltonian, Lagrangean, symmetries*). This also allows the application of a conceptual scheme to different physical situations (*thermodynamics, complex system theory, phase transitions, universality*).[64]
- Since mathematical structures build up an universe for themselves[65] they can be used for completing theoretical schemes or for envisaging new ones in the absence of direct empirical evidence (*displacement current in Maxwell's eqs. —from*

[62] To what extent does QM comply with this understanding of lawfulness is an outstanding question—between the incompleteness criticism (*Einstein*), the claim of universality of statistical laws (*Schrödinger*), the denial of the need for space–time description (Anschauung; *Heisenberg, Cassirer*), ...

[63] See, e.g., the exciting discussion in Penrose [40].

[64] The mathematical-symbolic schemes of classical physics, for instance, provide keys for developing symbols for the newer theories.

[65] See, e.g., Bourbaki [13].

symmetry considerations, conservation laws; string theories, super-symmetry—in the search for QG).

Such observations, however, also make clear that we must be able to use the mathematical schemes. Since physics is a quantitative science this means among others that we must be able to calculate. One speaks in this connections of *analytic* and *numerical* procedures.

A good part of the mathematical work of the previous centuries concerned the development of analytic procedures: special functions *(e.g. the well known Bessel functions)*, algorithms *(for solving equations, stochastic processes)*, ... They have provided important insights in physics (*planet trajectories in classical mechanics, simple atoms in quantum mechanics*). However they come to their limits: -With increasing complexity of the problems *(already the 3 body problem in Newtonian gravity has no general analytic solution; complex atoms in QM; phase transitions in QFT and Complex Systems, ...)*. -With the demand for high precision *(celestial mechanics, satellites, Rosetta mission, LIGO detector for gravitational waves, ...)*. Analytic studies today are especially used in models or approximations.

Numerical procedures also have a long history and in the mid of the twentieth century logarithm and Bessel function tables were still in use. However also computing machines started developing since the nineteenth century. And this development acquired in the last decades an extraordinary momentum concerning the capabilities of the machines, which also triggered an exploding field of applications, not least in science. Computers enter now the scene of natural sciences and this will be the subject of the second part.

2.9 Closing the Circle

Relating explanation to physical theories led us to derive a number of features of the former. We thus had to acknowledge a conditional character for the necessity, truth and reality which we can associate to explanation, relative to a (theoretically and empirically) well defined class of reproducible phenomena. We correspondingly have also seen explanation as an evolutionary process, both in its symbolic content and in its structure concerning the different procedures following from the relation to theory development, such as induction, abduction and deduction. That in connection to explanation mathematics is important, even at the meta-level of comparing theories and illuminating the lines of theory development.[66] And finally that an essential role is played by calculations which opens a window toward the role of computers and the revolution happening here.

We have considered explanation as provided in physical theories. We proceeded from physical theories to characterize it. We can accept that a particular theory provides a partial explanation for the physical world. Is explanation just the sum of

[66] This is of course a relevant issue in the philosophy of physics. For a nice example of using modern mathematical tools at the meta-theoretical level see, e.g., Barth [4].

what is provided in the particular theories or is there something more we can say about it? In fact there is also a reverse flow from explanation to physical theories. Explanation tries to produce a coherent picture and thus supports meta-theoretical perspectives. Such are for instance the unification and reduction as "regulative ideas"[67] in advancing theoretical knowledge, but also constructive principles "found on the way" such as "gauge symmetry"[68] or the cosmological principle. In that sense explanation synthesises the theoretical development and at the meta-theoretical level also reveals the features which may play a role in guiding the progress of the theories.

In a realistic perspective the same way we think *accomplished theories* and the explanations they provide have an ontological connection which ensures them a high level of success, we can think that the explanation accompanying the *progress of theories* also has access to an ontological order, because it acknowledges well defined steps and follows a well defined development "dynamics"—and appears successful. It is this ontological order immanent in the structure of the world which thus appears to *define, justify* and *guide* our steps and their proceeding from theory to theory. This order is revealed by the explanation provided in the theories which close over well defined classes of phenomena at each level of the physical knowledge; and by the explanation concerning the progress of physical knowledge where it suggests general principles and regulative ideas and also provides at each step reasons and hints for change.

3 Computers in Natural Sciences

3.1 Two Moments in the History of Computers

Historical descriptions are not common in essays, however I think that we should pay respect to the promoters of a development which is practically overrunning us today.

---

[67] Or "regulative principles"—not to be understood as constitutive (cf. I. Kant, The Critique of Pure Reason, Appendix).

[68] The name "gauge symmetry" is slightly misleading (cf footnote 37), nevertheless this concept proved to be very rich and extremely useful both in developing and in analysing theories.

Charles Babbage (1791-1871) | Ada Augusta Lovelace (1815-1859) | Replica of the Analytical Engine © Science Museum London

Fig. 4 Two pioneers. From commons.wikimedia.org. Left: Nat Library of Wales Catalog, Public Domain; middle: Science Museum Group, Public Domain; right: CC BY-SA 2.5

The present day developments were already foreseen mid of the nineteenth century. *Charles Babbage*: Mathematician, Philosopher, Inventor, he established together with John Herschel the Analytical Society, later the Royal Astronomical Society. He worked towards dissemination of the Leibniz differential calculus. He was Lucasian Professor of Mathematics (Cambridge), member of the Royal Society and of many Academies of Science. He promoted the British Society for the Advancement of Science. He designed the first programmable, efficient "computer", the "analytical machine", to be used for statistical calculations. Although never built in his lifetime this and further theoretical work is considered as fundamental for the development of computers.

He is often mentioned together with *Ada Augusta Lovelace*, born Byron, considered as "the first programmer". She received private mathematical instruction. She met Babbage as 17 years old and was fascinated by his work to which she kept contact over many years. She wrote programs for the analytic machine, and in 1843 in a paper translating work of Federico Luigi Menabrea (which provided a description of Babbage's analytical machine and programs for it) expressed many original, very interesting ideas, foreseeing among other things the use of computers for symbolic calculations. She denied, however, that computers may develop human-like intelligence (the *"Lovelace objection"*).

About 100 years later we witness the second impulse by which the computer era took its real start. *John von Neumann*, mathematician, physicist, computer scientist, made important contributions in quantum mechanics in formalising the theory and the measurement process, in mathematics, in functional analysis, in logics, in game theory and in computer science where he developed the computer architecture which bears his name. He worked in the Manhattan project to the realisation of the atomic and the hydrogen bomb and supported the concept of assured mutual destruction as the only way to ensure peace in the atomic era. He also devised the idea of self-reproducing automata to be sent in the Galaxy as witnesses of life on the earth.

Konrad Zuse (civil engineer, inventor, computer pioneer) devised a number of programmable computers: from the mechanical Z1 (1938) controlled by a perforated film, to Z3 (1941), the first functioning computer in the world, a "Turing complete", i.e. computationally universal machine, working with relays and freely programmable. He built the Z4 (1945), the first commercial computer which was subsequently used for many years at the ETH Zürich. He anticipated programming languages and the *John von Neumann* architecture. He was a passionate painter without having an art study and argued that he did not studied computer science either.

Practically at the same time we witness the work of *Alan Turing*—mathematician, computer scientist, logician, theoretical biologist, cryptoanalyst—who had fundamental influence on computer science. He formalized the concepts of algorithm and computation, devised the so called *universal Turing machine* and provided proofs for computability and for decision problems. In this connection one should also mention the so called *Turing-Church thesis*, which has many variations and interpretations in mathematics, computer science or philosophy. In one formulation the thesis says that physical computability can always be simulated by a Turing machine, which implies digitalisation. It is surely interesting to follow this discussion in connection with *mechanistic explanation;* this also connects directly to the question of the role of mathematics in physics opened in Sect. 2.6., but I shall not touch this here. One of the activities of Turing was building decoding machines which is said to have shortened

John von Neumann (1903-1954) | Konrad Zuse (1910-1995) | Alan Mathison Turing (1912-1954) | Replica of the decoding machine "Bombe" 1944

Fig. 5 The real start. From commons.wikimedia.org. Left: Attribution; middle-left © CC BY-SA 3.0; middle-right gemeinfrei; right © CC BY-SA 3.0

the war by up to 2 years. He also contradicted the *Lovelace objection* and devised in this connection the well known *"Turing test"* to differentiate between human and machine interlocutors (the tests until now are only slightly positive towards humans).

The ensuing development is dramatic. The "first electronic brain", ENIAC (1945) used vacuum tubes and achieved a speed of about 10 kHz which could be enhanced to 100 kHz or more by parallel processing. A 100 Byte magnetic core memory was added later. It occupied about 60 m^3 and needed 150 KW of electric power. A typical laptop today has a speed of about 2 GHz, also to be amplified by parallel processing (about 100,000 times ENIAC), a memory of 1 TByte = 1,000,000,000,000 Byte and a volume of about 1000 $cm^3 = 0.001\ m^3$. It takes less than 15 W of power. Big machines with parallel processing perform by factors of 1000 to 100,000 still better than that. The hardware development triggered software developments and a correspondingly increase of the application fields. Artificial intelligence and robotics start to catch up. This brief history may help to appreciate the future coming upon us.

3.2 A View on Computers in Sciences

Computers and explanation, Involvement in experiment, theory, understanding; Visualisation

Before entering the discussion on the role of computers in natural sciences we should remember that computers have long since been involved in epistemological and fundamental questions. We may quote the work of Turing in the foundations of mathematics where he uses the computer paradigm. Later this connection became even more explicit and involved philosophy, driven by the questions of Artificial Intelligence and Cognitive Science,[69] finding its way also in the theory of explanation.[70] This connection was strengthened by the Neural Network paradigm[71] which became the dominant paradigm in the present day Machine Learning field. Since we are interested, however, in the impact of computers on explanation in physics we shall not pursue this direction after having mentioned a few of the big names active here.

1993 the physicist Joseph Dreitlein from the University of Colorado in Boulder published a paper with the title *"The Unreasonable Effectiveness of Computer Physics"*,[72] with evident allusion to the 1960 paper of Wigner. He first raises the question, "what are the computers actually doing?" and then starts describing some of their amazingly successful interventions in physics. We shall come back to the first question in the last section. In the following we shall take this (already 27 years old!)

---

[69] For a few references; Churchland [16], Churchland and Sejnowsk [17], Boden [11].

[70] Thagard [50].

[71] McCulloch and Pitts [38].

[72] Dreitlein [19].

paper as a prompt to illustrate and discuss the present day involvement of computers in physics and the significance of this involvement for the question of explanation.

The role of computers in explanation comes together with the role of calculation in the latter as mentioned in Sect. 2.6. and related to the quantitative—mathematical foundation of physics. We thus find essential computer involvement in experiment and observation, in computer assisted theory development, in finding the right concepts for understanding and explanation. And in all fields and at all stages visualization, data processing, communication, became indispensable tools in research.

Many of these applications involve computer simulation: the computer emulates the dynamics implied by a theory or a theoretical model through the corresponding equations. This is sometimes also called "analysis from first principles" since it only involves the fundamental equations of the theory without appeal to further approximations or simplifying models. Simulations in statistical mechanics and quantum field theory employ furthermore stochastic methods to reproduce the microscopic fluctuations. A brief account on simulations is provided in the Sect. 3.3.

A further development involves symbolic manipulation (already foreseen by Ada Lovelace mid of the nineteenth century): this has been used in many areas and has proven to be essential, for instance in analytic studies using perturbation series (perturbation theory, Feynman graphs) of QFT, in analyses for GRT, in general symbolic programs such as Mathematica and in application from Machine Learning and AI. In the following, however, I shall concentrate on numerical applications in physics. Details and illustrations are given in Sect. 3.2.3.

3.2.1 Computer-Assisted Experiment and Observations.

I shall only mention some fields with paradigmatic applications:

(a) *On-line assignment:*

- Astronomical observations today are strongly dependent on eliminating disturbances, so for instance big telescopes work with a system of adjustments of small mirror elements to correct for atmospheric turbulence (adaptive optics). This happens by gauging the image on a known source and performing the corrections and this clearly must proceed in real time.
- High energy collision experiments produce events with tens of thousands of tracks (see Fig. 3 right—compare Fig. 2). One searches for a given signature—as the ones decoded on Fig. 2 (50 years ago this was done by eye ...)—and since the events are produced in fractions of seconds one needs to analyse them on-line in real time. This involves extremely fast, sophisticated algorithms for pattern recognition, etc.- besides beam control and general control of the collider. High energy physics experiments are unthinkable without powerful computers and software.

(b) *Calling for precision:*

- Cutting Edge Discovery Science needs high precision and handling of large amount of data. The LIGO experiment, e.g., which in 2015 produced first evidence of gravity waves needs to measure displacements at the level of 1/1000 the size of a proton (~1 fm = 10^{-15} m).
- Satellites and space missions, for instance the "New-Horizons" mission which started 2006 following the Voyager 1 and 2 missions and which after reaching Pluto 2015 at a distance of about 5 Lighthours entered the Kuiper belt with flyby of a number of asteroids. 2021 it is proceeding toward the end of the heliosphere aiming to reach the outer space space by 2035 at a distance of about 16 Lighthours (120 AE). The mission includes calculation of the orbit including acceleration near big planets and on board computing for flight control and data management (which clearly must proceed on board, including error detection and reboot, etc.). Its task is to describe and understand the structure of the solar system and its embedding in the interstellar space.

(c) *Empirical enhancement using simulations and impossible perspectives:*

- Milky Way has more than 10^{11} stars distributed in the arms of a barred spiral. Our view is from a rather periphery position in the plane of the Galaxy. On the basis of astronomical data computers can, however, help simulating any wanted perspective—see Fig. 6, Sect. 3.2.3a. This helps an intuitive assessment of the situation and of the structure of the Galaxy.
- The computer can simulate effects in the frame of a theory but not observed or even unobservable in nature. Such are problems in celestial mechanics (the design of the flight of space probes, the behaviour of planets and asteroids in "would be" situations—see Sect. 3.2.3b and Fig. 7), or the behaviour of matter at energies and temperature before they are reached experimentally

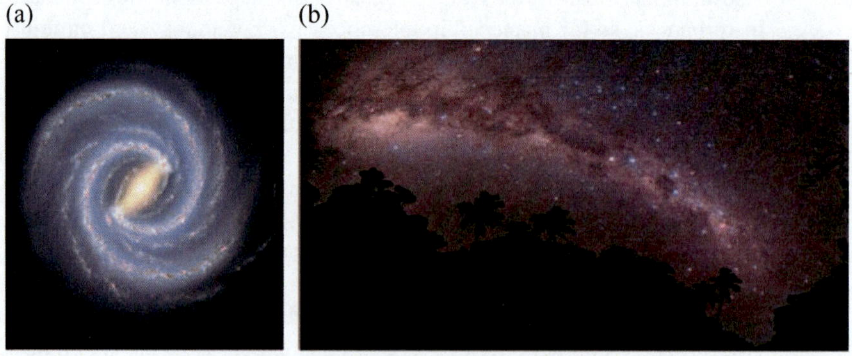

Fig. 6 **a** View of our Galaxy from above, artist reconstruction using astronomical input (left: NASA.gov, Public) and **b** a real "insider" view of Milky Way on a clear night far from other light sources (above: apod.nasa.gov, MangaiaMW_tezel)

(phase transitions in SM). We can thus obtain data about the predictions of the theory in situations or at parameter values which are not or not yet available experimentally. In cosmology one can simulate not directly observable situations, such as black hole collisions which are sources of gravitational waves, helping the interpretation of real observations or suggesting new effects.

3.2.2 Computer Assisted Theory Development

Again I shall only mention a few paradigmatic cases.

(a) *Asking questions and promoting hypotheses (research and curiosity).*

- The non-trivial behaviour of non-linear maps observed in computer simulations (e.g. the Mandelbrot set, the Feigenbaum diagram) led to hypotheses concerning classical chaos. Here visualisation plays a big role—see Sect. 3.2.3c and Fig. 8.
- Non-linear dynamics in complex systems puts many questions for which no analytic solutions can be given. One of the first computer involvements tried to study numerically the question, whether non-linearity ensures ergodicity[73] (*Fermi, Pasta and Ulam, 1953*); what they found was a non-ergodic, unexpected, quasi-linear behaviour. In the sequel of this work very many new insights in the physics of non-linear systems were obtained such as solitons or the interplay of regular and irregular motion, etc.
- Machine learning in the Neural Networks (NN) set up leads to problems which can be defined in the Complex Systems Theory (CS). Here also belongs training NNs for physical applications. One interesting question is whether statistical learning from unspecific reinforcement[74] is possible, that is if the assessment of the actions of an agent and the corresponding reward (positive or negative feedback) concerns not the result of each action but the end result of a series of actions. This is in fact the typical realistic situation in most contexts, think of playing checkers, or of an animal finding the path to water in a forest, or the NN training—the final result counts. We were not only able to answer the question positively but in a neural networks (NN) simulation[75] we also found a very interesting learning flow and an unexpected parameter dependence of the learning behaviour[76] (for details see Sect. 3.2.3d and Fig. 9).

---

[73] The capacity of a system to take all allowed configurations.

[74] Mlodinov and Stamatescu [39].

[75] Kühn and Stamatescu [35], Kühn et al. (Eds.) [36].

[76] Kühn and Stamatescu [37, 48].

(b) *Promoting and testing hypotheses and models.*

- QCD, the theory of the strong interactions responsible for the formation of nucleons and other *hadrons* from *quarks* and *gluons*, the fundamental fields of the theory, must ensure two important demands: The quarks should behave like non-interacting fields *(asymptotic freedom)* at high energy, while they should be strongly interacting and *confined* in hadrons at lower energy. In a certain representation of QCD belonging to the *Lattice Field Theory* approach (LFT, see Sect. 2.3) the second hypothesis is immanent, while for the first one the support comes from computer simulations.
- Decoding the structure of theories by using models. In LFT models which couple matter and electromagnetic or gluonic fields the effect of the matter fields can be evaluated from summing the contributions from closed paths on the lattice. In a certain example one considers a model accounting for the effect of a background electromagnetic field upon the matter but not for the back reaction of the matter upon the field. It is an example of separating different contributions and using lattice simulation and theoretical results in analysing the theory.[77]

(c) *Deriving post- and predictions from a theory.*

- The building of hadrons in QCD is not amenable to direct analysis and for a while only simple models could be used to obtain, say, the spectrum of hadrons. The development of computer simulations has been able to provide the hadron spectrum and further properties of the hadrons, model-free and without uncontrolled systematic approximations. This has been one of the first big successes of computer simulations in elementary particle physics.
- Besides a low temperature, "hadronic" phase where the quarks are bound in nucleons and other hadrons QCD is also expected to show a high temperature phase, where the hadrons "dissolve" into a "quark-gluon plasma". This theoretical expectation has been tested and supported by computer simulations, before being proven experimentally in high energy collisions at accelerators (at CERN and at the Brookhaven National Laboratory). See Sect. 3.2.3e and Fig. 10.

(d) *Developing and relating theories.*

- We can start with a theory defined at a small length scale (we speak of the UV—*ultraviolet*—scale, short distances). We can ask, what would this theory predict for the phenomena at much larger lengths (smaller energy, IR—*infrared* scale, long distances) and whether we can find concepts in IR and a succinct description for the corresponding phenomena. This is the "renormalization group" approach mentioned in Sect. 2.2 and can be

[77] Schmidt and Stamatescu [47].

followed in both directions: Finding an *effective* description at the IR level from the known UV theory; and developing hypotheses for a UV theory such that the known IR theory ensues. It typically involves further conditions (symmetries, "renormalizability", etc.) to define the searched-for theory. Computer simulations can be employed for both problems.

(e) *Finding the right concepts for understanding and explanation.*

- We mentioned solitons, turbulence, etc. which were studied in computer "experiments". Similar structures are sought in SME *(instantons, vortices)* trying to understand the mechanism of confinement and other expected features of the theory. They have been observed in computer simulations and their role in producing the above effects can be studied there.
- In the NN learning study (see Sect. 3.2.2.a) the peculiar fixed point and separatrix structure found suggests a concept for understanding the mechanisms of learning from unspecific reinforcement (see Sect. 3.2.3d and Fig. 9). The ensued analytic study was prompted by the numerical results.
- The computer simulations for non-linear maps and complex systems have also led to fruitful concepts: self-similarity, fractal structures, hierarchy of scales, etc. which helped understand such phenomena as *classical chaos* and provided thus new instruments for explanation.

3.2.3 Visualisation

In all scientific areas computers produce pictorial representations which offer a synthetic grasp of the theoretical structure and connections. Some few examples:

(a) *Impossible views.*

The view of our Galaxy seen from above gives a better intuition than a real picture, Fig. 6.

(b) *Abnormal situations in celestial mechanics.*

The planetary system is considered metastable,[78] but one may ask what could have happened for other initial conditions. In Fig. 7 are shown the orbits around the sun of the earth, the moon (undistinguished from that of the earth at the scale of the picture), and a "lost-mars" which would have been at some time very near to the earth and expelled after a near-encounter toward the left upper corner. This is a simple simulation of Newtonian interaction on a desktop as illustration for the "games" played innumerable times at another scale of precision in conducting planetary missions or controlling the movement of asteroids. The computer simulation allows to change the parameters to produce new effects.

---

[78] An interesting discussion of the stability properties of the solar system can be found in Petterson [41].

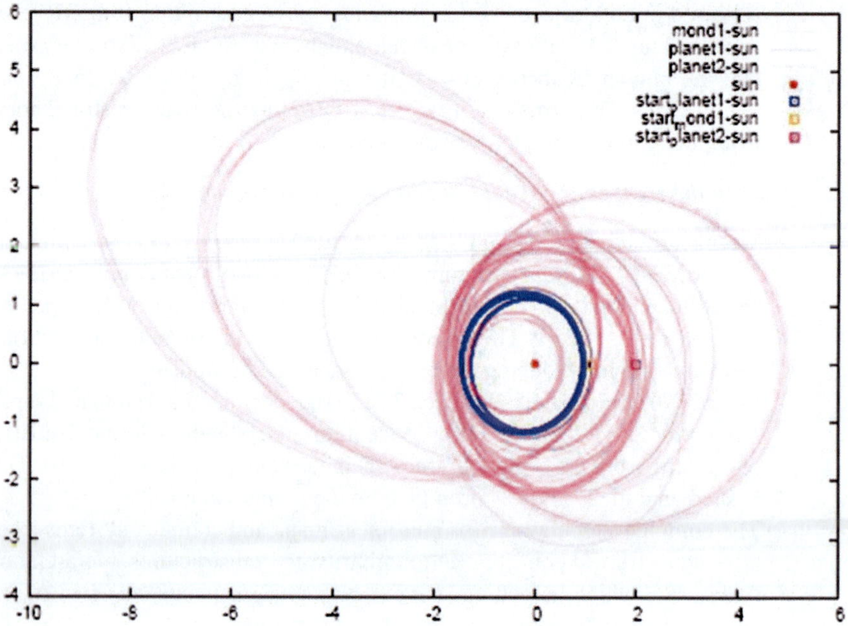

Fig. 7 4-body celestial mechanics: m(sun) > > m(earth) > m(mars) > m(moon), ecliptic initial conditions, Newtonian simulation, arbitrary ecliptic coordinates. Orbit and starting point of earth and moon are undistinguishable at the scale of the figure. © by the author

(c) *Classical chaos.*

The Mandelbrot-set M mentioned in Sect. 3.2.2.a is obtained from the non-linear mappingt $z_{n+1} = z_n^2 + c$ in the *complex z plane* and defined as the condition $c \in M$ if the iteration is bounded. The pictures allow an immediate grasp of the peculiar properties of the mapping: fractal structure and self-similarity at different scales, Fig. 8. This mapping is a (complex) deterministic process, but it shows unexpected

Fig. 8 *Left and middle*: The Mandelbrot set in the complex c plane and a zoom of a small part of the boundary. Black denotes the areas of convergence, dark blue signals fast divergence, the other colours indicate different speeds in reaching divergent behaviour. From commons.wikimedia.org, CC BY-SA 3.0. Right: The Feigenbaum diagram at given r, gemeinfrei, showing the emergence of classical chaos from increasingly rapid bifurcations of trajectories. From commons.wikimedia.org, Gemeinfrei

unstable behaviour which appears nevertheless to generate a complicated, reproducible organization making it a paradigm for "classical chaos". Another approach to classical chaos is described by the bifurcation behaviour of the solutions of the logistic equation $x_{n+1} = r.x_{n}.(1 - x_n), r, x \in \mathbb{R}$ when varying r—the Feigenbaum diagram which shows explicitly the mechanism of chaos generation.

(d) *Learning from unspecific reinforcement.*

In the learning process mentioned in Sect. 3.2.2.a the agent (or "student") is simulated as a neural network. At each trial the student is presented with a series of patterns which he must classify, and uses the so called Hebb rule to modify his internal structure (the *synapses*) according to the *local* stimulus/response *coincidences* between neurons based only upon his up to date knowledge (*blind association* = *self confidence* based on its up to date experience only). The teacher generates the patterns by a fixed stochastic rule and at the end of each trial tells the student which was the *proportion* of the good answers (i.e. in agreement with its—teacher's—own classification). Unless all the answers were correct the student *globally* contradicts *all* previous updates *equally* according to this *global* trial failure (*unspecific reinforcement* = *global feedback*). We called this learning algorithm AR *(association-reinforcement)* and an essential parameter is the strength of the reinforcement, r which tells how strong the feedback is compared with the *self-confidence*. Figure 9 shows the flow with the time of the student error ε_G and the normalization of its synapses Q and shows a peculiar structure with various learning curves depending on r and separated in two classes. For small values of r (lax teacher) the student always learns (the error drops exponentially, right hand curves), but slowly. Increasing r (rigorousness) the learning speed improves, but it reaches a threshold r_0 above which there is no

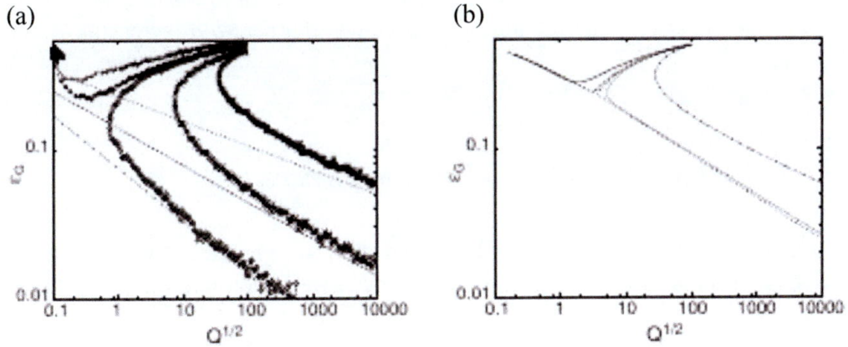

Fig. 9 Flow with time of the student error in the classification of new patterns (generalization error εG) vs normalisation of its synapses, Q. The figures show a flow starting with a candid student (upper middle point) and controlled by an attractive/repulsive fixed point further down which defines a separatrix dividing learning (right hand side flows) from no-learning behaviour (left hand side flows toward the attractive fixed point at εG = 0.5). The curves correspond to increasing rigorousness parameter r from right to left defining on which side of the separatrix the flow will go. The start parameters are slightly different in the two plots. **a** *Left*: computer simulation implementing the AR process. **b** *Right*: exact result from an analytic model representing the AR process. © by the author

learning: the flow turns up toward a fixed point of total confusion in the left upper corner ($\varepsilon_G = 0.5$, the answers are 50/50 guesses). The exponential decrease of the error for $r < r_0$ indicates that the student learned the (unknown to him) rule of the teacher's classification, the fastest for r just below the threshold. All these details are immediately seen from the diagram, Fig. 9. Thus we showed that global, unspecific reinforcement learning is possible, which was our hypothesis but we also obtained an unexpected result that increasing rigour while improving the performance reaches a threshold above which learning is killed. These exciting numerical results prompted an analytic study and in a certain setting the model was found amenable to analytic calculations allowing comparison with the numerical simulations.

A more involved simulation pertains the path of a "robot" on a table with obstacles learning its way to a certain place. It uses again the *association/reinforcement* rule where only the total result of a run is evaluated. A controlled random element in the behaviour of the robot is essential, allowing it to cope with changes or finding better solutions. This model can only be numerically simulated.[79]

(e) *Phase transitions*

The phase transition of QCD mentioned in Sect. 3.2.2.c can also be triggered by populating the vacuum with matter (quarks) which in the formulation of the theory is achieved by introducing a chemical potential μ. The transition can therefore be continued in the plane T—μ (temperature—chemical potential). The path integral is now genuinely complex which makes the simulation much more difficult. Some of the simulation methods redefine the path integral in the complex plane. The phase transition is signalised by the strong increase of certain quantities averaged over paths (the "order parameters") from practically 0 (cyan) in the hadronic ("confinement") phase to a large non-zero value in the quark-gluon-plasma phase (red). The contour maps of the order parameter describe the behaviour of the transition line in the T—μ plane and also suggest that the transition becomes sharper with increasing μ. See Fig. 10.

3.3 A Brief Account of Simulation in Physics

Simulation is essentially a computer emulation of physical processes, of their formalisation or of theoretical rules. There are a number of dichotomies applying to this game.

Concerning the "hardware": Analogue—Digital

The analogue computer is a physical device using known physical effects. Some early analogue computers used electric circuits and relied on the electrodynamic rules immanent in their functioning to ask and answer questions such as adding exponentials, etc. More involved applications, especially in biology ask how a system

---

[79] See Stamatescu and Kühn et al. (Eds.) [36].

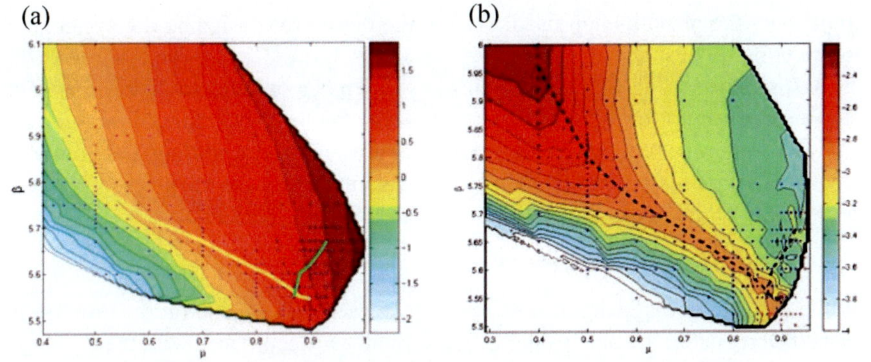

Fig. 10 Phase diagram of QCD with heavy quarks in the $T\,(\sim 1/\beta\,) - \mu$ plane. **a** *Left*:Order parameter. **b** *Right*: Susceptibility of the order parameter (essentially, its slope across the ridge) showing clearly the transition line. © by the author

of interconnected cells subject to an initial perturbation would behave as a whole: smooth, oscillatory, abrupt, etc. under varying the parameters of the system (e.g., the connections between the cells or the temperature). Hybrid analogue–digital systems use electronic models for the cells—so for neuron assemblies the so called Hodgkin-Huxley neuron realised with electric circuits and represented as a set of differential equations—to simulate the activity of biological neuron networks (see also Sect. 2.4).

The present day landscape is, however, dominated by the digital computer, whose history was hinted at in Sect. 3.1. Its involvement implicitly assumes that all our "encounters" with nature can be analysed in terms of human (binary) logics[80] (an instance of the Turing-Church thesis). Digital computers use rational numbers and binary operations to perform every real number operation. A digital computer can also be represented as a neural network (a network of artificial binary cells) and following the celebrated Charles Sherrington remark that "what the brain does is to take decisions" as taken over by McCulloch and Pitts[81] to introduce the neural networks, we appear to having closed the circle relating digital computers with human logics. Neuroscience and many directions in Artificial Intelligence followed this line and we witness a very rapid development.

Every knowledge which can be formalized can be implemented on digital computers (as already foreseen by Ada Lovelace mid of the nineteenth century and formally developed by Alan Turing 100 years later). Digital computers are of course "classical" devices, but they can also simulate quantum theory by implementing the corresponding rules *(Schrödinger equation, measurement rules,* etc.*)*. Notice that this is not what the quantum computers do: the latter do not simulate the theory but

[80] Non-standard logics can also be represented in binary logic.

[81] McCulloch and Pitts [38].

implement the phenomenon itself and use the theory for designing the system and interpreting the results.[82]

A simulation is an algorithm following a number of rules to produce at each step a set of values for the variables representing the basic quantities describing the analysed system (a "configuration"). The rules are designed according to the problem at hand. The configurations are then used to obtain solutions to the problem as limits or averages of the relevant quantities ("observables").

Concerning the algorithm: Deterministic—Stochastic

While the deterministic algorithms calculate at each step exactly (at the level of the machine precision) the new values of the variables in accordance with the equations of the theory, stochastic algorithms introduce a controlled random element ("noise") in their procedure which now involves "stochastic" variables. The noise is usually meant to take into account the physical fluctuations for instance in statistical or in quantum physics and it is represented in computation as "pseudo-random numbers" generated by some mathematical routine. There are different types of stochastic processes: non-linear or linear (the noise does or does not depend on the variables), Markovian or non-Markovian (each step depends only on the previous variable's configuration or on the whole preceding history), etc.

Concerning the problem set up: Classical—Quantum

Classical problems involve, for instance: Simple iterations such as the one described in Sect. 3.2.3c (the computation is simple but it needs to be repeated billions of times); Solving systems of differential equations, such as in Sect. 3.2.3b. Both use deterministic algorithms. Also implementing learning rules for evolving a neural network as in Sect. 3.2.3d is a classical problem, but it may use a stochastic algorithm to introduce uncertainties in behaviour of the learner (agent).

Simulations of classical statistical mechanics systems at equilibrium also use stochastic processes to produce statistical *ensembles*. One proceeds from the so-called *partition function* whose expression describes the ensemble of configurations with their weights—the Boltzmann factors, which are positive definite and can be interpreted as probabilities. Since the number of configurations (i.e., situations of the system) may be enormous the algorithms performs "importance sampling" according to these probabilities using a *real stochastic process*. The collection of configurations so obtained is then used to obtain averages of interesting quantities, study phase transitions, etc.

Also simulations in quantum physics use stochastic algorithms to produce ensembles of configuration. A typical quantum theoretical problem is *Quantum Chromodynamics* (QCD), a quantum field theory, see Sect. 3.2.3e. Here we start from the path integral formulation in which, however, the weights of the paths are now complex

[82] There are a number of approaches in this field which mix analogue (cold atoms, optical lattices, etc.) and digital features in various combinations.

and cannot be used as probabilities for importance sampling. The trick is to use certain mathematical features of these theories which allow to rotate the time in the complex plane from the real to the imaginary axis (*Wick rotation*). In a second step we *discretize* the fields on a space-(imaginary) time lattice. We obtain in this way a so called *Euclidean Lattice Field Theory* (LFT) which is essentially a *Statistical Mechanics* system at equilibrium with real, positive Boltzmann factors and for which we can use the same methods as above. In the end we must interpret the results back in real time.

However even in the Euclidean formulation some physical question, such as unbalanced background charges introduce imaginary parts in the action. Moreover, there are problems such as non-equilibrium dynamics which require real time evolution and must be studied on a Minkowski lattice without performing the Wick rotation. In such cases refined algorithms are called for since the Boltzmann factors are now complex. The most studied procedure is to redefine the variables as complex numbers and extend the analysis to their complex planes. A far developed approach which has provided results for realistic situation (e.g., QCD for non-zero baryon density) uses a *complex stochastic algorithm*, the Complex Langevin Equation.[83]

As can be seen, simulations build up a domain for themselves. This includes setting up the theory in an adequate form, developing algorithms, etc.[84] These studies are called "from first principles" since they directly implement the basic relations of the theory. The role of simulations is to ensure computability for complex theories which is one major factor in explanation. Besides producing post- and predictions for empirical tests of the theories (*deductive effort)* they also hint at many *right concepts* and lead to insights into the theory and contribute to the *abductive effort* by suggesting and testing hypotheses. Simulations are very important in evolving explanation as related to the particular theories as well as to their progress. With the rapid development of the computers we may generally expect synergy effects between application and theoretical development in computer science with further impact on sciences.

3.4 The New Computing

Machine learning; Neuromorphic computing; Quantum computer; Entanglement and decoherence

The tremendous development in both hard- and software has opened the way for a new definition of the role of computers in science. Here a brief account of the present day involvement of computers should be provided: the new computing. This regards both the hardware and the tasks and methods. The main features of this development can be subsumed in three notions: miniature and flexible design, neural networks

[83] Cf. Scherzer et al. [46] and references therein. For a review see Berger et al. [6].

[84] Of course one uses simulations also for processes, hypotheses, etc. as far as these can be formalized.

and complex architectures, machine learning and autonomous development. Many present approaches are expanding research from the end of the last century; it is, however, its present momentum in opening new perspectives and applications which may justify the term "new computing".

For an example, I shall briefly mention one comprehensive program which shows many of the present lines of research: the European Human Brain Project (https://www.humanbrainproject.eu). As already the title suggests, the program has a strong connection to human biology and cognition. It is distributed over many research groups. It involves fundamental and applicative research and is set up to provide interactions and synergy effects at various scales concerning.

Neuroinformatics: Brain atlas, data sets, 2D-3D viewer,

Brain Simulation: Simulation of neurons and collectives, validate models,

Analysis of high level brain functions: *From pattern recognition to cognition,*

High Performance Analytics and Computing: Large scale data mining, large scale simulations,

Medical Informatics: Bridge brain-science, clinical research and patient, improve diagnosis,

Neuromorphic Computing: Simulations on special architectures inspired from biological brains,

Neurorobotics: Link simulated brain to robotic bodies, test brain and robot models,

Ethical and Societal Assessment: Accompanying studies and discussion,

with influence from and upon related areas including machine learning and physics simulations. Just for an impression, brain models have been developed at the scale of a rodent brain, and tests for brain function models on neuromorphic computers allow a speed up by a factor of 1000 over biological processes.[85] The Human Brain Project above gives an impression on the horizon of the present developments.

A recent development which is interesting also for the question of explanation takes place in the frame of simulations for systems with many degrees of freedoms, which include QFT defined as LFT (see Sect. 3.3). Such models typically show phase transitions, see e.g. Sects. 3.2.2c and 3.2.3e which we can observe in computer simulations and describe using *order parameters*—observables distinguishing between the different phases. This behaviour is visible in the configurations produced in the simulation and the order parameter is in most cases suggested theoretically. In that sense the computer helps *testing* a hypothesis. Models with a more complicated structure may show more phase transitions and the building of hypotheses for *order parameters* may be ambiguous. Using machine learning methods in a refined neural networks set-up and an adequate way of "asking" the computer may itself *propose* a hypothesis and may provide access to the "*right concepts*", represented here by the order parameters. Such methods are also used in data mining, pattern finding,

[85] Wunderlich et al. [56] and further publication of the Neuromorphic Computer Group at the University of Heidelberg. Neuromorphic computers are "hybride" (analog-digital) machines with some millions of Hodkgin-Huxley neurons in variable architectures.

etc. This is an example of how computer science (here: *machine learning*) enters the problem of explanation in physics beyond calculation and simulation.[86]

The above examples concern classical computers. A rapid development presently pertains also quantum computers[87] which in some sense and to various degrees can be seen as analogue computers. By now their development followed the classical binary paradigm with the equivalent of the neurons being here two-states quantum objects (qubits) which could be brought in entangled states. Equivalents of the logical gates *(and, or, not,..)* are then defined and as long as we can shield the systems from *decoherence* effects due to the environment we can use the *entanglement* in specially designed algorithms for non-classical calculations. The effort until now has been to show that q-computers can outperform classical ones on some selected problems such as factorization large integers (interesting for cryptography) where the corresponding algorithms for classical computers would be much less efficient. The general superiority is still under debate. But further developments which would take advantage of the full quantum zoo could open new horizons. This may affect the full architecture—e.g., by redefining the "qubits" on higher complex manifolds and re-designing the "logical gates" accordingly—definitely breaking with the classical paradigm. Genuine analogue quantum simulation may represent a far reaching perspective leading to unforeseeable developments both in applications and in connection with explanation.

3.5 What Are Computers Actually Doing?

We have seen computers being involved in essential ways in practically all domains of physics research in all fields. They help or permit performing and interpreting evolved experiments, they allow extrapolating our empirical basis beyond its natural limits. They are involved at all stages of theoretical developments: inductive, in data handling, deductive, in obtaining post- and predictions from the theory, abductive, suggesting hypotheses by producing unexpected effects. In the fields of the fundamental interactions, of complex systems, of astronomy and cosmology, etc. Thereby computer simulations became a powerful tool in all these enterprises. So what can we say about the computers commitment in explanation?

Dreitlein [19] quotes two typical points of view of scientists concerning the involvement of computers in science. Computers can be seen as:

- producing final data from initial data according to some given rules, or
- reproducing the "code of nature".

Concerning the first perspective we must remember that computers are *not supposed to produce new physics,* but they can help us understanding the hints from nature by helping us to obtain empirical information and by bringing us on the track

[86] For an introduction and review see Carleo et al. [14].

[87] For partial assessments see e.g. Altman et al. [1], Grambling and Horowitz [26].

of effects inherent but hidden in the observational data. Moreover, the computers also directly support explanation by allowing to test our theories and "decode" them and find or suggest relevant concepts. And computers help us at an important step in the theory development by allowing to test and develop hypotheses on the way to new or improved explanation.

Hence: *Computers, while not directly explaining nature, provide vital knowledge and tools for explanation, assisting the scientist in an interactive game in his work to explain the world.*

But we don't know what the future developments, e.g. in Artificial Intelligence will bring (may *Turing's* refutation of *Lovelace'* objection hold?), and whether in a (far?) future computers may be able to learn more about the universe than we do ... And this brings us to the second perspective. This is of course a speculation, nevertheless let us see what it might mean:

- In a trivial sense it just says that computers reproduce our logical thinking, which is (at least partially?) compatible with nature since it may have been shaped by the history of our interaction with the world; thus computers help us extend our logical capacities (see Sect. 3.3).
- But in a less trivial interpretation it says that *computers may help us overcome the limitations of our classical logical thinking and access new thinking horizons* (may *quantum computing* be a suggestion in this direction?) bringing also *new dimensions* in explanation.

Whatever the answer, we may expect a lot of development—and fast.

References

1. E. Altman et al., Quantum Simulators: Architectures an Opportunities. PRX Quantum 2, 017003 (2021), arXiv:1912.06939v2 [Quantum-ph]A (2019)
2. R. Ariew, Pierre Duhem (2020). https://plato.stanford.edu/archives/fall2020/entries/duhem/
3. J. Barnes, *Aristoteles: Eine Einführung* (Reclam, Stuttgart, 1992)
4. L. Barth, A mathematical framework to compare classical field theories (2019). arXiv:1910.08614
5. J.S. Bell, *Speakable and unspeakable in quantum mechanics: collected papers in quantum philosophy* (Cambridge Univ. Press, Cambridge, UK, 2004)
6. C.E. Berger et al., Complex Langevin and other aproaches to the sign problem in quantum many-body physics. Phys. Rep. **892**(2021), 1–54 (2021). https://doi.org/10.1016/j.physrep.2020.09.002
7. A. Bird , E. Tobin, Natural kinds (2018). https://plato.stanford.edu/archives/spr2018/entries/natural-kinds/
8. M. Bitbol et al. (eds.), *Constituting Objectivity* (Springer, Berlin, Heidelberg, New York, 2009)
9. Blanchard et al. (Eds.) *Decoherence, Theoretical, Experimental and Conceptual Problems,* Springer, Berlin, Heidelberg, New York w York (1998)
10. Blum, W. et al. (Eds.), *Werner Heisenberg. Gesammelte Werke,* Piper, München (1989)
11. M. Boden (ed.), *The philosophy of Artificial Intelligence* (Oxford University Press, Oxford, 1990)
12. D. Bohm, B.J. Hiley, *The Undivided Universe: Ontological Interpretation of Quantum Theory* (Chapman & Hall, Routledge, 1993)

13. N. Bourbaki, (group work pseudonym), (1939–1980), *Eléments de mathématique,* Hermann, Chemnitz
14. Carleo, G. et al.Machine Learning and the physical sciences. *Rev. Mod. Phys. 91.045002* (2019). arXiv:1903.10563 [physics.comp-ph]
15. E. Cassirer, *Zur Modernen Physik*, Wissenschaftliche Buchgesellschaft, Darmstadt (1987)
16. P.M. Churchland, *A Neurocomputational Perspective: The Nature efMind and the Structure of Science* (MIT Press Cambridge, MA, 1989)
17. P.S. Churchland, T.J. Sejnowski, *The Computational Brain* (MIT Univ. Press, Cambridge, MA, 1992)
18. J.T. Cushing, *Philosophical Concepts in Physics* (Cambridge Univ. Press, Cambridge, UK, 1998)
19. J. Dreitlein, The unreasonable effectiveness of computer Physics. Found. Phys. **23**, 923 (1993)
20. P. Duhem, *The Aim and Structure of Physical Theory* (Princeton University Press, Princeton, NJ, 1954)
21. B. Falkenburg, *Particle Metaphysics*, Springer, Berlin, Heidelberg, New York (2007)
22. M. Ferrari, I.-O. Stamatescu (eds.), *Symbol and physical knowledge* (Springer, Berlin, Heidelberg, New York, 2002)
23. R. Feynman, P., Leighton, R. B., Sands, M., *The Feynman Lectures on Physics* (Addison-Wesley, Reading, MA, 1964)
24. C.I. Gerhardt, (Ed.), *G.W. Leibniz: Sämtliche Schriften und Briefe*, Berlin Verlag, Berlin (1849)
25. T. Görnitz, B. Görnitz, *Von der Quantenphysik zum Bewusstsein*, Springer, Berlin, Heidelberg, New York (2016)
26. E. Grambling, M. Horowitz, *Quantum Computing: Progress and Prospects*, The National Academy Press, https://doi.org/10.17226/25196.Grumbling, E (2019)
27. H. von Helmholtz, *Philosophische Vorträge und Aufsätze* (Akademie Verlag, Berlin, 1971)
28. C. Hempel, *Aspects of Scientific Explanation* (Free Press, New York, NY, 1965)
29. K. Hepp, C. Koch, Nature **440**, 611 (2006)
30. H. Hertz, *Die Prinzipien der Mechanik* (Akademische Verlaganstalt, Leipzig, 1894)
31. E. Joos, H.D. Zeh, The emergence of classical properties through interaction with the environment. Z. Phys. B **59**, 223–243 (1985)
32. E. Joos , et al., *Decoherence and the Appearance of a Classical World in Quantum Theory,* 2nd Edn. Springer, Berlin, Heidelberg, New York (2003)
33. C. Kiefer, *Quantentheorie*, S. Fischer, Frankfurt Main (2002)
34. C. Kiefer, *Der Quantenkosmos*, S. Fischer, Frankfurt Main (2008)
35. R. Kühn, I.-O. Stamatescu, J. Phys. A Math. Gen. **32**, 5479 (1999)
36. R. Kühn et al. (eds.), *Adaptivity and Learning* (Springer, Berlin, Heidelberg, 2003)
37. R. Kühn , I.-O. Stamatescu, *Biol. Cybernetics 97 (2007) 99* and *101 (2009) 40*
38. W.S. McCulloch, W.A. Pitts, Bull. Math. Biophysics **5**, 1151 (1943)
39. L. Mlodinow, I.-O. Stamatescu, Internat. J. Comput. Inform.Sci. **14**, 201 (1985)
40. R. Penrose, *The emperor's New Mind* (Oxford University Press, Oxford, 1989)
41. I. Petterson, *Newton's Clock – Chaos in the Solar System,* Henry Holt and Co, New York, NY (1993)
42. B. Rettler, A.M. Bailey, Object (2017).https://plato.stanford.edu/win2017/entries/object/
43. W.C. Salmon, *Causality and Explanation* (Oxford University Press, Oxford, 1998)
44. E. Scheibe, *Die Philosophie der Physiker*, C.H. Beck (2007)
45. E. Scheibe, G. Süssman (Hrsgb.), *Einheit und Vielfalt*, Goettingen (1973)
46. M. Scherzer et al. Deconfinement phase transition line with the Complex Langevin equation up to mu/T ~5. Phys. Rev. D **102**, 014515, arXiv:2004.05372 (2020)
47. M. G. Schmidt, I.-O. Stamatescu, Matter determinants in background fields using random walk world line loops on the lattice. Mod. Physics Letters A **18**(22), https://doi.org/10.1142/S0217732303011204, arXiv:hep-lat/0209120 (2004)
48. E. Seiler, I.-O. Stamatescu (eds.), *Approaches to Fundamental Physics* (Springer, Berlin, Heidelberg, New York, 2007)
49. M. Tegmark, The mathematical universe. Found. Phys. **38**(2), 101–150 (2008)

50. P. Thagard, *Computational Philosophy of Science* (MIT Press, Cambridge, MA, 1988)
51. G. 't Hooft, *The Cellular Automaton Interpretation of Quantum Mechanics*, Springer Open (2016)
52. C.F. von Weizsäcker, *Aufbau der Physik* (Springer, Berlin, Heidelberg, New York, 1985)
53. A. Wachter, *Relativistic Quantum Mechanics* (Springer, Berlin, Heidelberg, New York, 2011)
54. E.P. Wigner, The unreasonable effectiveness of mathematics in the natural sciences. Pure Appl. Math. **13**, 1–14 (1960)
55. J. Woodward, Scientific Explanation (2021). https://plato.stanford.edu/archives/spr2021/entries/scientific-explanation/
56. T. Wunderlich et al., Demonstrating advantages of neuromorphic computing: a pilot study. Frontiers in Neurophysics (2019). https://doi.org/10.3389/fnins.2019.00260
57. H.D. Zeh, On the interpretation of measurement in quantum theory. Found. Phys. **1**, 69–76 (1970)
58. W.H. Zurek, Environment induced superselection rules. Phys. Rev. **D 26**, 1862–1880 (1982)
59. W.H. Zurek, Decoherence and the transition from quantum to classical. Physics Today ***44**, 36–44* (1991)

Searching for Dieter Zeh

Peter Byrne

I never met Dieter Zeh in the flesh, nor did I ever talk to him by phone, Skype, or Zoom. Ours was exclusively an email relationship for more than a decade, and it worked.

It began in 2006, when I embarked upon writing a biography of the late physicist, Hugh Everett III, who, a half century previous, had proposed a controversial solution to the measurement problem in quantum mechanics. I reached out by email to Zeh, who had proposed a solution to the measurement problem similar to Everett's. In 1970, Zeh introduced a phenomenon later known as decoherence, the physical process by which entangling quantum systems take on macroscopic shapes. Zeh's theoretical schema is intellectual bound to Everett's, although the two scientists never met nor communicated.

Here is the back story.

Everett 27, published his doctoral dissertation on the "relative states" formulation of quantum mechanics in 1957. For many years, his famous thesis was unknown to all but a thin strata of physicists and philosophers with an interest in quantum foundations. Everett's motivation was to explicate a paradox presented when measuring the properties of a quantum "superposition," in which, for example, a carbon atom simultaneously occupies multiple locations. For reasons that are not understood, interacting with or observing a superposed quantum object via a macroscopic measuring instrument typically results in but one of the object's many positions registering as a discrete value. This is strange, because the bedrock quantum mechanical formula, Schrödinger's wave equation, tracks all of the evolving, superposed microscopic properties of a quantum system, treating each element in a superposition as equally real, equally probable.

Everett asked, what happens if we allow Schrödinger's ruthlessly linear wave function to guide the evolution of *macroscopic* objects? His idea was to treat the

P. Byrne (✉)
Peter Byrne Ink, 101 H Street, STE O, Petaluma, CA 94952, USA
e-mail: pbyrne@sonic.net

C. Kiefer (ed.), *From Quantum to Classical*, Fundamental Theories of Physics 204,
https://doi.org/10.1007/978-3-030-88781-0_13

wave formula as the blueprint of all that is, as describing the evolution of all of physical reality as completely quantum mechanical, as deterministic, as smoothly uninterrupted by discontinuities or indeterminacies. The consequence of treating the Schrödinger formula as seamless, as continuous and universally applicable elicits a conception of a universe of, as Everett put it, "equally real" universes.

Picture *everything* as an entirely quantum structure, as governed by an infinitely complicated and continuously evolving and entangling and non-discontinuous, *non-collapsing* "universal wave function" encompassing all that is, ever was, or will be. Everything that can happen does happen somewhere inside this *configuration* space, which is not the same as our spacetime, and, in the larger scheme of things, *probability* is meaningless. Macroscopic objects, including human brains, are describable as quantum superpositions smeared across multiple, tree-like branches of a constantly evolving, universe of branching universes as they interfere with each other, overlap, and ineluctably foliate, writing parochially self-descriptive wave functions incorporated into the master equation describing an infinite and non-denumerable set of sub-universes of a mutating, over-arching superposition of universes, or multiverse, in modern parlance.

According to Everett, from an observer's point of view, every quantum subsystem is constantly "splitting" into myriads of copies of itself as it interacts with its environment. Each copy of a mouse brain or a planet is quantum mechanically *correlated* to a different outcome of the same event, if there are alternatives. Sentient objects, the observers, believe in the laws of probability, of chance, of, in the case of quantum theory, the Born rule, a postulate that alchemizes snippets of non-probabilistic wave functions into probabilistic statements with predictive accuracy when viewed from *inside* "our" branch by practitioners of the quantum arts, who believe, quite reasonably, in the laws of probability.

Much has been written about Everett's interpretive theory, both laudatory and critical and fantastical. Notably, the interpretation is underdetermined; it predicts the universe we see, but so do its rival theories. In its popular conception, the "Many Worlds" theory is philosophically disturbing to One Worlders who crave ontological uniqueness. But it attracts those who do not mind the idea of infinitely uncountable copies of themselves riding a wave function roller coaster of the ecstatic and the dreadful. Win billions of dollars in a lottery! Get killed by an asteroid! Or gain *quantum immortality* as a increasingly worm-like, Voldemortian creature shaking off fatal diseases and bodily decay in probabilistic spasms of surviving against ever-shrinking odds as anything that can physically happen, does happen, whether you like it or not.

On its more practical side, the Everett approach is important to developing a theory of quantum gravity. It conceives a way of including the observer as part of the wave function of the universe, as opposed to having to be situated outside the universe, which is, to date, impossible. The theory is useful: it combines cosmology and nuclear physics. String theorists love it.

An inveterate empiricist, Everett did not speculate in his writings on the ontological reality of the universes which his logic had spawned. And neither, really, did Zeh. Based for most of his career at the University of Heidelberg, Zeh upheld

Everett's formulation because it buttressed his own theory. "Decoherence without Many Worlds is an incomplete story," Zeh told me.[1]

In 1973, Everett's relative states solution to the measurement problem was renamed the Many Worlds theory by physicist Bryce DeWitt, who strongly promoted the idea that the universes contain trillions of versions of ourselves living out diverging histories. Many Worlds has gained considerable traction in physics, and in Western culture, too. But Everett never wrote another jot of quantum mechanics after his dissertation was published. He left theoretical physics behind, choosing to carve out instead a brilliant career in the monied world of military operations research. Everett, a mathematical and computational polymath, was not only a top secret-cleared nuclear war planner, he also created the first relational database during the Cold War, and a famous optimization algorithm that bears his name.

Addicted to alcohol, cigarettes, rich foods, and extramarital fornications, Everett, age 51, died of a heart attack in 1982. Although he did not live to know it, his theory has had a profound effect on the foundations of quantum mechanics. It is a contender on the short list of approaches purporting to explain how our classical, macroscopic world emerges from the microscopic quantum substratum.[2] Everett's interpretation of quantum mechanics was written when he was studying physics at Princeton University under John Wheeler, who enthusiastically supported the relative states theory until *his* mentor, Neils Bohr, dismissed it as heretical and stupid. That story is amply told elsewhere.[3]

Zeh became one of the fiercest champions of Everett's theory, which he had discovered while researching his seminal 1970 paper, "On the Interpretation of Measurement in Quantum Theory." In this path-opening work, Zeh, as did Everett, claimed to be able to solve the measurement problem in quantum mechanics by abolishing the empirically nonsensical postulate of a non-linear wave function collapse in favor of an eternally evolving, non-collapsing wave function spanning the observable universe and beyond.[4]

Zeh and Everett share a vision of a universally operative Schrödinger equation which documents the evolution of the atomic elements forming the observer and the object observed *relative* to each other and the quantum environment. Everett and Zeh both explain the appearance of calculable probabilities as the subjective experiences of observers unaware that their seemingly unique, self-conscious observer states are constantly splitting, replicating in wherever configuration space exists in

[1] Zeh to Byrne, April 5, 2008.

[2] The short list includes Everett's relative states theory, hidden variables theories, such as Bohmian mechanics, and several collapsing wave function theories. Everett's universal wave function is a key component of the Landscape model in string theory, according to Leonard Susskind and other physicist-cosmologists.

[3] Byrne [1], Osnaghi et al. [2].

[4] Wave function collapse was postulated as a physical action by John Von Neumann in 1932. Bohr was not a fan of the collapse postulate, and the philosophical stance of the Copenhagen Interpretation with which Bohr is associated is often misunderstood as "anti-realist". See Barad [3] for an erudite and convincing defense of Bohr's philosophical stance from the point of view of what Barad terms "agential realism".

order to preserve the linearity of the ever evolving correlations of each observer state with all of the also splitting, replicating quantum subsystems in their environments. In other words, the elements composing any superposition correlate to a span of *equally real* branching universes. The goings on inside the infinite set of all universes is describable, in theory, by a *universal wave function*. Conscious observers go about the business of life in each branch according to what appear to be objectively sound probabilistic laws. And there *are* probabilistic laws in the branches in which there are such laws.[5] But probability is non-applicable inside the interfering amplitudes composing the universal wave function that encompasses all occurrences. And how could it be otherwise? A special operation, the Born rule, must be exerted to draw forth a probability statement from the otherwise merrily propagating non-probabilistic wave forms. Probabilities, for Everett and Zeh are frequencies of certain measurements *within* a *typical* branch, not *across* branches.[6]

But what is the nature of the physical mechanism by which probability-laden branches emerge from the non-probabilistic quantum soup? Ruminating on this question led Zeh to decoherence. But was Everett there before him?

Although the pace of our many conversations about Everett's interpretation declined after the biography was published in 2010, Zeh and I emailed every now and then up until 2017, shortly before he exited our branch of the wave function, leaving behind a rich set of quantum entanglements, including his friends, colleagues, and an historically important body of work. When Claus Kiefer asked me to write a chapter for this book about Zeh, I was flattered. But not being a scientist (I am a journalist), I was perplexed as to how to go about it.

I turned on my computer, searching for Zeh's emails, only to discover to my dismay that all except the most recent exchanges had vanished! A search through dusty boxes stuffed with research revealed *printouts* of most of our emails. What cyberspace taketh, paper preserves!

And then there are the books. My bookshelves contain a small cross section of popular and academic books on quantum foundations. And many of them have "Zeh, Dieter" listed in their indexes. First, we shall take a brief tour of these books, to find out who is this man, Dieter Zeh, as recorded there. Then, we will explore how he patiently taught me, a non-mathematical journalist, a little bit of quantum theory from thousands of miles away in Heidelberg. We will learn how Zeh came to terms with Everett's adumbration of decoherence. Lastly, I will relate a funny, yet poignant tale about how Zeh's work was still meeting opposition from science journals decades after decoherence was mainstreamed into the physics discourse.

---

[5] Is there a probability of branches without probability? Maybe, but the point is that as far as the universal wave function and its unsquared amplitudes is concerned, probability is latent or emergent. Maybe branches with no observers or "maverick" branches have no probability laws. But then they might not qualify as branches. As Everett remarked, language is ever a problem in quantum mechanics.

[6] Discussing the language used in this section with me, Jeffrey A. Barrett comments, "For Everett, one gets the *appearance* of probabilistic events from the statistical distribution of results in a typical branch (in his special norm-squared coefficient sense of typical)".

I am grateful to have known Dieter.

1 Indexing Zeh

One thing I have learned about the typical scientist is that when presented with a book on their specialty, the first, and perhaps the only action s/he takes is to thumb through the index looking for their own name, and, if spotted, perusing those pages with satisfaction or outrage, depending. It amuses me to imagine Zeh flipping through the indices of a few of these books, the ones, in fact, that I have in my office. It is not purely for laughs, though, as it reveals something about his place in the pantheon of remarkable physicists. Zeh's imagined responses are in ***"bold"***.

The Wave Function—Essays on the Metaphysics of Quantum Mechanics by David Albert and Alyssa Ney is an overview of the epistemological and philosophical issues with quantum theory. It devotes an entire chapter to the Everett Interpretation. The index indicates that Zeh is mentioned on page 209 in a chapter by David Wallace.[7] Alas, he is not there! Wojciech Zurek and several of Zeh's collaborators appear in a footnote on the development of decoherence theory, however.

What can I say? Next.

The Quantum Labyrinth by Paul Halperin explores the lives and thoughts of Richard Feynman, John Wheeler, and Hugh Everett III. Zeh gets two short paragraphs, the first noting his intellectual debt to Everett. Halperin frames Zeh's vision of Everett's branching universes as the mind of the observer forking into "many minds" at the instant of entanglement with a quantum system, without explaining the richness of Zeh's profoundly physical, non-idealist concept. The second paragraph begins, "Zeh helped develop another idea related to the many minds concept, called decoherence. Wojciech Zurich, a student of Wheelers ... was another of its prominent developers."[8] And those few sentences are the only place that the words "Zeh" or "decoherence" appear in the entire book.

Which must be boring book. Yawn. Next.

Murray Gell-Mann's *The Quark and the Jaguar* introduces a quantum theory of decohering universes at great length. It spends parts of two sentences vaguely crediting Zeh, Zurek, and Erich Joos with making "valuable contributions" to the theory of decoherence.[9]

Well, Murray was never one to share the glory. Next.

Jeffrey A. Barrett's *The Quantum Mechanics of Minds and Worlds* explicates Everett's theory, but the only mention of Zeh is in a footnote. Barrett wrote,

---

[7] Ney and Albert [4].

[8] Halpern [5].

[9] Gell-Mann [6].

"Everett did not think decoherence effects were necessary to explain our determinate experience."[10]

A footnote! Next.

Max Jammers' *The Philosophy of Quantum Mechanics* lists Zeh as an Everett "sympathizer" in a footnote in a section on "Many Worlds Theories."[11] The book was written long before the phenomenon of decoherence was experimentally validated in the mid-1990s, as it has been many times since then.

Jammer read my paper when it was published! Beeindruckend, wie cool ist das!"[12]

In 2004, Cambridge University Press published an anthology, *Science and Ultimate Reality.* Max Tegmark observes, "Everett's work had left two crucial questions unanswered: first of all, if the world actually contains bizarre macrosuperpositions, then why don't we perceive them? The answer came in 1970, when Dieter Zeh showed that the Schrödinger equation itself gives rise to a type of censorship effect. This effect became known as decoherence, and was worked out in great detail by Wojciech Zurek, Zeh, and others over the following decades."[13]

Tegmark goes on to say that decoherence solves a crucial question in quantum mechanics, that of how a preferred measurement basis (our universe with its consistent history and causality) physically emerges from the universal wave function as it spreads through Hilbert or configuration space, streaming diverging realities.

There is also a chapter on decoherence written by Zeh which treats the vastly entangling universal wave function as an It rather than a Bit.[14] Zeh's last paragraph reads, "However you turn it: *In the beginning was the wave function.* We may have to declare victory of the Schrödinger over the Heisenberg picture."[15]

A book well worth buying and reading, several times.

Other books on Everett or decoherence generously credit Zeh with pioneering the concept of decoherence as a physical phenomenon. *The Many Worlds of Hugh Everett III, Multiple Universes, Mutual Assured Destruction, and the Meltdown of a Nuclear Family* allocates ample space to the work of Zeh and Zurek in formulating the existence of decoherence as assisted by Everett's relative state formulation.[16] In *The Everett Interpretation of Quantum Mechanics: Collected Works of Everett 1955–1980 with Commentary*, co-author Jeffrey A. Barrett commented, "Zeh (1970) was the first to suggest adding decoherence considerations to Everett's pure wave mechanics to explain the dynamical stability of relative states. ... It was arguably Zeh's and Zurek's idea to understand decoherence as the mechanism for the *physical*

---

[10] Barrett [7].

[11] Jammer [8].

[12] Translation: "Wow, how cool is that!".

[13] Barrow et al. [9].

[14] Zeh [10].

[15] Ibid, p. 119.

[16] Byrne [11].

emergence of quasi-classical worlds from the objective global state. … [T]hey were adding interpretive tools to Everett's theory."[17]

Ah, that is much better than a footnote, Barrett! I agree with you!

2 Dark Days of Decoherence

In October 2015, Zeh invited physicist-author Adam Becker to his house in Neckargemund, Germany for tea and conversation. Becker's *What is Real?* relates Zeh's decades-long journey advocating for decoherence and the Everett interpretation, often as a voice in the wilderness. When Zeh began questioning the postulate of wave function collapse, his Nobel Prize winning thesis advisor, Hans Jenson, warned him, "that further work on the subject [of quantum foundations] would extinguish his academic career."[18]

Zeh had tenure and he persisted in his advocacy for the new theories in which he believed, but he did suffer some academic consequences. Most notably, he had a hard time getting his initial paper on decoherence published in a scientific journal. Eventually, Zeh reached out to Eugene Wigner at Princeton University, who was himself no fan of the collapse model. Wigner was an instrumental force in getting Zeh's (1970) paper published in the inaugural edition of *Foundations on Physics*.

Early advocates of Zeh's decoherence model included John Bell, John Wheeler, and, of course, Zurek (Wheeler's doctorand), who published a paper on decoherence in 1991 in *Physics Today*, in which he credited Zeh as the forerunner of decoherence theory.

Zeh told Becker that during "the dark days of decoherence" he and his students were denied jobs and promotions and even workplace collegiality.[19] But as the theory of decoherence developed over over the years, it gained experimental traction. It is now largely uncontroversial and accepted, although subject to interpretations that differ from the Zeh-Everett approach, including wave function collapse. As we shall learn, the darkness of challenging the status quo of wave function collapse still shades mainstream physics, especially when decoherence is interpreted as validating Everett's interpretation.

3 The Email Trail

When I contacted Zeh in 2006, I knew that he was highly respected by the physicists and philosophers who delved into the forbidding realm of quantum foundations. Not content to rest upon his laurels as a decoherence pioneer, Zeh was always pushing

[17] Barrett and Byrne [12].

[18] Becker [13].

[19] Ibid, p. 209.

boundaries. I found him actively agitating against the bodily existence of elementary particles, which he viewed as an untenable consequence of the Heisenberg picture of quantum mechanics with its crudely fashioned "observables" which were, in his view, not actual corpuscles, but wave packets entangling with the purely quantum environment and decohering entire universes at light speed like a giant "zipper", as Zeh so colorfully put it. He also called the decoherence process 'delocalization' and 'dislocalization', which at first confused me as I had thought decoherence was a *localizing* process. In our conversations, I did not have the nerve to admit to my confusion about the terms, but I learned what he meant by reading "Quantum Discreteness is an Illusion," a paper which he was trying to get published, and for which he was getting flack from journal editors. Zeh explained, "The term 'dislocalization' was suggested to me by John Wheeler in place of my original 'delocalization' to characterize the transformation of local superpositions into entangled (nonlocal) ones—in contrast to objects that are merely extended in space."[20] Like with the Everett worlds, Zeh's physically instantiated dislocalization process links or entangles an observer state, or any object state interacting with its quantum surroundings, with an unfolding branch of the universal wave function, a branch selected by decoherence out of all the gateways seeded in the local superposition.

I read many of Zeh's papers (skipping over the equations, which might as well be Martian to me, frankly) and asked him scores of questions. He was generous with his time and never patronizing, despite my lack of scientific training or expertise. He never called me thick-headed, although I had the impression that there were the odd branches of the wave function where he might have muttered it. I believe he was a bit piqued by my somewhat presumptuous suggestion that *Everett had somehow discovered the phenomenon of decoherence* … years before Zeh did … and he was initially adverse to my suggestion that Everett may have, in some sense, derived the Born Rule from pure wave mechanics.

Our email correspondence began in June 2006 with a discussion of how Everett's theory relates to decoherence. Zeh wrote, "I feel that Everett was outstanding in apparently being the first physicist who dared to regard the nonlocality that is part of the wave function (nonlocal entanglement) as an essential part of REALITY—several years before Bell's discussion!"[21]

More than a year later, he wrote, "What most physicists do not realize is that my 1970 paper was predominantly meant as an argument for Everett, whose interpretation I had independently [arrived at] by these decoherence arguments."[22]

Prior to 1970, Zeh became aware of Everett's 1957 "short" dissertation as it was published by *Reviews of Modern Physics*. The short thesis was a much-changed version of the longer thesis which Everett had submitted to Wheeler. The "long" thesis laid out a mathematical argument that eliminated the measurement problem. It was severely edited and rewritten by Wheeler and Everett (1) to make an argument more relevant to general relativity than to solving the measurement problem, and

---

[20] https://arxiv.org/pdf/0809.2904.pdf, p. 7.

[21] Zeh to Byrne, June 16, 2006.

[22] Zeh to Byrne, September 19, 2007.

(2) to complete Everett's doctoral requirements without irritating his dissertation committee by discarding the widely accepted notion of wave function collapse.

The long thesis was not published until 1973.

In our first two years of conversations, Zeh and I discussed the long thesis; newly discovered correspondence between Everett and Bryce DeWitt and others; newly discovered transcripts of a January 1957 conference on quantum gravity at which Wheeler had presented the gist of Everett's unedited dissertation, shortly before it was cut, edited, and published a few months later; and the transcript of an impromptu discussion about his theory by Everett at a conference of world-leading physicists at Xavier University in 1959.

Zeh opined that because Everett was not aware of the role of entanglement, he could not have come up with the idea of decoherence. But to my neophyte eye, the new material, and many passages in the long thesis made it clear that Everett was aware of entanglement: he just called it the "correlation" of quantum systems. I could not imagine that Everett was unaware of the *actuality* of decoherence even though he did not name it as such. And I was puzzled as to how the mathematically brilliant Everett could have fooled himself that he had derived a probability measure equivalent to the Born Rule, if he had not. Plus, Wheeler had seemed to enthusiastically agree with him about the efficacy of the probability measure, although later he threw Everett's theory under a Copenhagen-driven bus to placate Bohr, who had rejected Everett's interpretation as a violation of quantum indeterminism.

In May 2008, after reading the new materials, Zeh wrote to me, "I now agree that he 'deduced' something in analogy to phase space volume as a measure of probability, but: (1) there is no phase space in quantum theory; the norm of a wave function, which appears in the Born rule, is only similarly conserved under the exact (deterministic) dynamics, (2) this is not a deduction in a strict sense, but a plausibility (or dynamical consistency) argument, so the last sentence of the thesis is misleading."[23]

Here is the last sentence of Everett's long thesis before it was edited and rewritten by him and Wheeler as more of an argument for quantizing gravity than for solving the measurement problem, which had been Everett's original intention: "Aside from any possible practical advantages of the theory, it remains a matter of intellectual interest that the statistical assertions of the usual interpretation do not have the status of independent hypotheses, but are deducible (in the present sense) from the pure wave mechanics that starts completely free of statistical postulates."[24]

The Born rule is a robust and practical postulate. Was Everett's supposed derivation of it from first principles somehow better? Explanatory, as well as predictive? A stroke of genius?

Zeh continued in the May 2008 email, "Before applying a measure, in quantum theory you first have to present arguments that the 'branches' become autonomous, and thus form 'independent worlds'. This is in my opinion most convincingly done

[23] Zeh to Byrne, May 9, 2009.

[24] Barrett and Byrne [14].

by means of decoherence—*but Everett may have had some preliminary ideas of such a concept*."[25]

At the Xavier University conference, Everett remarked, "I can therefore assert that the probabilistic interpretation of quantum mechanics can be deduced quite as rigorously from pure wave mechanics as the deductions of statistical physics." His mathematical choice of measure was the Lebesque measure, which he liked because, he said, "it is preserved under the transformation of phase space."[26] Everett felt justified in using the classical phase space measure to track events in one of the branching quantum worlds, the classical, probabilistic universe which we inhabit. And he never deviated from holding onto his assertion that he had, in effect, *derived* an equivalent to the Born rule, the postulate which works so well in our branch, as opposed to having to *postulate* a new probability measure for the branching universes.[27]

Zeh told me, "Apparently he hoped to find a justification for the probability interpretation (and perhaps Born's rule) in the mathematics. This would be the typical mistake of a young scientist. ... The main problem of measurement is precisely the transition from entangled superpositions to (apparent) ensembles. So it would not suffice to point out some formal similarities. *Where does he explicitly use the word 'entanglement'?* There is no doubt that he was well aware of the difference, since he regarded the wave function as describing reality. ... However, one must also expect that he was in some state of confusion, which is normal when you start thinking about a completely novel idea that has strange consequences. Just compare it with the inconsistent claims of the founding fathers of quantum theory!"[28]

Zeh warned me, "Mathematics can provide useful tools for physics, but you cannot derive physics from mathematics."[29]

Rooting through Everett's papers stored in cardboard boxes in the dirt-floored Los Angeles basement of Everett's musician son, Mark Everett, I had found the unpublished, handwritten drafts of Everett's long dissertation with his extensive notes and revisions. In mid-2008, I Xeroxed these several hundred pages and mailed copies in large boxes to several physicists and philosophers who were Everett fans, including Zeh.

I was hoping that these experts would troll through the (Martian) technicalities and spot young Everett miraculously discovering decoherence and unequivocally deriving the Born rule—thereby putting paid to his many critics. After reading the materials, the experts hosed my hopes. There was plenty of wonderful new contextual material, they said, but no evidence that Everett had discovered decoherence or derived the Born rule in his drafts, including a very early draft which he had titled, "Wave Mechanics Without Probability."

---

[25] Zeh to Byrne, May 9, 2008. Emphasis added.

[26] Barrett and Byrne [15].

[27] On June 11, 1957 Everett wrote an eloquent letter to the distinguished physicist, E. T. Jaynes, presenting his use of the Lebesque measure in light of Jayne's work. Barrett and Byrne [16].

[28] Zeh to Byrne, June 10, 2008. Emphasis added.

[29] Zeh to Byrne, October 4, 2008.

Zeh was much impressed, however, with the example Everett used in that draft of a "splitting, intelligent amoeba" as a metaphor for the observer state splitting at every quantum interaction.[30] The amoeba metaphor did not make it into any of the published versions of Everett's theory. It was replaced by bifurcating cannonballs, which may have appealed to the militarist in Everett, at the expense of poetic power.

Discussing the handwritten drafts and early typed versions of Everett's thesis, Zeh asked me if Everett had ever used the word "entanglement" in his drafts, which he did not. Everett called that phenomenon the "correlation" of quantum subsystems. Zeh then asked me if Everett ever used the word "environment" in his drafts and notes.[31] I replied that Everett wrote of quantum subsystems being "strongly correlated to our environment" in three of the drafts, but the word "environment" did not make it into the published versions.

The next day, Zeh qualified a previous comment, "I had seen the pages you mention, but overlooked the word 'environment'. The word 'mistake' that I used was a bit misleading. What I meant is that he is using an inappropriate strategy in this respect. … The fact that HE does not use the word 'entanglement', does not mean too much by itself, since it was not very well known at that time. Although I knew Schrödinger's 'Verschraenkung', in the 70s, I translated it with 'quantum correlation,' which I clearly distinguished from statistical correlations, however. The difference became clear in Bell's theorem—but it was also evident from many-particle states in applied QT (atoms etc.). Nonetheless, it was often confused with statistical correlation—even by Schrödinger. Everett's measure of correlations measures only probability correlations!"[32]

In other words, Everett's phase space measure was appropriate for measuring the quantum probabilities we experience from the peculiar point of view of our decohered branch, but his statistical measure is not derived from the existence of infinitely many entangled branches (the universal wave function itself) and does not answer the burning question of how can there be probabilities in a multiverse in which everything that is possible happens.[33]

Then comes the kicker in June 2008: Zeh observes that in his drafts and the long thesis, "*HE seems to come very close to decoherence*, although I am not sure he is fully aware of the entanglement concept. Even where he mentions correlations with the environment, he seems to mean classical correlations. Entanglement (and the resulting decoherence) is about complex amplitudes (phase relations), while HE always refers to probability correlations (which are positive numbers). This may be the reason why he claims to 'deduce' the probability interpretation, a claim that is otherwise hard to understand. One may even speculate if this was the reason for

---

30 Barrett and Byrne [17].

31 Zeh to Byrne, June 10, 2008.

32 Zeh to Byrne, June 11, 2008.

33 Maybe there could be such a measure, in theory, because not *everything* is possible? But there is no way to measure the possible against the impossible … the road to madness wends through quantum mechanics …

Wheeler's later preference for a fundamental information concept underlying QT ('quantum information').

> In any case, this passage appears interesting enough to make it available to science historians in an appropriate form. For me it is just exciting trying to follow his early attempts. I don't know how I had reacted 40 years ago if I had seen it! His most important insight certainly is the splitting observer, illustrated by the amoeba. ... In hindsight it appears all so obvious that you don't understand why people refuse to get convinced.[34]

A few weeks later, Zeh wrote, "One may also assume that Everett would have one day discovered the full implications of decoherence, but when he refers to the environment he seems to mean exclusively its macroscopic entanglement with the object—not the essential entanglement of all microscopic (thermal) variables.[35] It is only this fact which gives rise to 'classical' properties (namely those being stable against decoherence) ...

> Pointing out the existence of a MATHEMATICAL structure of a 'measure' is certainly far from giving a proof of its physical meaning. HE clearly shows that the wave function 'splits' into dynamically autonomous branches, but this can be understood as an apparent indeterminism (the basis of a probabilistic process) only if you POSTULATE that the observer is a (local) part of an autonomous branch (the amoeba example!) And next you need a measure or weighting, which he can justify (not derive) only by assuming that (a) the probabilities in each branching event are independent of the history (defined within the present branch), and (b) the sum of probabilities of all branches is constant. This is plausible, but nonetheless an assumption. He seems quite clear about this at [the Xavier conference] with Dirac and others.[36]

In August, Zeh wrote, "I think Everett (like many others) may have meant various things when he talked about probabilities. At some points he just may mean a non-unique (indeterministic) evolution, in other cases the precise measure that describes the observed frequencies of outcomes in series of measurements. (Even mathematical theories of probability differ—for example in referring to time or not.) The only thing he never means in my opinion is the psychology of an 'agent'. ...[37]

> Everett made the correct assumption about splitting observers in order to maintain an effective indeterminism (but he did NOT derive it), and he demonstrated that the correct measure (probability weighting) follows from some plausible assumptions (such as conservation of

---

[34] Zeh to Byrne, June 11, 2008. Emphasis added.

[35] Zeh seems to be missing that Everett's conceptualization of correlation was essentially what we now call entanglements between the microscopic systems composing macroscopic objects. Everett did, however, have an information theoretic approach to modeling correlation/entanglement. That differs from Zeh's approach to modeling, which is more grounded in nuclear physics. The meaning of correlation between quantum systems is discussed at length in Everett's drafts, "Quantitative Measure of Correlation," and "Probability in Wave Mechanics, and the long thesis, "Theory of the Universal Wave function. See Barrett and Byrne [18].

[36] Zeh to Byrne, July 17, 2008.

[37] Zeh to Byrne, August 19, 2008. Zeh was opposed to the proposals of David Wallace, Simon Saunders, and David Deutsch that the Born rule can be justified in Everettian quantum mechanics because a "rational agent" would make bets as if the probabilities calculated by the rule were true. In our correspondence, Zeh several times denigrated that notion as psychology, not physics.

probability AND independence from pre-history)—but other possibilities are NOT LOGICALLY excluded. The empirical fact that we are not living in a maverick component[38] is not trivial. So he may have "nailed" it—but he did not deduce it without further assumptions or postulates.[39]

And a month later, "*I think one can now conclude that Everett 'derived' Born's probabilities (the weights) not strictly, but in the sense that they are the only reasonable choice.*[40] This was not quite clear from his RMP article."[41]

In October 2008, Zeh concluded, "It is difficult to find out what HE really meant. I may have too much of my own interpretation in mind, but I guess he did think of an objective 'weight' given by the norm. However, since for him objective reality was represented by the complete global wave function (for which the weights of the components do have meaning in their superposition!) probabilities could be meaningful only in a 'subjective' sense (that is, for 'our' specific branch—although they are then 'objective' for all observers in the same branch). It was natural for him to assume that 'we' are living in a 'normal' branch (one with non-negligible weight).

In this sense he would have derived the Born rule; I cannot imagine any other concept of an empirical rule than that found to apply in our branch—hence it is 'subjective' in this theory of a multiverse. So if the Born rule applies, it trivially also 'appears' to apply. It can evidently not hold for all branches—and not even for their majority IF you define this majority by simply counting them without using a weight![42] (They do hold for their majority if defined by means of the weights).[43]

To add more context to this discussion, Everett scholar Jeffrey A. Barrett has recently written, "Note that Everett did not seek to deduce quantum probabilities *over branches*. The probability associated with each branch is always only one. Rather, he wanted to derive the appearance of the standard quantum probabilities from the statistics of records *within a typical branch*. ... Importantly, Everett did not derive the standard *probabilistic predictions* of quantum mechanics regarding chance events, and he knew it. He clearly and repeatedly insisted that there were no probabilities in pure wave mechanics. .... Rather than claim that his theory predicted the same chance events as predicted by the nonlinear collapse dynamics, Everett

---

[38] I.e. a branch where the usual physical laws do not apply.

[39] Zeh to Byrne, August 19, 2008.

[40] Emphasis added. Is Zeh echoing the Oxford Everettian's assumption that using the Born rule is a rational choice? Sadly, I never thought to ask him.

[41] Zeh to Byrne, September 18, 2008. "RMP" refers to the Review of Modern Physics which published Everett's severely edited thesis in 1957.

[42] Zeh completely discounted DeWitt's and Neil Graham's version of a probability measure on the Everett branches because it relied on counting separate worlds and was, in his view, a circular argument. See Barrett and Byrne [8] for discussions of this issue, as well as Barrett [19].

[43] Zeh to Byrne, October 10, 2008.

showed that it would appear to a typical relative observer that there had been such events."[44]

Back to Zeh: "I have never held the position that only one branch contains conscious observers … Let me explain that since my early interests in philosophy I always held the position that only subjective awareness is uncountably real, while all physical reality must in principle remain a 'heuristic fiction' (thinking of Descartes' demon). This included the consciousness of others. So I distinguished between two concepts of a 'subjective/objective' relation: that referring to individual observers in a branch world and that referring to individual branches. Their combination leads to the concept of individual 'versions' of an observer. So I remember that I once called the denial of the other Everett branches 'solipsism in Hilbert space' (I should better have said 'in configuration space').[45]

In August 2009, I asked Zeh to review several draft chapters of my Everett biography. And for the first time, he chastised me. He was concerned that I was giving to much play to DeWitt's easily popularized Many Worlds version of Everett, which treats all of the universes as ontologically real, as "schizophrenia with a vengeance," as he described the many worlds theory, passing it off as Everett's schematic, which was not accurate, although it generated a lot of controversy in physics and made Everett into an ATM for science fiction writers.

Zeh said that DeWitt and David Deutsch and others insisted on using the static Heisenberg picture of quantum mechanics to situate Everett's relative states theory and that was the wrong approach. Everett's time evolving universal wave equation was modeled in the Schrödinger picture. Despite the common misperception that the Schrödinger and Heisenberg models are equivalent to each other, they are not! Zeh told me, repeatedly.

> You should in my opinion have emphasized the important fact that [David] Deutsch [who is a proponent of the Many Worlds theory as ontologically strong] is representing DeWitt's—not Everett's—Many Worlds. He has often said that he is using the Heisenberg picture (not the wave function itself as describing reality, as Everett did). Therefore, his 'worlds' are trajectories in classical configuration space—similar to Feynman paths. So he does not need decoherence to separate different branches of the wave function … Perhaps [Deutsch] does not even see the difference. It becomes particularly obvious when he claims that a quantum computer is calculating in parallel worlds. However, then every time-dependent wave function or other superposition would already represent a continuum of different worlds. The point of Everett's branching is its irreversibility that would prevent any quantum computer from working. A quantum computer has to live in one quasi-classical Everett 'world' (a partial wave function).[46]

Zeh was also concerned that I devoted an entire chapter to the Oxford University interpretation of Everett, as conceived of by Deutsch and the philosophers, Simon

[44] Barrett [20].

[45] Zeh to Byrne, May 23, 2009.

[46] Zeh to Byrne, August 18, 2009.

Saunders and David Wallace. Zeh loved that the Oxford Everettians deployed decoherence to "solve" the preferred basis problem,[47] but he could not abide their reliance upon "psychology," i.e., assuming that the Born rule is a good measure of probability inside a decohered Everett world since a supposedly "rational" agent would place bets on future outcomes in accord with mod squared amplitude measure, the Born rule.[48]

"I also think that the philosophers are just confusing the issue," Zeh told me. "Their probability problem is a red herring. You have to postulate a probability measure for the branches according to what we observe—and that is it. ... It is true that Everett was ahead of me by ten years (although I did not know his work when I started), but I was ahead of Deutsch by 15 years and ahead of Saunders by 25 years (and they definitely knew both Everett's and my work). However, these Oxford people seem to cite only each other. I have never seen anything new to me written by them (except for the questionable stuff), while I have addressed many of these questions, always advocating for Everett, long before them."[49]

In the spring of 2009, Zeh submitted a paper titled "Quantum Discreteness is an Illusion" to a prestigious scientific journal. It was written for non-experts, as well as experts. The very readable paper was yet another stage in Zeh's advocacy for differentiating the Schrödinger picture from the Heisenberg picture. He relied heavily on his interpretation of the Everett theory to argue for the non-existence of ontologically discrete atomic particles that mysteriously "jump" around. For Zeh, Nature *is* the wave function. Particles are heuristic fictions. There are no so-called "quantum jumps," nor "wave/particle duality," nor wave collapses caused by ghostly "minds".

The paper was peer reviewed and revised, and then re-reviewed and rejected. The final reviews were mixed, but one reviewer who disagreed with its premises, said it was still worth publishing as an example of a trend in physics. Another reviewer said there was already too much published on the topic, even though the paper was good.

Writing to Zeh, the journal editor rejected the paper, explaining, "As you can see, the reviews vary quite widely. On the one hand, I think it is safe to say that some people feel (as I do) that many worlds interpretation is useless. I have never seen the point of it, except that people feel a great urge to want the quantum world to act like the classical world. ... And I have never seen a single interpretation that leads to any experimental consequence."[50]

Zeh was disappointed, perhaps enraged, commenting to me, 'It may be a good idea to collect any further material of this kind. The Copenhagen lobby seems to be as strong as ever."[51]

---

[47] In a nutshell: the problem of how the ever-splitting worlds maintain consistent histories and physical laws.

[48] Zeh to Byrne, August 18, 2009.

[49] Ibid.

[50] Zeh emailed the editor's note and the referee reports to me on May 21, 2009.

[51] Zeh to Byrne, May 21, 2009.

When I asked the journal editor to explain his aversion to Everett's and Zeh's work, he expressed outrage that Zeh had shared his and the reviewer's communications with a journalist. He then asked me to please not report on the incident in my upcoming Everett biography. I complied, because it was not relevant to the book. He later assigned a harsh critic of Everett's theory to review the biography for the journal, presuming, one can logically suppose, that the book would be trashed.

In this universe, the review was laudatory.

In 2010, in a strange symmetry, "Quantum Discreteness is an Illusion" was published by *Foundations in Physics*, the same journal that had bravely published Zeh's first paper on decoherence in 1970–40 years prior.

Our last email conversation was on August 23, 2017. Dieter sent me the *23rd version* of his previously published paper, "The Strange (hi)story of particles and waves".[52] It is a wonderful read, in many ways summing up his scientific personality and his contributions to the history of science. And I was not surprised by the final sentence of his email: "P.S. Emotions (and tradition) still prevent most physicists from accepting Everett."[53]

References

1. P. Byrne, *The Many Worlds of Hugh Everett III, Multiple Universes, Mutual Assured Destruction, and the Meltdown of a Nuclear Family* (Oxford University Press, 2010)
2. S. Osnaghi, F. Freitas, O. Freire, The origin of the Everettian Heresy, in *Studies in History and Philosophy of Science Part B: Studies in History and Philosophy of Modern Physics*, vol. 40, no. 2, May 2009
3. K. Barad, *Meeting the Universe Halfway—Quantum Physics and the Entanglement of Matter and Meaning* (Duke University Press, 2007)
4. A. Ney, D. Albert, *The Wave Function—Essays on the Metaphysics of Quantum Mechanics* (Oxford University Press, 2013), p. 209
5. P. Halpern, *The Quantum Labyrinth* (Hachette Book Group, 2018), pp. 249–250
6. M. Gell-Mann, *The Quark and the Jaguar* (W. H. Freeman & Co., 1994), pp. 138, 149
7. J.A. Barrett, *The Quantum Mechanics of Minds and Worlds* (Oxford University Press, 1999), p. 221n
8. M. Jammer, *The Philosophy of Quantum Mechanics* (Wiley, 1974), p. 519
9. J.D. Barrow, P.C.W. Davies, C.L. Harper, *Science and Ultimate Reality* (Cambridge University Press, 2004), p. 473
10. H.D. Zeh, The wave function: it or bit?, in J.D. Barrow, P.C.W. Davies, C.L. Harper (eds.) *Science and Ultimate Reality* (Cambridge University Press, 2004), pp. 103–199
11. P. Byrne, *The Many Worlds of Hugh Everett III, Multiple Universes, Mutual Assured Destruction, and the Meltdown of a Nuclear Family* (2012), pp. 102, 261, 364–366, 368–370
12. J. Barrett, P. Byrne, *The Everett Interpretation of Quantum Mechanics: Collected Works of Everett 1955–1980 with Commentary* (Princeton University Press, 2012), pp. 46–47
13. A. Becker, *What is Real?* (Basic Books, 2018), p. 198

[52] https://arxiv.org/abs/1304.1003.

[53] Zeh to Byrne, August 23, 2017. Zeh was posting on a physics blog *Not Even Wrong* in March 2018, not long before he passed. His parting shot: "In my opinion there are no arguments but only emotions against Everett! So please take some time to study my website!".

14. J. Barrett, P. Byrne, *The Everett Interpretation of Quantum Mechanics: Collected Works of Everett 1955–1980 with Commentary* (Princeton University Press, 2012a), p. 196
15. J. Barrett, P. Byrne, *The Everett Interpretation of Quantum Mechanics: Collected Works of Everett 1955–1980 with Commentary* (Princeton University Press, 2012b), pp. 267–279
16. J. Barrett, P. Byrne, *The Everett Interpretation of Quantum Mechanics: Collected Works of Everett 1955–1980 with Commentary* (Princeton University Press, 2012c), pp. 261–264
17. J. Barrett, P. Byrne, *The Everett Interpretation of Quantum Mechanics: Collected Works of Everett 1955–1980 with Commentary* (Princeton University Press, 2012d), p. 69
18. J. Barrett, P. Byrne, *The Everett Interpretation of Quantum Mechanics: Collected Works of Everett 1955–1980 with Commentary* (Princeton University Press, 2012)
19. J.A. Barrett, *The Conceptual Foundations of Quantum Mechanics* (Oxford University Press, 2019)
20. J.A. Barrett, *The Conceptual Foundations of Quantum Mechanics* (Oxford University Press, 2019), pp. 158–159